Networked Nonlinear Stochastic Time-Varying Systems

Networked Nonlinear Stochastic Time-Varying Systems

Analysis and Synthesis

Hongli Dong, Zidong Wang and Nan Hou

CRC Press is an imprint of the
Taylor & Francis Group, an **informa** business

First edition published 2022
by CRC Press
6000 Broken Sound Parkway NW, Suite 300, Boca Raton, FL 33487-2742

and by CRC Press
2 Park Square, Milton Park, Abingdon, Oxon, OX14 4RN

ISBN: 978-1-032-03878-0 (hbk)
ISBN: 978-1-032-03880-3 (pbk)
ISBN: 978-1-003-18949-7 (ebk)

DOI: 10.1201/9781003189497

Typeset in CMR10
by KnowledgeWorks Global Ltd.

This book is dedicated to the Dream Dynasty, consisting of a group of diligent people who have enjoyed intensive research into analysis and synthesis for networked nonlinear stochastic time-varying systems ......

Contents

Preface

Nowadays, with the development of computers and communication network technology, the complexity of networked systems tends to increase rapidly. In the areas of engineering, computer science and mathematics, a growing attention has been paid to the analysis and synthesis issues for nonlinear stochastic time-varying systems with different network-enhanced complexities. Specifically, the complexities contain time-delays, saturations, nonlinearity, stochasticity, time-varying parameters, randomly occurring phenomena, state-multiplicative noises, fading channels, redundant channels, measurement quantizations, actuator and sensor failures and stochastic inner couplings, etc, the ignoring of which may deteriorate the working effects of the designed filters, fault estimators and reliable controllers. Thus, it has practical significance to investigate the filtering, fault estimation and reliable control problems for systems with such network-enhanced complexities and avoid the negative influences from the network-induced phenomena on the system performance.

In this book, we cope with the filter design, fault estimation and reliable control problems for different classes of nonlinear stochastic time-varying systems with network-enhanced complexities (e.g. state-multiplicative noises, stochastic nonlinearities, stochastic inner couplings, channel fadings, redundant channels, measurement quantizations, actuator and sensor failures, mixed time-delays and state saturations). Particularly, the randomly occurring phenomena under consideration contain randomly occurring sensor saturations, randomly occurring nonlinearities, randomly occurring uncertainties, randomly occurring faults, randomly occurring gain variations and randomly varying topologies. The content of this book involves three parts. In the first part, the finite-horizon filtering, fault estimation and reliable control are studied for networked control systems with state-multiplicative noises, channel fadings, randomly occurring sensor saturations, randomly occurring nonlinearities and uncertainties, measurement quantizations, mixed time-delays, state saturations, actuator and sensor failures and randomly occurring faults. Sufficient conditions or necessary and sufficient conditions are established which ensure that the designed filters, fault estimators and reliable controllers satisfy the H_∞ performance constraint or the covariance constraint or the $P(k)$-dependent constraint via proposing new techniques. In the second part, the approaches and skills acquired in the former part are adopted to design the distributed state estimators, distributed fault estimators and distributed filters for three classes of nonlinear time-varying systems with randomly switching topologies, redundant channels and randomly occurring gain variations and nonlinearities under the Round-Robin protocol over sensor networks. In the third part, the problems of variance-constrained H_∞ state estimation, partial-nodes-based state estimation and recursive filtering are considered for nonlinear time-varying complex networks with randomly varying topologies, random coupling strengths, stochastic inner coupling and measurement quantization under the random access protocol.

The book is arranged as follows. In Chapter 1, research backgrounds of the filter design, fault estimation and reliable control problems are reviewed for nonlinear stochastic time-varying systems with network-enhanced complexities, and the research contents of the following chapters are outlined. In Chapter 2, in terms of the developed event-triggered framework, the finite-horizon H_∞ filtering problem is dealt with for discrete time-varying systems where fading channels, randomly occurring nonlinearities and multiplicative noises are taken into account, and the variance-constrained H_∞ control problem is investigated for a class of discrete time-varying systems with randomly occurring saturations, stochastic nonlinearities and state-multiplicative noises. In Chapter 3, the time-varying reliable H_∞ output feedback controllers are designed for a class of discrete time-varying systems where randomly occurring uncertainties, randomly occurring nonlinearities and measurement quantizations are considered. In Chapter 4, for two classes of nonlinear stochastic time-varying systems with randomly occurring faults, the finite-horizon H_∞ fault estimation problem is discussed via the recursive linear matrix inequality method and the coupled backward recursive Riccati difference equation method. In Chapter 5, the set-membership filtering problem is dealt with for a class of time-varying state-saturated systems with mixed time-delays under the communication protocol. In Chapter 6, the distributed state estimation problem is examined for a kind of nonlinear time-varying stochastic systems through sensor networks with randomly switching topologies as well as redundant channels, the non-fragile distributed fault estimation problem is solved for a class of time-varying systems subject to randomly occurring nonlinearities and randomly occurring gain variations, furthermore, the filtering problem is focused on for nonlinear systems over sensor networks whose topologies are changeable subject to Round-Robin protocol in the finite-horizon case. In Chapter 7, the variance and H_∞ performance constrained state estimators are designed for a class of nonlinear time-varying complex networks with randomly varying topologies, stochastic inner coupling and measurement quantization, the robust finite-horizon state estimation problem is also investigated for a class of time-varying complex networks under the random access protocol through available measurements from only a part of network nodes. In Chapter 8, the recursive filtering problem is investigated for a class of time-varying complex networks with state saturations and random coupling strengths under an event-triggering transmission mechanism. In Chapter 9, this book is summarized and future research topics are discussed.

List of Figures

List of Tables

Acknowledgments

We would like to express our sincere thanks to those who have been devoted to the research in this book. Particular thanks are given to: Professor Steven X. Ding from University of Duisburg-Essen, Duisburg, Germany; Professor Huijun Gao from Harbin Institute of Technology, Harbin, China; Professor Bo Shen from Donghua University, Shanghai, China; and, Professor Derui Ding from University of Shanghai for Science and Technology, Shanghai, China. Special thanks are given to Fei Han, Hongyu Gao, Haijing Fu and Fan Yang for their considerable help in the editorial and proofreading work.

The accomplishment of this book was supported in part by the National Natural Science Foundation of China under Grants 61873058, 61873148, 61933007 and 62073070, the Natural Science Foundation of Heilongjiang Province of China under Grants ZD2019F001, the Fundamental Research Funds for Provincial Undergraduate Universities of Heilongjiang Province under Grants GJQHB201801 and KYCXTD201802 and the Alexander von Humboldt Foundation of Germany.

Symbols

$\mathbb{R}^n$ — The n-dimensional Euclidean space.

$\mathbb{R}^{n\times m}$ — The set of all $n \times m$ real matrices.

$X \geq Y$ ($X > Y$) — $X - Y$ is positive semi-definite (positive definite), where X and Y are real symmetric matrices.

M^T — The transpose of the matrix M.

$M^{\dagger} \in \mathbb{R}^{n\times m}$ — The Moore-Penrose pseudo inverse of $M \in \mathbb{R}^{m\times n}$.

$\mathbf{0}_n$ (or simply $\mathbf{0}$) — The n-dimensional zero matrix.

0 — The zero matrix of compatible dimensions.

I_n or simply I — The n-dimensional identity matrix.

$\text{diag}\{A_1, A_2, \cdots, A_m\}$ — A diagonal matrix with $A_1, \cdots, A_m$ in the diagonal.

$\mathbb{E}\{x\}$ — Expectation of the stochastic variable x.

$\mathbb{E}\{x|\, y\}$ — Expectation of x conditional on y.

$\text{Prob}\{\cdot\}$ — The occurrence probability of the event "·".

$\text{Prob}\{\cdot|\cdot\}$ — The conditional probability of the event "·".

"$*$" — An ellipsis for terms induced by symmetry in symmetric block matrices.

$\otimes$ — The Kronecker product.

$\mathbf{1}_n$ — $[1, 1, \ldots, 1]^T \in \mathbb{R}^n$.

$l[0, N]$ — The space of vector functions over $[0, N]$.

$l_2[0, N)$ — The space spanned by square summable and vector-valued functions $\epsilon = \{\epsilon_k\}_{0\leq k\leq N}$ whose norm is defined by $\|\epsilon\|_2 = (\sum_{k=0}^{N} \|\epsilon_k\|^2)^{(1/2)}$.

$l_2[0, \infty)$ — The space of square summable sequences.

$l_2[0, N]$ — The space of square summable sequences over $[0, N] := \{0, 1, 2, \cdots, N\}$.

$l_2([0, N], \mathbb{R}^n)$ — The space of nonanticipatory square-summable n-dimensional vector-valued functions.

tr(A) — The trace of a matrix A.

$\|A\|$ — The norm of a matrix A defined by $\|A\| = \sqrt{\text{trace}(A^T A)}$.

$\|A_k\|^2_{[0,N]}\mathbb{E}\{\sum_{k=0}^{N} \|A_k\|^2\}$.

$(\Omega, \mathscr{F}, \text{Prob})$ A probability space, where Prob, the probability measure, has total mass 1.

$\|x\|$ The Euclidean norm of a vector x.

$\mathbb{N}^0$ The set $\{0, 1, \ldots\}$.

$\mathbb{Z}_1$ The set of zero and all negative integers.

N_0 A positive integer.

$l_2([0, N], \mathbb{R}^m)$ The space of square-summable m-dimensional vector-valued functions on the interval $[0, N]$.

min{a,b} The minimum value in variables a and b.

List of Acronyms

ASFs	actuator and sensor failures
CC	covariance control
CNs	complex networks
FCs	fading channels
FDI	fault detection and isolation
LMI	linear matrix inequality
MTDs	mixed time-delays
MQs	measurement quantizations
NCSs	networked control systems
NTVSs	networked time-varying systems
PNB	partial-nodes-based
RAP	random access protocol
RCSs	random coupling strengths
RCs	redundant channels
RDEs	Riccati difference equations
RLMI	recursive linear matrix inequality
RMSE	root-mean-square errors
RNNs	recurrent neural networks
ROFs	randomly occurring faults
ROGVs	randomly occurring gain variations
ROMDs	randomly occurring multiple delays
RONs	randomly occurring nonlinearities
ROSSs	randomly occurring sensor saturations
ROUs	randomly occurring uncertainties
RSTs	randomly switching topologies
RVSDs	randomly varying sensor delays
RVTs	randomly varying topologies
SICs	stochastic inner couplings
SMNs	state-multiplicative noises
SNs	sensor networks
WTOD	Weighted Try-Once-Discard
RR	Round-Robin

1

Introduction

In the last decades, the progress of computers and communication network technology has a vital impact on the applications of control systems. However, due to the increasing complexity of the controlled plants, the control systems start to develop toward a more decentralized and intelligent direction, accordingly, the traditional point-to-point structure of control systems has been replaced by the private or public networks which can undertake the main transmission task through the components including sensors, controllers and actuators that distributed in different areas. Generally, when there exists a closed-loop control via a communication channel, the controlled systems can be called networked control systems (NCSs). NCSs have many merits over traditional control systems, such as less modularity, easy maintenance, low cost and high reliability. In virtue of its practicability and validity, NCSs are implemented in various fields, for example, car automation, experimental facilities, medical industry, domestic robots, space exploration, aircraft, automobiles, military systems and so on. Particularly, it is worth noting that virtually most of the practical engineering systems are time-varying, stochastic and nonlinear due to that the system parameters are definitely dependent on time and there exist random influence factors from the outer environment/network burden. However, such time variations of parameters, stochasticity and nonlinearity may result in considerable complexities in the system analysis and synthesis. For this reason, the networked time-varying systems (NTVSs) have become an increasingly hot research field.

During the analysis process of NCSs in real networked environments, there is another issue that cannot be ignored, that is, the frequently encountered and unavoidable incomplete information phenomenon which would seriously degrade the system performance. Two sorts of factors would bring about this phenomenon strikingly. First, the inherent characteristics of the NCSs such as large scale, multiple components, high complexity, intermittent sensor breakdowns, network jamming or data transmission in noisy surroundings and wide distribution may lead to higher probabilities of the occurrence of insufficient measurements in networked systems than in traditional systems. Second, since the network is a dynamic system, its quality of service depends on many factors, for example, bandwidth capacity, cable quality and so on. If the phenomena such as cable aging, interface failures, limited bandwidth and

DOI: 10.1201/9781003189497-1

network congestion arise, they will bring negative influences on the application service of networked systems.

It is known to all, as the fundamental problems, filter/estimator/controller design problems have received considerable attention in signal processing and control engineering. On another research aspect, it is familiar to us that our methods and technologies used in classical or modern control system design are almost under the ideal situation that all system actuators and sensors are in good working conditions. Nevertheless, a number of control systems designed by utilizing traditional techniques are unable to satisfy the required performance in the presence of actuator faults, sensor faults and component faults. Even worse, faults would appear at any time in the actual operation of the system, which may generate abnormal production and even make the entire production process stop. Consequently, we must improve the reliability and security of the system to ensure the production process to operate in a safe, reliable and efficient manner. Based on the issues above, the essential problem to be settled first is estimating the shapes and sizes of faults. After that, in order to maintain the ideal performance, the reliable control is necessary especially in safety-critical systems which should tolerate failures in system components, such as military space mission, aircrafts, nuclear reactors, etc.

Sensor networks (SNs) have already become an ideal research area for control engineers, mathematicians and computer scientists to manage, analyze, interpret and synthesize functional information from real-world SNs. For distributed estimation/filtering problems, each individual sensor in a sensor network locally estimates/filters the system state not only from its own measurement but also its neighbouring sensors' measurements according to the given topology. The possible complexity of such a topology poses many challenges for scientists and engineers, and it is difficult to analyze these networks thoroughly with currently available estimation/filtering algorithms. Therefore, there is an urgent need to research on modelling, analysis of behaviours, systems theory, estimation and filtering in SNs. Moreover, it should be noted that the phenomena of incomplete measurements and nonlinearities might occur more seriously and ubiquitously in the practical complex networks (CNs) due to their complicated inter-node coupling and communication topologies. As a result, it is of great significance to intensively investigate the impact from incomplete measurements and nonlinearities on the behaviour of CNs.

In this chapter, on one hand, the background of filtering/fault estimation/reliable control is introduced for nonlinear stochastic time-varying systems with different kinds of network-enhanced complexities including fading channels, missing measurements and redundant channels, randomly occurring gain variations, randomly occurring sensor saturations, state saturations, measurement quantizations, stochastic nonlinearities, actuator and sensor failures, randomly occurring faults, randomly occurring nonlinearities, state-multiplicative noises, the event-triggered mechanism, communication protocol, randomly occurring uncertainties, communication time-delays, randomly varying topologies, stochastic inner couplings, random coupling strengths and so on. The models of nonlinear stochastic time-varying systems include the general NCSs, SNs and CNs. Problems of event-triggered H_∞ filtering/control, set-membership/recursive filtering, H_∞ state/fault estimation, partial-nodes-based state estimation, recursive fault estimation, variance-constrained control/estimation as well as reliable control are analyzed. On the other hand, the outline of the book is shown explicitly.

1.1 Background

1.1.1 Nonlinear Stochastic Time-Varying Systems

In recent decades, it has been known that virtually most of the practical engineering systems are time-varying as the complexity of industrial systems grows rapidly and the parameters of the real-time systems are definitely dependent on time. Then, the time-varying nature has gradually become an indispensable means of reflecting the fast changes in system dynamics, however, such time variations of parameters may result in considerable complexities in the system analysis and synthesis. For this reason, the networked time-varying systems have become an increasingly hot research field. It is worth noting that, when dealing with such time-varying systems, one would be more interested in their transient performances over a finite period than the traditional steady-state behaviours over the infinite horizon.

It is known that nonlinearity is a ubiquitous feature in a large class of practical systems and, if not properly coped with, the nonlinearity would inevitably deteriorate the operating efficiency or even cause the instability of the plant. On the other hand, as the network scale increases, the randomly occurring phenomena (ROP) have emerged. ROP refer to those phenomena that appear intermittently in a random way based on certain probability law, such as randomly occurring nonlinearities [42, 151], randomly occurring uncertainties [79], randomly occurring sensor delays [42] and randomly occurring sensor saturations [186]. For several decades, nonlinearity and stochasticity serve as two of the most active research topics in systems that have found successful applications in a variety of engineering systems such as automotive engines, robot manipulators, aircraft and electrical motors. Thus, it is meaningful to take nonlinear stochastic time-varying systems into consideration and carry out the relative analysis and synthesis.

SNs have gradually been undergoing a quiet revolution in all aspects of the hardware implementation, software development and theoretical research. The inherently asynchronous sensor network is comprised of numerous sensor nodes with computing and wireless communication capabilities, where the nodes are spatially distributed to form a wireless ad hoc network and every node has its own notion of time. Specifically, SNs do possess their own characteristics due mainly to the large number of inexpensive wireless devices (nodes) densely distributed and loosely coupled over the region of interest. The past decade has seen successful applications of SNs in many practical areas ranging from military sensing, physical security, air traffic control, to distributed robotics and industrial and manufacturing automation. Accordingly, theoretical research on SNs has gained an increasing attention from multiple disciplines including engineering, computer science and mathematics. For example, plenty of fundamental questions have been addressed about the connections between sensor network topology and dynamic properties including stability, controllability, robustness and other observable aspects [9, 10, 12, 93, 94, 126, 160]. However, some major problems have not been fully investigated, such as the behaviour of stability, estimation and filtering for SNs with incomplete/imperfect/stochastic topology under the communication protocol, as well as their applications in, for example, distributed signal processing.

The complex network consists of many network nodes, in which the outer coupling refers to the interaction of different nodes between each other and the inner coupling represents the inner connection strength inside a single node. The past two decades have witnessed an ever-increasing research interest in CNs due primarily to their explicit insights in describing a wide range of practical systems with complex structures, network topology evolution, junction and node diversity, dynamic complexity and multiple merging of complexity. Up to now, a rich body of literature has been reported on dynamics analysis problems for CNs, see [3, 13, 26, 34, 70, 99, 117, 135, 141, 146, 192, 196]. Among others, the state estimation problem has drawn particular research attention as the network states are often immeasurable directly from incomplete network outputs especially for large-scale networks [1, 85, 91, 103, 113]. For example, the state estimation problem has been investigated in [34] for the discrete time-delay nonlinear CNs with randomly occurring sensor saturations (ROSSs) and randomly varying sensor delays (RVSDs).

1.1.2 Network-Enhanced Complexities

In this subsection, we will first analyze the factors that influence the complexity of the networked systems, and then we will point out several challenging problems.

Fading Channels

Due to the rapid development of network technologies, network-induced phenomena such as packet dropouts [140, 183], communication delays [56] and signal quantization [82] have been well studied for filtering and control problems of networked systems. However, the network-induced channel fading problem has received little attention despite its practical significance in wireless mobile communications. Generally speaking, the main causes for fading effects are the multi-path propagation and the shadowing from obstacles, which are widely regarded as a kind of channel unreliability described by a random process reflecting the random changes of amplitude and phase of the transmitted signal. If not dealt with adequately, the phenomenon of network-induced channel fading would inevitably deteriorate the filtering performance of systems under investigation. To date, some pioneering work has appeared in the literature concerning NCSs with fading channels, see [32] and the references therein.

Missing Measurement and Redundant Channels

Due to the wireless communication nature and the low cost of sensors, the communication within a network is often unreliable and the missing measurement (if the measurement is missing entirely, then it is called packet loss or packet dropout) is among the most frequently encountered phenomena that should be taken into serious consideration in the distributed state estimation problem. It has been recognized that the dropout rate of the packet transmissions has a negative impact on the estimation performance and, to this end, a few algorithms have been proposed to ensure certain robustness of the distributed state estimators against the random packet dropouts obeying a known probability distribution law (see [44], [102]). Another active way of improving the communication quality is to introduce multiple/redundant channels for the signal transmission, therefore decreasing the overall probability of packet dropouts. In fact, the redundant channel communication protocol has been proposed in [62] to offer some suboptimal choices for information transmission when the primary channel fails and, since then, such a protocol has been widely exploited when the reliability (rather than resource) becomes the main concern (see [37], [221]).

Randomly Occurring Gain Variations

In the course of physical implementation, it is often the case that the digital filters/estimators suffer from a certain degree of parameter drifts/variations/perturbations. Most existing design methods, however, are dependent on the essential assumption that the gains/parameters of the estimator can be accurately realized. Obviously, ignorance of such parameter drifts would probably cause performance degradation or even instability to the underlying systems. As such, it becomes a significant issue to design *non-fragile* filters/estimators so as to tolerate the possible gain variations. Recently, a rich body of literature has appeared on the non-fragile filter/estimator design problems. For instance, in [22,202], non-fragile Kalman filters and non-fragile H_∞ filters have been designed, respectively, to mitigate the effects of finite word length on filter implementation. In [215], non-fragile distributed H_∞ filtering problems have been investigated for a class of discrete-time

Takagi-Sugeno systems over SNs. In addition, due to unpredictable network-induced fluctuations and circumstance changes, the variations of the estimator parameters may occur in a random way, and such a phenomenon is referred to as the randomly occurring gain variations (ROGVs). Accordingly, it is of crucial importance to investigate the influence of ROGVs on the estimator/filter/controller performance.

Randomly Occurring Sensor Saturations

Due to physical and safety constraints, the sensor saturation is probably one of the most commonly encountered phenomena in practical control systems that can severely degrade the system performance or even lead to unstable behaviours. So far, considerable research attention has been paid to the filtering and control problems for systems with sensor saturation, see [58,65,194,201] and the references therein, where the saturation has been implicitly assumed to occur definitely, i.e., sensor always undergoes saturation. Such an assumption, however, is not always true. For example, in a networked environment, the sensor saturations may occur in a probabilistic way where the saturation amplitude/intensity may be randomly changeable. Such kind of randomly occurring sensor saturations may result from network-induced intermittent sensor failures, sensor aging or sudden environment changes [34].

State Saturations

As is well known, the state saturation is a typical kind of nonlinearities that occurs frequently in engineering practice due to some unavoidable physical constraints (e.g. limits on power, capacity and amplitude) of the system devices. Saturations could have a notable impact on the dynamical behaviours that would degrade the system performance or even jeopardize the system stability. For state-saturated system, all the system states are constrained within a given range and the conventional system analysis/synthesis approach is *no longer* applicable. As such, dedicated efforts ought to be sought to look into the influence from the saturations. Recently, the filtering problems for state-saturated systems have been investigated in [35, 154, 190, 191]. For instance, the H_∞ filtering problem has been investigated in [35] for a class of state-saturated discrete systems subject to randomly occurring nonlinearities and successive packet dropouts. In [190, 191], the recursive filtering problems have been dealt with for systems with state saturations over wireless SNs where fading/missing measurements, quantization effects and randomly occurring nonlinearities have all been taken into special account.

Measurement Quantizations

In NCSs, quantizer is a device or algorithmic function for the sake of processing the signals which require to be quantized before transmission [15, 40]. Quantization is a process which converts a real-valued signal into a piecewise constant one taking on a finite set of values by the quantizer. There are several reasons which would impact the system behaviour caused by quantization. For one thing, if the signal is out of the range of the quantizer, then the control law designed for the ideal case may lead to instability of the system. The other one is the deterioration of performance near the equilibrium: it is worthwhile to mention that if the distinction between the current and the desired values of the state is small, and higher precision is necessary, then asymptotic convergence performance of the system is unacquirable in this condition. Recently, some methods have been demonstrated in [48, 53, 142, 210, 211] to deal with the quantization problems. Up to now, much efforts have been devoted to solve the filtering and fault detection problems for networked systems with signal quantization, and some effective algorithms have been reported in [82, 96, 209]. For example, in [96], the fault detection problem has been addressed for NCSs with Markovian packet dropouts as well as quantizations.

Stochastic Nonlinearities

Stochastic nonlinearities are often found in NCSs where the nonlinearities are induced by randomly fluctuated network loads due mainly to the communication limitations. Particularly, as pointed out in [127, 208], the statistical description of stochastic nonlinearities could cover several classes of well-studied nonlinearities in stochastic systems such as (1) linear system with state- and control-dependent multiplicative noise; (2) nonlinear systems with random vectors dependent on the norms of states and control input; and (3) nonlinear systems with a random sequence dependent on the sign of a nonlinear function of states and control inputs. Nevertheless, despite the above advantages of such a nonlinearity description, the analysis and synthesis problems for time-varying systems with stochastic nonlinearities need to be further investigated, which motivates some work in this book.

Actuator and Sensor Failures

It is noticeable that, in almost all the aforementioned literature, the components of the control systems have been implicitly assumed to be fully reliable. This assumption is, unfortunately, not always true since the failures of control components (e.g. sensors and actuators) often occur in practical applications due to a variety of reasons, for example, the abrupt changes of working conditions, the erosion caused by sever circumstance, the internal component constraints, the intense external disturbance and the aging of sensors or actuators, etc. Therefore, it is of both practical significance and theoretical importance to design a reliable controller against possible actuator and sensor failures such that the essential performance of the controlled system can be guaranteed. In fact, in the past two decades, the problem of reliable controller design has attracted much research attention and many approaches have been proposed in the literature including Hamilton-Jacobi equation approach [105,112], robust pole region assignment technique [67,223], algebraic Riccati equation approach [158, 203] and linear matrix inequality approach [24, 121]. Despite the fruitful results on *time-invariant systems* over an *infinite horizon*, it is worth pointing out that the *finite-horizon* reliable control problem for *time-varying* systems has not been thoroughly investigated yet, not to mention the case complicated further by nonlinearity and stochasticity.

Randomly Occurring Faults

It is worth mentioning that, in the existing literature concerning finite-horizon fault estimation problems, it has been implicitly assumed that the occurred fault signals are instantaneous, that is, the actuator/sensor faults occur in a deterministic way. Such an assumption, unfortunately, is not always true. For example, in a networked control system, due to the bandwidth limitation of the shared links as well as the unpredictable variation of the network conditions, a number of network-induced intermittent phenomena (including electromagnetic interference, severe packet loss, data collision or temporary failure of the sensors/actuators) could be regarded as different kinds of faults when the reliability becomes a concern. Obviously, in terms of the random nature of the network load, these kinds of intermittent faults could be better modelled as randomly occurring faults (ROFs) whose occurrence probability can be estimated via statistical tests. In other words, the network-induced ROFs are typically time-varying and would act in a probabilistic fashion.

Randomly Occurring Nonlinearities

Nonlinear control has been a mainstream of research topics due primarily to the fact that nonlinearity is a ubiquitous feature in a large class of practical systems and, if not properly coped with, the nonlinearity would inevitably degrade the system performance or even lead to the instability of the controlled systems. As discussed in [39, 40, 187], in today's popular networked systems such as the internet-based three-tank system for leakage fault diagnosis, the occurrence of nonlinearities is often of *random nature* resulting from sudden environment changes, intermittent transmission congestion, random failure and repairs of components, etc. Accordingly, the so-called randomly occurring nonlinearities (RONs) have started to gain some research interest and several initial results have been reported on the filtering problems subject to *additive noises*, see e.g. [33, 150].

State-Multiplicative Noises

Note that many plants may be modelled by systems with *state-multiplicative noises* and some characteristics of nonlinear systems can be closely approximated by models with multiplicative noises rather than by linearized models. To be specific, the mutually uncorrelated zero-mean Gaussian sequences can be exploited to regulate the phenomena of state-dependent noises and depict the probabilistic ways that the parameter system matrices enter the system. Furthermore, it is recognized that the introduction of state-multiplicative noises into the system model could better reflect the engineering practice in networked environments. However, in the context of nonlinear finite-horizon H_∞ filtering and control for the stochastic systems with exogenous disturbance, the results on state-multiplicative noises have been very few, and this constitutes partial motivation for the book.

The Event-Triggered Mechanism

In NCSs, an important issue is how to transmit signals more effectively by utilizing the available but limited network bandwidth. To alleviate the unnecessary waste of communication/computation resources that often occurs in conventional time-triggered signal transmissions, a recently popular communication schedule called event-triggered strategy has been proposed in [45, 64, 72, 83, 131, 132, 162, 180, 214]. The triggering mechanism refers to the situation where the measurement output is transmitted to a remote controller/filter only when certain conditions are satisfied. In other words, a constant measurement signal is maintained until a specified event condition is violated in an event generator. In comparison with the conventional time-triggered communication, a notable advantage of the event-triggering scheme is its capability of reducing redundant transmissions while preserving the guaranteed system performance. In recent years, increasing attention has been drawn on the event-triggered techniques for stochastic systems and many important results have been reported in the literature, see [73, 84, 125, 174]. However, it should be pointed out that, many established results referring to event-triggering schemes are in the framework of continuous-time systems and, when it comes to the discrete time-varying systems with a relative error-based event-triggering criterion, the corresponding results have been scattered.

Communication Protocol

In a typical resource-constrained environment, frequent data transmissions among elements would inevitably lead to data congestion/collision caused by the limited bandwidth of the transmission channels. In order to mitigate the side effects (e.g. data congestion and collision) from the network-induced phenomena and reduce the occupation rate of network resource, an efficient way adopted widely in industry is to employ the communication protocol so as to regulate which node can access the network to transmit signal at each time step. By introducing certain protocol (data scheduling strategy), only one node (or one component) is allowed to transmit data at each transmission instant for the large-scale complex network (or general NCSs). Then, the communication quality/accuracy/efficiency is improved and the undesired network-induced phenomena are further alleviated such as packet dropouts and communication delays.

It is noteworthy that SNs are composed of abundant sensor nodes scattered around the target system. Thus, the amount of measurements from the sensors is voluminous. Considering the limited bandwidth in the networked system, such huge pressure on the data transmission may cause packet dropout, nonlinearity, channel fading, and then result in negative effects on the filtering (or estimation) performance. In order to tackle this challenge, similarly, the communication protocol is utilized.

Some well-known scheduling protocols include the Round-Robin (RR) protocol [171,197], the Weighted Try-Once-Discard (WTOD) protocol [46,108] and the random access protocol [116]. To be specific, the Round-Robin protocol conforms to a novel equal communication rule for measurements to arrange when the information could be delivered from the sensor nodes to the filter/estimator. As one of the most popular communication rules, the Round-Robin mechanism has found widespread utilizations in computing, communication, the load adjustment of server and so on [119,228]. The WTOD protocol is regarded as the most effective one since the 'most needed' node is always selected to transmit according to certain prescribed selection principle, see e.g. [229]. By utilizing the random access protocol, one node (or one component or one sensor) is chosen stochastically to access the communication network to transmit data at each time step [110, 227].

Randomly Occurring Uncertainties

As is well known, modelling errors (usually parameter uncertainties) constitute an important kind of complexities for system modelling that has a great impact on the subsequent system analysis and design [74]. In today's networked systems, it is quite common that the network load is randomly fluctuated and the signal transmission suffers from unpredictable networked-induced phenomena owing to limited bandwidth. As such, the occurrence of the parameter uncertainties in a networked environment is often of random nature resulting from abrupt phenomena such as modification of the operating point of a linearized model of nonlinear systems, random failures and repairs of the components, changes in the interconnections of subsystems and sudden environment disturbances, etc. Very recently, in [79], the concept of randomly occurring uncertainties (ROUs) has been introduced to reflect the probabilistic fashion of the network-induced modelling error and the corresponding sliding mode control problem has been considered. However, when it comes to the finite-horizon reliable H_∞ control problem involving ROUs for nonlinear time-varying stochastic systems, the related results are very few and the situation is even worse when measurement quantizations, actuator and sensor failures are also taken into account.

Communication Time-Delays

In nowadays popular NCSs, due to the limited bandwidth of communication channels, frequent data exchanges/transmissions would unavoidably lead to network-induced phenomena such as packet dropouts, communication delays and signal quantization. Communication time-delay is one of the main factors contributing to the complexities of analyzing networked systems. In the past few decades, the effects from various time-delays on the complex systems have been thoroughly investigated and a rich body of literature has been available, see e.g. [34] for the H_∞ state estimation problem with RVSDs [134] for the robust adaptive synchronization with multiple time-varying coupled delays, [124] for the pinning control approach with mixed time-delays, and [224] for the distributed adaptive perimeter control scheme with delayed state interconnections reflecting data processing and communication delays. The types of communication time-delays include, but are not limited to, randomly occurring multiple delays, mixed time-delays (the constant delay and the distributed delay) and time-varying delays.

Randomly Varying Topologies

It is now well recognized that the coupling among different nodes of the CNs has a great impact on the performance of the CNs [179]. The topology structure of CNs is configured to set the physical layout and determine how the various nodes are connected via the transmission media. Usually, the topology of CNs is characterized by the so-called outer-coupling configuration matrix. In practical engineering, the connection topology of CNs might be randomly changeable as a result of the diversities in the linking structures and spatio-temporal evolution (e.g. the link failures/construction or new addition/elimination of network nodes in wearable SNs and virus spreading networks). Consequently, it is necessary to consider the impacts from the unfixed topologies when handling the dynamical behaviours of the CNs. Up to now, several frequently studied time-varying topologies include the switching topology [100, 159], the randomly varying topologies [38], the switching disconnected topology [28] and the semi-Markov jump topology [152]. As is known, the Markovian jumping parameters could be well suited to model such randomly varying topologies, since MJPs can describe the sudden occurrence of random failures and repairs of the components, changes in the interconnections of subsystems, unpredicted environment changes and so on [111,136,177].

Stochastic Inner Couplings

In addition to the topology that describes the connection between neighbouring nodes, the inner coupling matrix characterizing the inner linking strength should also be taken into consideration. The coupling strengths, however, are sometimes partially unknown and stochastically fluctuated due to the unavoidable modelling errors in either the structure or the coupling scheme of the CNs [206]. It should be emphasized that three special types of couplings (i.e. the stochastic coupling, the uncertain coupling and the quantized jumping coupling) have been investigated, respectively, in [163,175,196]. In [146], the uncertainties have been considered in the inner coupling matrix of the CNs by means of an interval matrix method. A thorough literature review has revealed that, despite its practical insights, the state estimation problem for CNs with both stochastic inner coupling and randomly varying topologies has not been adequately studied yet, and this constitutes a crucial research topic to be investigated.

Random Coupling Strengths

The couplings among nodes for CNs are usually described by constant matrices, see e.g. [118,133]. It has been pointed out in [176] that, in engineering applications, the coupling strengths might be subject to random yet limited fluctuations due primarily to the influence of environmental factors and, therefore, it makes practical sense to research into the issue of *random coupling strengths* for CNs, which has been recently paid some initial attention. For example, in [130], the randomly coupled network of Kuramoto oscillators has been studied where the coupling strengths are governed by a uniform random distribution within a specified domain. In [99], a state estimator has been designed for CNs with random coupling strengths obeying the similar distribution as in [130], where the selection method of coupling strengths is similar to that of [130].

1.1.3 Filter Design, Fault Estimation and Reliable Control

In recent years, theoretical and practical research on time-varying networked systems has gained an increasing attention from multiple disciplines including engineering, computer science and mathematics. Lying in the core part of the area are the filtering, state/fault estimation and reliable control problems that have recently been attracting growing research interests. In particular, an urgent need has arisen to understand the effects of network-enhanced complexities on filtering, state/fault estimation and reliable control performance in networked systems.

Robust/H_∞ Filter Design

The past several years have witnessed the rapid progress and extensive applications of filtering in the real world, such as in spacecraft, navigation, digital image processing, remote sensing technology, signal denoising, target tracking and industrial monitoring, where the Kalman filter is widely deployed. It is worth mentioning that the Kalman filtering approach can be executed only under the assumption that all noise terms and measurements own known distributions and an accurate knowledge of the underlying linear system model is available. However, it is hard to provide the ideal condition under the effects of measurement noises, modelling errors, parameter uncertainties and external disturbances. In this case, the robust/H_∞ performance of the networked systems has been paid adequate research attention due to its engineering significance. To mention a few, in [52], considering the linear time-varying systems, the robust non-fragile filtering problem has been addressed with multiple packet dropouts and a locally optimal filtering algorithm has been established. Specially, in the minimum mean-square error sense, a globally optimal filtering scheme has been proposed in [97] based on the result in [52]. In view of the missing measurements and quantization effects, the effective H_∞ filtering scheme has been presented in [183]. In addition, the filtering problems for nonlinear time-varying networked systems have come into our vision in parallel to the linear systems. For example, in [33] and [66], the H_∞ filters have been constructed respectively for discrete time-varying networked systems with randomly occurring nonlinearities and fading measurements and with time-varying delays, and a further result of a probability-guaranteed H_∞ finite-horizon filtering method has been reported in [79] for a class of time-varying nonlinear systems with sensor saturations.

Recursive Filter Design

In recent years, due to the inevitable nonlinearity problem of practical systems, the analysis and synthesis of nonlinear systems have become a very active research topic and some results have been published [78, 81]. Note that the traditional Kalman filtering theory may not satisfy the required performance in the case that the system model is nonlinear even along with uncertainties. For the sake of improving its abilities of handling nonlinearities and uncertainties, many optional methods have been explored in the literature including the robust extended Kalman filter [81, 89], H_∞ filtering [33,66,79]), etc. Except for the methods above, the recursive filtering approach which can deal with this kind of problems has stirred an increasing research interest, and some latest results can be given in [52, 78, 81, 82] and the references therein. In [81], the recursive filter has been designed for time-varying nonlinear systems encountering probabilistic sensor delays and finite-step correlated measurement noises. Furthermore, the recursive filter has been constructed for nonlinear time-varying networked systems with multiple missing measurements or quantization measurements. For example, the recursive filtering problem has been developed in [78] for the nonlinear system with random parameter matrices, correlated noises and multiple fading measurements.

In recent years, the state estimation problems for CNs have received considerable research attention, where the adopted methodologies mainly include the linear matrix inequality approach for time-invariant CNs and recursive matrix equation approach for time-varying CNs, see [99,137,146,178] for some representative works. In the context of time-varying CNs, the recursive filtering problem has recently attracted a particular research interest due mainly to its advantages in online implementations. One of the challenging tasks that we would have to face in the recursive filtering problem is how to analyze the exponential boundedness of the filtering error in the mean square. However, when mentioning parameterizing an upper bound on the filtering error covariance matrix and subsequently minimizing through designing the filter gain at each time step, there have been few relevant results, which need to be further investigated.

Distributed Filter Design

Over the past ten years or so, SNs have proven to be a persistent focus of research attracting an ever-increasing attention in the areas of systems and communication. A typical sensor network is composed of a large number of spatially distributed autonomous sensor nodes, where each sensor has wireless communication capability as well as some level of intelligence for signal processing and for disseminating data [9, 10, 12, 93, 94, 126, 160]. The development of SNs is originally motivated by military applications such as distributed localization, power spectrum estimation and target tracking problems. With recent intensive research in this area, SNs have a wide-scope domain of applications in areas such as environment and habitat monitoring, health care applications, traffic control, distributed robotics and industrial and manufacturing automation [9, 10, 12, 23, 161].

As one of the most fundamental collaborative information processing problems, the distributed filtering or estimation problem for SNs has gained particular concerns from many researchers and a wealth of literature has appeared on this topic, see e.g. [17–19, 43, 50, 104, 148, 156, 164, 212] and the references therein. For distributed filtering problems, the information available on an individual node of the sensor network is not only from its own measurement but also from its neighbouring sensors' measurements according to the given topology. As such, the main difficulty in designing distributed filters lies in how to cope with the complicated coupling issues between one sensor and its neighbouring sensors and how to reflect such couplings in the filter structure specification.

It is worth noting that most reported results concerning the distributed filtering/estimation algorithms are for linear and/or deterministic systems. Since nonlinearities are ubiquitous in practice, it is necessary to consider the distributed filtering problem for target plants described by nonlinear systems. On the other hand, distributed filtering in a sensor network inevitably suffers from the constrained communication and computation capabilities that would degrade the network performances. The occurrence of incomplete information in SNs is more complex and severer due primarily to the network size, communication constraints, limited battery storage, strong coupling and spatial deployment.

Since the results of filter design for nonlinear stochastic time-varying NCSs with incomplete information are rarely reported compared with the linear ones, it provides guiding references for future research.

Set-Membership Filter Design

The set-membership filter can provide a set of ellipsoidal sets that contains all possible actual states with 100% confidence, where such design is in accordance with the engineering requirement. Obviously, distinct from other existing filtering strategies, the set-membership filtering scheme stands out for two prominent features: (1) it requires a hard bound constraint rather than the statistical description on the system noises; and (2) it is capable of providing a set of ellipsoidal sets that contains all possible actual states with 100% confidence. In view of the aforementioned advantages, the set-membership filtering has broad range of applications including robot localization and map building, neural network training, fault diagnosis and path tracking. So far, significant advances have been made in the analysis and synthesis of the set-membership filters [11, 69, 200]. For example, the problem of set-membership filtering has been considered in [200] for discrete-time systems with nonlinear equality constraint among their state variables.

Partial-Nodes-Based Estimator Design

In engineering practice, it is sometimes impractical (or even impossible) to *collect* the measurements from *all* of the nodes in CNs for a variety of reasons such as the limited transmission ability of sensors, harsh working environments, large spatial distribution of the network nodes, and budget constraints. In this case, the measurements from only *partial* nodes are available for the dynamics analysis (e.g. state estimation) of the overall network, and the resulting partial-nodes-based (PNB) analysis problem is of great importance for CNs and has recently gained much research attention [106, 114, 115]. For instance, in [106], on the basis of the extended Kalman filtering scheme, the recursive state estimator has been designed for an array of stochastic CNs with measurement signals from merely a part of nodes, which can optimize the upper bound of the filtering error covariance.

Fault Estimation for Time-Varying Systems

In practical engineering, faults are undesirable deviations of system parameters from normal states, which are caused by unexpected model uncertainties, time delays, disturbances, perturbations and noises. As the unacceptable deviation will prevent the control system from achieving the desired performance, the fault detection and isolation (FDI) issues are of great significance and have received a wide range of attention, see, e.g. [29, 47, 77, 88, 120, 129, 143, 189, 205, 209, 225]. However, the difficulty encountered is to get the accurate size of the fault from an FDI scheme [109]. Therefore, the fault estimation issue is introduced to derive the size and shape of the faults and reconstruct the fault signals so as to perform the required fault detection automatically [86].

The key of fault estimation is to design a fault estimator. Regard the input signal and measurable output signal as the inputs of the estimator, and the output signal is the reconstructed fault estimation signal. Moreover, the reconstructed signal can be used as the input signal of the fault-tolerant controller to improve the fault-tolerant performance of the system. Compared with the fault diagnosis method based on residual generation, the fault estimation one can better represent the severity of the fault. In recent years, many researchers in the fields of FDI have been devoted to explore effective fault estimators to obtain the on-line accurate fault information, and plenty of important results have been published. There exist some common methods of fault estimation, including adaptive observer [98], neural network [226], T-S observer [153], robust Kalman filter [82], sliding mode observer approach and H_∞ optimization method [145, 198, 199, 217, 218], and some other excellent fault estimation approaches have been shown in the literature [129, 139, 207, 219, 220].

In regard to the fault estimation problem for time-varying systems, some efforts have been made to design fault estimators on a finite time-horizon accompanying with incomplete information. For example, the estimation problems of randomly occurring faults over a finite horizon have been handled in [39] for systems with different sources of disturbances via the recursive Riccati difference equation approach, while the recursive matrix inequality method has been utilized in [41] to deal with the estimation problem. As is well known, the nonlinearities and uncertainties may make the system modelling complex, therefore, it is necessary to tackle them carefully when analyzing the complex dynamical systems under increasing performance requirements [7, 31, 32, 39, 123]. For instance, the fault signal has been estimated in [39] for a class of time-varying systems with stochastic nonlinearities, and it can be seen that a novel performance requirement against different sources of disturbances has been introduced. Especially mentioned, for the purpose of receiving less conservative results as well as computationally attractive algorithms, the Krein-space approach has been introduced to deal with the finite-horizon fault estimation problems and this method has been proven to be an effective tool to solve the filtering problems with the performance index described by an indefinite quadratic form. To list a few, in [225], the H_∞ fault filter has been constructed for linear discrete time-varying systems by using the Krein-space theory. Moreover, in [144], the measurement delays have been taken into consideration, and a finite-horizon H_∞ fault estimator has been obtained with the help of the Krein-space theory. The robust H_∞ fault estimation problem has been addressed in [145] in the framework of Krein spaces. Unfortunately, up to now, the fault estimation problem for nonlinear stochastic time-varying systems has gained very little research attention despite its practical importance.

Reliable Control for Time-Varying Systems

In practical control systems especially NCSs, intolerable system performance will appear due to a variety of reasons (changes of working conditions, sensors or actuators aging, zero shift, electromagnetic interference and network disturbance) [213]. Therefore, it is practically crucial to design a controller to remain the stability of the system and ensure the system to run properly even the failure exists. Reliable control refers to that no matter there is a failure or not, a controller can always be designed to keep the system stable and meet certain performance requirements. In recent years, the development of control system is more and more complex, and thus enhancing the reliability has attracted public concerns.

Over the past several decades, different methods have been presented so as to satisfy various performance requirments. In [173], a methodology has been developed for the design of reliable centralized and decentralized control systems, which meets the H_∞ performance under the conditions not only when all control components are operational, but also when there are sensor or actuator outages in the centralized case or control channel outages in the decentralized case. After that, a procedure has been put forward for the design of reliable linear-quadratic state-feedback controller which guarantees the system stability and a known quadratic performance bound [172]. In addition, a method based on the Hamilton-Jacobi inequality approach has been presented for the design of reliable nonlinear control systems and the H_∞ performance has been taken into account in [204]. In terms of the linear matrix inequality approach, a method has been proposed for designing a reliable fuzzy controller in [25] in the sense of asymptotic stability, and the received controller is suitable for the systems whose control components are operating well or in fault.

Along with the methods proposed, a rich body of relevant literature subject to incomplete information or faults issues have been published, see e.g. [57,166,167,216]. The reliable H_∞ control problem has been investigated for discrete-time linear time-varying systems in view of the admissible infinite distributed delays and possible actuator failures in [188], ensuring that the closed-loop system is exponentially stable with a given disturbance attenuation level γ. In [166], the proposed reliable controller has been designed for NCSs which undergo probabilistic actuator fault, measurements distortion, random network-induced delay and packet dropout. Lately, as a newly emerged research topic, the finite-horizon reliable H_∞ output feedback control problem has been raised for a class of discrete time-varying systems in [40]. The main idea is to design a time-varying output feedback controller over a given finite horizon such that, in the simultaneous presence of randomly occurring uncertainties, randomly occurring nonlinearities and measurement quantization, the closed-loop system achieves a prescribed performance level. In [107], the reliable control problem has been settled for nonlinear systems with stochastic actuators fault and random delays via a T-S fuzzy model approach. It is worth mentioning that a novel state and sensor fault observer has been proposed in [57] to estimate system states and sensor faults at the same time, where the considered sensor fault may be in any form, even in the unbounded one. To the best of our knowledge, the reliable control for nonlinear time-varying complex systems with randomly occurring incomplete information has a few results, which still remains as a challenging research topic.

Variance-Constrained Control/Estimation for Time-Varying Systems

In the past few decades, there has been a surge of research interest in the stochastic control problem since stochastic modelling has been successfully applied in many fields. A large body of literature has been devoted to the stochastic control or filtering problem for different systems such as polynomial stochastic systems [6,8,75,76], linear retarded stochastic systems [60], Markovian jumping systems [68], switched stochastic systems [169,

193], discrete-time stochastic systems with state-dependent noises [222] and nonlinear stochastic systems [128, 168]. Among various stochastic control schemes, the covariance control (CC) theory has gained particular research attention due primarily to the fact that the performance requirements of many engineering control systems are naturally expressed as the upper bounds on the steady-state variances [4, 59, 90, 184]. It has been shown that the CC approach is ideally suited to handle the multi-objective design problems where the multiple objectives include, but are not limited to, variance constraints, H_2-norm specification, H_∞ performance index and pole placement [122]. The CC theory was originally developed for linear systems and has been recently extended to nonlinear stochastic systems [43, 122, 184]. It is worth pointing out that most results concerning the CC theory have focused on the steady-state behaviours for time-invariant systems over an infinite horizon. However, virtually almost all real-time control processes are time-varying especially when the noise inputs are nonstationary [36, 150, 183]. In such cases, it would make more sense to consider the covariance control problems for time-varying systems over a finite-horizon in order to provide a better transient performance.

On the other hand, in engineering-oriented state estimation problems, it is quite common that the performance requirements are expressed in terms of the upper bounds on the estimation error covariance [5, 16]. Rather than the optimal estimation minimizing the error covariance, the variance-constrained estimation provides a more practical yet flexible approach that aims to design acceptable estimators such that the corresponding error covariance satisfies the prespecified upper bound constraints [43]. Up to now, a lot of research attention has been paid to the variance-constrained problems, see, for instance, [20, 21, 30, 43] and the references therein. To be more specific, in [20, 21], the passivity-constrained fuzzy control and sliding mode fuzzy control problems have been discussed for nonlinear ship steering systems and continuous stochastic nonlinear systems subject to individual state variance constraints. The robust H_∞ finite-horizon filtering problem has been studied in [43] for a class of uncertain nonlinear discrete time-varying stochastic systems with multiple missing measurements, where both estimation error covariance constraint and a prescribed H_∞ performance requirement have been taken into consideration. As is well known, time-varying systems exist widely in practical engineering and, therefore, it is of vital importance to investigate the finite-horizon estimation problem for time-varying systems. Unfortunately, to the best of the authors' knowledge, in spite of their practical significance, the H_∞ state estimation problems for time-varying CNs have not been thoroughly investigated yet, let alone the consideration of the variance constraints on the estimation errors.

1.2 Outline

The organizational structure of this book is presented in Fig. 1.1 and the outline of this book is shown as follows.

- In Chapter 1, the research background is firstly introduced, which mainly involves nonlinear stochastic systems, SNs, CNs and problems of filtering, estimation and reliable control, then the outline of the book is listed.
- In Chapter 2, the event-triggered H_∞ filtering and variance-constrained H_∞ control problems are investigated for discrete time-varying systems. Based on the relative error with respect to the measurement signal, the event-triggered scheme is proposed with introducing an event indicator variable, which is to determine whether the measurement output is transmitted or not. A set of time-varying filters is designed for a class discrete time-varying systems with fading channels, randomly occurring nonlinearities and multiplicative noises, such that the influence from the exogenous disturbances onto the filtering errors is attenuated at the given level quantified by an H_∞-norm in the mean square sense. By utilizing stochastic analysis techniques, sufficient conditions are established to ensure that the dynamic system under consideration satisfies the H_∞ filtering performance constraint, and then a recursive linear matrix inequality (RLMI) approach is employed to design the desired filter gains. Moreover, another set of time-varying output feedback controllers is designed for a class of discrete time-varying systems with randomly occurring saturations, stochastic nonlinearities and state-multiplicative noises, such that, over a finite horizon, the closed-loop system achieves both the prescribed H_∞ noise attenuation level and the state covariance constraints. A recursive matrix inequality approach is developed to derive the sufficient conditions for the existence of the desired finite-horizon controllers, and the analytical characterization of such controllers is also given.
- In Chapter 3, we study the finite-horizon reliable H_∞ output feedback control problem for a class of discrete time-varying systems, where ROUs, RONs as well as measurement quantizations are considered. With the purpose of reflecting the reality, the deterministic actuator failures and probabilistic sensor failures are taken into account. Unlike the quantized actuator failure by a deterministic variable varying in a given interval, the quantization method of sensor failure is controlled by an individual random variable, whose value range is $[0,1]$. According to the known conditional probability of the Bernoulli distributed white sequences, both the nonlinearities and the uncertainties are entered into the system in a random manner. The main purpose of the problem is to design a time-varying output feedback controller in a given finite horizon, at the same

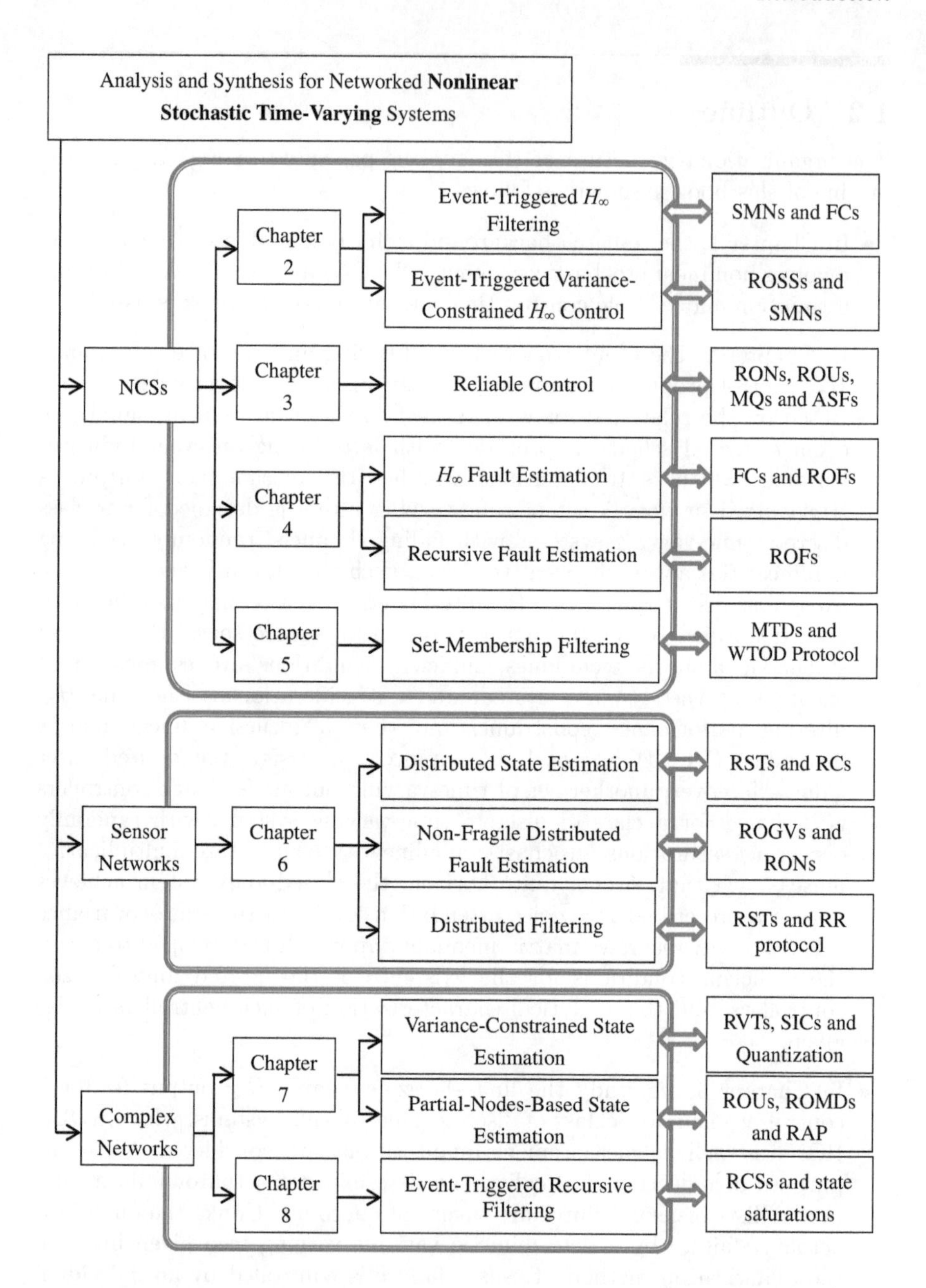

FIGURE 1.1
The structure of the book (the abbreviations are defined in the List of Acronyms).

time ROUs, RONs, actuator and sensor failures as well as measurement quantizations are all present, such that the closed-loop system can achieve the required performance level of H_∞-norm. The expected output feedback controller is designed through intensive stochastic analysis and recursive linear matrix inequality method.

- Chapter 4 is concerned with the finite horizon estimation of randomly occurring fault with fading channels. Firstly, the H_∞ estimation of ROFs with fading channels is discussed. The system model (dynamical plant) is subject to Lipschitz-like nonlinearities and a set of Bernoulli distributed white sequences is utilized to describe faults occurring in a random way. By utilizing the stochastic analysis techniques and a recursive linear matrix inequality approach, sufficient conditions for ensuring the previous performance constraints are obtained for fault estimation. Moreover, the finite horizon estimation problem of randomly occurring faults is investigated for a class of nonlinear systems whose parameters are all time-varying. Different from other results, statistical means are used to describe the stochastic nonlinearities entering the system, which can include several classes of well-studied nonlinearities. By using the completing squares method and stochastic analysis techniques, necessary and sufficient conditions are established for the existence of the desired finite horizon H_∞ fault estimator whose parameters are then obtained by solving coupled backward recursive Riccati difference equations (RDEs).

- Chapter 5 is to deal with the set-membership filtering problem for a class of time-varying state-saturated systems with mixed time-delays under the communication protocol. Under the WTOD protocol, only the sensor node with the largest measurement difference is allowed to access the shared communication network at each transmission instant. The purpose of the problem addressed is to design a set of set-membership filters such that, in the simultaneous presence of mixed time-delays, state saturation, WTOD protocol and bounded noises, the filtering error dynamics is confined to certain ellipsoid regions. A sufficient condition is derived to guarantee the existence of the desired set-membership filters by means of the solutions to a set of recursive linear matrix inequalities. Subsequently, an optimization problem subject to certain inequality constraints is put forward to acquire the minimized ellipsoid in the sense of matrix trace. A simulation example is presented to illustrate the effectiveness of the proposed filter design scheme.

- In Chapter 6, several distributed state estimation problems over SNs are studied. Firstly, we consider the distributed estimation problem for a class of time-varying nonlinear stochastic systems over SNs with redundant channels and randomly switching topologies. The switch topologies of sensor network are described by Markovian jumping parameters.

Considering the realities of data transmission, the quality of network communication is improved by using redundant channels. The purpose of the problem is to design a class of distributed state estimators so that the state estimation error dynamics satisfies the given average H_∞ performance constraint. Sufficient conditions are obtained through the stochastic analysis, and the desired state estimators are achieved by calculating a set of RLMIs. Secondly, the non-fragile distributed fault estimation problem for a class of time-varying systems subject to RONs and randomly occurring gain variations is also studied, and a set of parallel results is derived. Thirdly, the H_∞ filtering problem is focused on for nonlinear systems over SNs whose topologies are changeable subject to Round-Robin protocol in the finite horizon case. The Round-Robin communication strategy is employed to save the limited bandwidth and reduce the network resource consumption.

- In Chapter 7, we consider a new state estimation problem for nonlinear time-varying complex networks, where randomly varying topologies (RVTs), stochastic inner coupling and measurement quantization are turned over. The so-called stochastic disturbances in the inner coupling of complex networks are modelled by using a Gaussian random variable. Considering the distortion of signals phenomenon during transmission, the incomplete measurements and measurement quantization are used. A sufficient condition is first established in terms of a set of recursive linear matrix inequalities under which the expected estimation error and the desired H_∞ performance requirements are guaranteed. Then, the estimator gain parameters are derived. Besides, the robust finite-horizon state estimation problem is investigated for a class of time-varying complex networks under the random access protocol through available measurements from only a part of network nodes. The random access protocol is adopted to orchestrate the data transmission at each time step based on a Markov chain. The underlying complex networks are subject to randomly occurring uncertainties, randomly occurring multiple delays as well as sensor saturations. Sufficient conditions are provided for the existence of such time-varying partial-nodes-based H_∞ state estimators via stochastic analysis and matrix operations. Finally, two simulation examples prove the effectiveness of the estimator design methods.

- In Chapter 8, the recursive filtering problem is investigated for a class of time-varying complex networks with state saturations and random coupling strengths under an event-triggering transmission mechanism. The coupled strengths among nodes are characterized by a set of random variables obeying the uniform distribution. The event-triggering scheme is employed to mitigate the network data transmission burden. The purpose of the problem addressed is to design a recursive filter such that, in the presence of the state saturations, event-triggering communication mechanism and random coupling strengths, certain locally optimized upper bound is guaranteed on the filtering error covariance. By using the stochastic analysis technique, an upper bound on the filtering error covariance is first derived via the solution to a set of matrix difference equations. Next, the obtained upper bound is minimized by properly parameterizing the filter parameters. Subsequently, the boundedness issue of the filtering error covariance is studied. Finally, a numerical simulation example is provided to illustrate the effectiveness of the proposed algorithm.

- Chapter 9 summarizes the book and points out some possible research directions related to the work done in this book.

2

Event-Triggered Multi-objective Filtering and Control

In this chapter, we consider the event-triggered H_∞ filtering problems for discrete time-varying systems. First, the event-triggered H_∞ filtering problem is investigated for discrete time-varying systems with multiplicative noises. Based on the relative error with respect to the measurement signal, a scheme is proposed in order to determine whether the measurement output is transmitted or not. Some uncorrelated random variables are introduced, respectively, to govern the phenomena of state-multiplicative noises, randomly occurring nonlinearities as well as fading measurements. The purpose of the addressed problem is to design a set of time-varying filter such that the influence from the exogenous disturbances onto the filtering errors is attenuated at the given level quantified by a H_∞-norm in the mean square sense. It is shown that a RLMI approach is employed to design the desired filter gains. Second, based on the relative error with respect to the measurement signal, an event indicator variable is introduced and the corresponding event-triggered scheme is proposed. A set of unrelated random variables is exploited to govern the phenomena of randomly occurring saturations, stochastic nonlinearities and state-dependent noises. Finally, two illustrative examples are provided to demonstrate the effectiveness and applicability of the proposed filter design schemes.

2.1 Event-Triggered H_∞ Filtering with Fading Channels

In this section, a general event-triggered framework is developed to deal with the finite-horizon H_∞ filtering problem for discrete time-varying systems with fading channels, randomly occurring nonlinearities and multiplicative noises. By utilizing stochastic analysis techniques, sufficient conditions are established to ensure that the dynamic system under consideration satisfies the H_∞ filtering performance constraint, and then a RLMI approach is employed to design the desired filter gains.

DOI: 10.1201/9781003189497-2

2.1.1 Problem Formulation

In this section, we consider a discrete time-varying nonlinear stochastic system. The plant under consideration is of the following state-space model:

$$\begin{cases} x(k+1) = \big(A(k) + \sum_{i=1}^{r} w_i(k)A_i(k)\big)x(k) + \alpha(k)g(k, x(k)) \\ \qquad\qquad + D_1(k)v(k) \\ \tilde{y}(k) = C(k)x(k) + D_2(k)v(k) \\ z(k) = L(k)x(k) \end{cases} \tag{2.1}$$

where $x(k) \in \mathbb{R}^{n_x}$ represents the state vector; $\tilde{y}(k) \in \mathbb{R}^{n_y}$ is the process output; $z(k) \in \mathbb{R}^{n_z}$ is the signal to be estimated; $w_i(k) \in \mathbb{R} \sim \mathcal{N}(0,1)$; $v(k) \in l_2([0,N], \mathbb{R}^{n_v})$ is the disturbance/measure noise, in which $l_2([0,N], \mathbb{R}^{n_v})$ is the space of square summable mean vector-value functions $v(k)$ in interval $[0,N]$ with the norm $\|v(k)\|_{[0,N]} = \sqrt{\sum_{k=0}^{N} \mathbb{E}\{\|v(k)\|^2\}}$; $A(k)$, $A_i(k)$, $C(k)$, $D_1(k)$, $D_2(k)$ and $L(k)$ are known, real, time-varying matrices with appropriate dimensions.

The nonlinear vector-valued function $g : [0,N] \times \mathbb{R}^{n_x} \to \mathbb{R}^{n_x}$ is continuous, and satisfies $g(k,0) = 0$ and the following sector-bounded condition:

$$\big[g(k,x) - g(k,y) - \Phi(k)(x-y)\big]^T \big[g(k,x) - g(k,y) - \Psi(k)(x-y)\big] \leq 0 \tag{2.2}$$

for all x, $y \in \mathbb{R}^{n_x}$, where $\Phi(k)$ and $\Psi(k)$ are real matrices with appropriate dimensions.

The variable $\alpha(k)$ in (2.1), which accounts for the randomly occurring nonlinearity phenomena, is a Bernoulli distributed white sequences taking values on 0 or 1 with

$$\text{Prob}\{\alpha(k) = 1\} = \bar{\alpha}, \quad \text{Prob}\{\alpha(k) = 0\} = 1 - \bar{\alpha} \tag{2.3}$$

where $\bar{\alpha} \in [0,1]$ is a known constant.

In this section, we consider an unreliable wireless network medium utilized for the signal transmission. In this case, the fading channels become a concern and the actual measured output $y(k)$ is described by

$$y(k) \;=\; \sum_{s=0}^{l_k} \beta_s(k)\tilde{y}(k-s) + D_3(k)\xi(k) \tag{2.4}$$

where $l_k \triangleq \min\{l, k\}$ with l being the given path number, $\xi(k) \in l_2[0,N]$ is an external disturbance, and $\beta_s(k)$ $(s = 0, 1, \cdots, l_k)$ are the channel coefficients that are mutually independent random variables taking values on the interval $[0,1]$ with $\mathbb{E}\{\beta_s(k)\} = \bar{\beta}_s$ and $\text{Var}\{\beta_s(k)\} = \nu_s$.

For simplicity, we set $\{\tilde{y}(k)\}_{k\in[-l,-1]} = 0$, $C(k)_{k\in[-l,-1]} = 0$ and $\big[v^T(k) \quad \xi^T(k)\big]_{k\in[-l,-1]} = 0$.

Remark 2.1 *The Rice fading model (2.4), which is capable of accounting for channel fading, time-delay and data dropout simultaneously, has been widely utilized in the area of signal processing and remote control. Also, it can be seen from (2.1) that both the parameter system matrices $A_i(k)$ $(i = 1, 2, \ldots, r)$ and the nonlinear function $g(k, x(k))$ enter the system in probabilistic ways depicted by the random variable $w_i(k)$ and $\alpha(k)$, separately. As such, the system model described in (2.1)–(2.4) could better reflect the engineering practice in networked environments.*

For the purpose of reducing data communication frequency, the event generator is constructed which uses the previously measurement output to determine whether the newly measurement output will be sent out to the filter or not. In this section, such an event generator function $f(.,.)$ is defined as follows:

$$f(\sigma(k), \delta) \triangleq \sigma^T(k)\Omega\sigma(k) - \delta y^T(k)\Omega y(k) \tag{2.5}$$

where $\sigma(k) \triangleq y(k_i) - y(k)$ with $y(k_i)$ being the measurement at the latest event time k_i and $y(k)$ is the current measurement. Ω is a symmetric positive-definite weighting matrix and $\delta \in [0, 1)$ is the threshold.

The execution (i.e. the transmission of the measurement output to the filter) is triggered as long as the condition

$$f(\sigma(k), \delta) > 0 \tag{2.6}$$

is satisfied. Therefore, the sequence of event-triggered instants $0 \le k_0 \le k_1 \le \cdots \le k_i \le \cdots$ is determined iteratively by

$$k_{i+1} = \inf\{k \in \mathbb{N} | k > k_i, f(\sigma(k), \delta) > 0\}. \tag{2.7}$$

Accordingly, any measurement data satisfying the event condition (2.6) will be transmitted to the filter.

Remark 2.2 *Different from the traditional filtering problems, in this section, the event trigger is adopted in order to reduce the data communication frequency and network bandwidth usages. With the event trigger applied here, unnecessarily frequent transmission could be avoided when the change rate of the measurement signals is relatively small. Obviously, the set of event instants is only a subset of the time sequences, i.e. $\{k_0, k_1, k_2 \ldots, \} \in \{0, 1, 2, \ldots\}$. Note that, when $\delta = 0$, all the measurement sequences would be transmitted, and the problem addressed reduces to the traditional filtering one.*

For system (2.1), the following time-varying filter structure is proposed:

$$\begin{cases} \hat{x}(k+1) = A(k)\hat{x}(k) + \bar{\alpha} g(k, \hat{x}(k)) \\ \qquad\qquad - K(k)\Big(y(k_i) - \sum_{s=0}^{l} \bar{\beta}_s C(k-s)\hat{x}(k-s)\Big) \\ \hat{z}(k) = L(k)\hat{x}(k) \end{cases} \tag{2.8}$$

where $\hat{x}(k) \in \mathbb{R}^{n_x}$ is the estimate of the state $x(k)$, $\hat{z}(k) \in \mathbb{R}^{n_z}$ represents the estimate of the output $z(k)$ and $K(k)$ is the filter gain matrix to be designed.

By letting $e(k) \triangleq x(k)-\hat{x}(k)$, $\eta(k) \triangleq \begin{bmatrix} x^T(k) & e^T(k) \end{bmatrix}^T$, $\tilde{z}(k) \triangleq z(k)-\hat{z}(k)$, $\varpi(k) \triangleq \begin{bmatrix} v^T(k) & \xi^T(k) \end{bmatrix}^T$, $\bar{g}(k) \triangleq \begin{bmatrix} g^T(k,x(k)) & g^T(k,x(k)) - g^T(k,\hat{x}(k)) \end{bmatrix}^T$, $\tilde{\alpha}(k) \triangleq \alpha(k) - \bar{\alpha}$ and $\tilde{\beta}_s(k) \triangleq \beta_s(k) - \bar{\beta}_s$, we have the following augmented system to be investigated:

$$\begin{cases} \eta(k+1) = \mathcal{Y}_l(k) + \left(\sum_{i=1}^{r} w_i(k)\bar{A}_i(k) + \tilde{\beta}_0(k)\bar{C}_2(k) \right)\eta(k) \\ \qquad + \sum_{s=1}^{l} \tilde{\beta}_s(k)\bar{C}_2(k-s)\eta(k-s) + \tilde{\beta}_0(k)\bar{D}_2(k)\varpi(k) \\ \qquad + \sum_{s=1}^{l} \tilde{\beta}_s(k)\bar{D}_2(k-s)\varpi(k-s) + \tilde{\alpha}(k)S_1\bar{g}(k) \\ \tilde{z}(k) = \bar{L}(k)\eta(k) \end{cases} \tag{2.9}$$

where

$$\begin{aligned} \mathcal{Y}_l(k) &\triangleq \bar{A}(k)\eta(k) + \bar{\alpha}\bar{g}(k) + \sum_{s=1}^{l} \bar{\beta}_s\bar{C}_1(k-s)\eta(k-s) + \bar{K}(k)\sigma(k) \\ &\quad + \sum_{s=1}^{l} \bar{\beta}_s\bar{D}_2(k-s)\varpi(k-s) + \bar{D}_1(k)\varpi(k), \\ S_1 &\triangleq \begin{bmatrix} I & 0 \\ I & 0 \end{bmatrix}, \ \bar{D}_1(k) \triangleq \begin{bmatrix} D_1(k) & 0 \\ D_1(k) + \bar{\beta}_0 K(k)D_2(k) & K(k)D_3(k) \end{bmatrix}, \\ \bar{K}(k) &\triangleq \begin{bmatrix} 0 \\ K(k) \end{bmatrix}, \ \bar{D}_2(k-s) \triangleq \begin{bmatrix} 0 & 0 \\ K(k)D_2(k-s) & 0 \end{bmatrix}, \\ \bar{L}(k) &\triangleq \begin{bmatrix} 0 & L(k) \end{bmatrix}, \ \bar{C}_2(k-s) \triangleq \begin{bmatrix} 0 & 0 \\ K(k)C(k-s) & 0 \end{bmatrix}, \\ \bar{A}(k) &\triangleq \mathrm{diag}\{A(k), A(k) + \bar{\beta}_0 K(k)C(k)\}, \\ \bar{A}_i(k) &\triangleq \mathbf{1}_2 \otimes \begin{bmatrix} A_i(k) & 0 \end{bmatrix}, \ \bar{C}_1(k-s) \triangleq \mathrm{diag}\{0, K(k)C(k-s)\}. \end{aligned}$$

Our objective of this section is to design a time-varying filter of the form (2.8) such that, for the given positive scalar γ, the dynamic system (2.9) satisfies the following filtering performance requirement:

$$J \triangleq \mathbb{E}\left\{ \sum_{k=0}^{N-1} \left(\|\tilde{z}(k)\|^2 - \gamma^2\|\varpi(k)\|_U^2 \right) \right\} - \gamma^2 \sum_{i=-l}^{0} \mathbb{E}\left\{\eta^T(i)V_i\eta(i)\right\} < 0$$
$$(\forall\{\varpi(k)\}, \eta(i) \neq 0) \tag{2.10}$$

where U and V_i are given positive definite weighted matrices. $\|\varpi(k)\|_U^2 \triangleq \varpi^T(k)U\varpi(k)$.

2.1.2 Design of Filter Gain

In this subsection, let us investigate both the event-based filtering performance analysis and filter design problem for system (2.9). The following lemma will be utilized to acquire the main results.

Lemma 2.1 *[14] (Schur Complement) Given constant matrices $\mathcal{S}_1, \mathcal{S}_2, \mathcal{S}_3$ where $\mathcal{S}_1 = \mathcal{S}_1^T$ and $0 < \mathcal{S}_2 = \mathcal{S}_2^T$, then $\mathcal{S}_1 + \mathcal{S}_3^T \mathcal{S}_2^{-1} \mathcal{S}_3 < 0$ if and only if*

$$\begin{bmatrix} \mathcal{S}_1 & \mathcal{S}_3^T \\ \mathcal{S}_3 & -\mathcal{S}_2 \end{bmatrix} < 0 \quad or \quad \begin{bmatrix} -\mathcal{S}_2 & \mathcal{S}_3 \\ \mathcal{S}_3^T & \mathcal{S}_1 \end{bmatrix} < 0. \tag{2.11}$$

Firstly, we propose the following event-based filtering performance analysis results for a class of nonlinear time-varying systems with multiplicative noises and fading channels.

Theorem 2.1 *Consider the discrete time-varying nonlinear stochastic system described by (2.1)–(2.4). Let the disturbance attenuation level $\gamma > 0$, the positive definite weighted matrices $U > 0$, $V_i > 0$ $(i = -l, -l+1, \ldots, 0)$, the event weighted matrix $\Omega > 0$, the scalar $\delta \in [0,\ 1)$ and the filter gain matrix $\{K(k)\}_{k\in[0,\ N-1]}$ in (2.8) be given. For the augmented system (2.9), the performance criterion (2.10) is guaranteed for all nonzero $\varpi(k)$ if there exist families of positive scalars $\{\lambda(k)\}_{k\in[0,\ N-1]}$, positive definite matrices $\{P(k)\}_{k\in[0,\ N]} > 0$ and $\{Q(i,j)\}_{i\in[-l,\ N], j\in[1,\ l]} > 0$ satisfying*

$$\begin{aligned} &\Gamma(k) \\ = \ &\bar{\Gamma}(k) \\ &+ \begin{bmatrix} \Gamma_{11}(k) & * & * & * & * & * \\ \delta\bar{\beta}_0(\Lambda_\beta \bar{\mathcal{C}}_l(k))^T \Omega \bar{C}(k) & \Gamma_{22}(k) & * & * & * & * \\ \Gamma_{31}(k) & \Gamma_{32}(k) & \Gamma_{33}(k) & * & * & * \\ \delta\bar{\beta}_0(\Lambda_\beta \bar{\mathcal{D}}_l(k))^T \Omega \bar{C}(k) & \Gamma_{42}(k) & \Gamma_{43}(k) & \Gamma_{44}(k) & * & * \\ \lambda(k)\mathcal{U}_1(k) & 0 & 0 & 0 & -\lambda(k)I & * \\ 0 & 0 & 0 & 0 & 0 & -\Omega I \end{bmatrix} \\ < \ &0 \end{aligned} \tag{2.12}$$

with the initial condition

$$\gamma^2 V_0 - P(0) > 0, \ \ \gamma^2 V_{-i} - \sum_{j=i}^{l} Q(-i,j) > 0 \ (i = 1, 2, \ldots, l) \tag{2.13}$$

where

$$\mathcal{U}_1(k) \ \triangleq \ I \otimes (\Phi(k) + \Psi(k))/2, \ \bar{\Gamma}(k) \triangleq \left[\bar{\Gamma}_{ij}(k)\right]_{\{i=1,2,\ldots,6; j=1,2,\ldots,6\}},$$

$$
\begin{aligned}
\mathcal{U}_2(k) &\triangleq I \otimes (\Phi^T(k)\Psi(k) + \Psi^T(k)\Phi(k))/2, \\
\bar{Q}(k,l) &\triangleq \text{diag}\{Q(k-1,1), Q(k-2,2), \cdots, Q(k-l,l)\}, \\
\bar{\Gamma}_{11}(k) &\triangleq \bar{A}^T(k)P(k+1)\bar{A}(k) - P(k) + \nu_0 \bar{C}_2^T(k)P(k+1)\bar{C}_2(k) \\
&\quad + \sum_{i=1}^{r} \bar{A}_i^T(k)P(k+1)\bar{A}_i(k) + \sum_{j=1}^{l} Q(k,j), \\
\bar{\Gamma}_{21}(k) &\triangleq (\Lambda_\beta \bar{\mathcal{C}}_{1l}(k))^T P(k+1)\bar{A}(k), \\
\bar{\Gamma}_{22}(k) &\triangleq (\Lambda_\beta \bar{\mathcal{C}}_{1l}(k))^T P(k+1)\Lambda_\beta \bar{\mathcal{C}}_{1l}(k) - \bar{Q}(k,l) \\
&\quad + (\bar{\Lambda}_\gamma \bar{\mathcal{C}}_{2l}(k))^T \bar{P}(k+1)\bar{\Lambda}_\gamma \bar{\mathcal{C}}_{2l}(k), \\
\bar{\Gamma}_{31}(k) &\triangleq \bar{D}_1^T(k)P(k+1)\bar{A}(k) + \nu_0 \bar{D}_2^T(k)P(k+1)\bar{C}_2(k), \\
\bar{\Gamma}_{32}(k) &\triangleq \bar{D}_1^T(k)P(k+1)\Lambda_\beta \bar{\mathcal{C}}_{1l}(k), \\
\bar{\Gamma}_{33}(k) &\triangleq \bar{D}_1^T(k)P(k+1)\bar{D}_1(k) + \nu_0 \bar{D}_2^T(k)P(k+1)\bar{D}_2(k), \\
\bar{\Gamma}_{41}(k) &\triangleq (\Lambda_\beta \bar{\mathcal{D}}_{2l}(k))^T P(k+1)\bar{A}(k), \ \bar{P}(k+1) \triangleq I_l \otimes P(k+1), \\
\bar{\Gamma}_{42}(k) &\triangleq (\Lambda_\beta \bar{\mathcal{D}}_{2l}(k))^T P(k+1)\Lambda_\beta \bar{\mathcal{C}}_{1l}(k) + (\bar{\Lambda}_\gamma \bar{\mathcal{D}}_{2l}(k))^T \bar{P}(k+1) \\
&\quad \times \bar{\Lambda}_\gamma \bar{\mathcal{C}}_{2l}(k), \\
\bar{\Gamma}_{43}(k) &\triangleq (\Lambda_\beta \bar{\mathcal{D}}_{2l}(k))^T P(k+1)\bar{D}_1(k), \\
\bar{\Gamma}_{44}(k) &\triangleq (\Lambda_\beta \bar{\mathcal{D}}_{2l}(k))^T P(k+1)\Lambda_\beta \bar{\mathcal{D}}_{2l}(k) + (\bar{\Lambda}_\gamma \bar{\mathcal{D}}_{2l}(k))^T \bar{P}(k+1) \\
&\quad \times \bar{\Lambda}_\gamma \bar{\mathcal{D}}_{2l}(k), \\
\bar{\Gamma}_{51}(k) &\triangleq \bar{\alpha}P(k+1)\bar{A}(k), \ \bar{\Gamma}_{52}(k) \triangleq \bar{\alpha}P(k+1)\Lambda_\beta \bar{\mathcal{C}}_{1l}(k), \\
\bar{\Gamma}_{53}(k) &\triangleq \bar{\alpha}P(k+1)\bar{D}_1(k), \ \bar{\Gamma}_{54}(k) \triangleq \bar{\alpha}P(k+1)\Lambda_\beta \bar{\mathcal{D}}_{2l}(k), \\
\bar{\Gamma}_{61}(k) &\triangleq \bar{K}^T(k)P(k+1)\bar{A}(k), \ \bar{\Gamma}_{62}(k) \triangleq \bar{K}^T(k)P(k+1)\Lambda_\beta \bar{\mathcal{C}}_{1l}(k), \\
\bar{\Gamma}_{55}(k) &\triangleq \bar{\alpha}^2 P(k+1) + \bar{\alpha}(1-\bar{\alpha})S_1^T P(k+1)S_1, \\
\bar{\Gamma}_{63}(k) &\triangleq \bar{K}^T(k)P(k+1)\bar{D}_1(k), \ \bar{\Gamma}_{64}(k) \triangleq \bar{K}^T(k)P(k+1)\Lambda_\beta \bar{\mathcal{D}}_{2l}(k), \\
\bar{\Gamma}_{65}(k) &\triangleq \bar{\alpha}\bar{K}^T(k)P(k+1), \ \bar{\Gamma}_{66}(k) \triangleq \bar{K}^T(k)P(k+1)\bar{K}(k), \\
\Gamma_{11}(k) &\triangleq -\lambda(k)\mathcal{U}_2(k) + \bar{L}^T(k)\bar{L}(k) + \delta(\bar{\beta}_0^2 + \nu_0)\bar{C}^T(k)\Omega\bar{C}(k), \\
\Gamma_{22}(k) &\triangleq \delta(\Lambda_\beta \bar{\mathcal{C}}_l(k))^T \Omega \Lambda_\beta \bar{\mathcal{C}}_l(k) + \delta(\bar{\Lambda}_\gamma \bar{\mathcal{C}}_l(k))^T \Omega \bar{\Lambda}_\gamma \bar{\mathcal{C}}_l(k), \\
\Gamma_{31}(k) &\triangleq \delta\big((\bar{\beta}_0^2 + \nu_0)\bar{D}(k) + \bar{\beta}_0 \bar{D}_3(k)\big)^T \Omega \bar{C}(k), \\
\Gamma_{32}(k) &\triangleq \delta(\bar{\beta}_0 \bar{D}(k) + \bar{D}_3(k))^T \Omega \Lambda_\beta \bar{\mathcal{C}}_l(k), \\
\Gamma_{33}(k) &\triangleq -\frac{\gamma^2}{l+1}U + \delta\big(\bar{\beta}_0 \bar{D}(k) + \bar{D}_3(k)\big)^T \Omega \big(\bar{\beta}_0 \bar{D}(k) + \bar{D}_3(k)\big) \\
&\quad + \delta\nu_0 \bar{D}^T(k)\Omega\bar{D}(k), \\
\Gamma_{42}(k) &\triangleq \delta(\Lambda_\beta \bar{\mathcal{D}}_l(k))^T \Omega \Lambda_\beta \bar{\mathcal{C}}_l(k) + \delta(\bar{\Lambda}_\gamma \bar{\mathcal{D}}_l(k))^T \Omega \bar{\Lambda}_\gamma \bar{\mathcal{C}}_l(k), \\
\Gamma_{43}(k) &\triangleq \delta(\Lambda_\beta \bar{\mathcal{D}}_l(k))^T \Omega (\bar{\beta}_0 \bar{D}(k) + \bar{D}_3(k)), \\
\Gamma_{44}(k) &\triangleq -\frac{\gamma^2}{l+1} I_l \otimes U + \delta(\Lambda_\beta \bar{\mathcal{D}}_l(k))^T \Omega \Lambda_\beta \bar{\mathcal{D}}_l(k)
\end{aligned}
$$

$$
\begin{aligned}
& +\delta(\bar{\Lambda}_\gamma \bar{\mathcal{D}}_l(k))^T \Omega \bar{\Lambda}_\gamma \bar{\mathcal{D}}_l(k), \\
\bar{\mathcal{C}}_{1l}(k) &\triangleq \mathrm{diag}\{\bar{C}_1(k-1), \bar{C}_1(k-2), \ldots, \bar{C}_1(k-l)\}, \\
\bar{\mathcal{D}}_{2l}(k) &\triangleq \mathrm{diag}\{\bar{D}_2(k-1), \bar{D}_2(k-2), \ldots, \bar{D}_2(k-l)\}, \\
\bar{\mathcal{C}}_{2l}(k) &\triangleq \mathrm{diag}\{\bar{C}_2(k-1), \bar{C}_2(k-2), \ldots, \bar{C}_2(k-l)\}, \\
\bar{\mathcal{D}}_l(k) &\triangleq \mathrm{diag}\{\bar{D}(k-1), \bar{D}(k-2), \ldots, \bar{D}(k-l)\}, \\
\bar{\mathcal{C}}_l(k) &\triangleq \mathrm{diag}\{\bar{C}(k-1), \bar{C}(k-2), \ldots, \bar{C}(k-l)\}, \\
\bar{C}(k-s) &\triangleq \begin{bmatrix} C(k-s) & 0 \end{bmatrix}, \ \Lambda_\beta \triangleq \begin{bmatrix} \bar{\beta}_1 I & \bar{\beta}_2 I & \cdots & \bar{\beta}_l I \end{bmatrix}, \\
\bar{D}_3(k) &\triangleq \begin{bmatrix} 0 & D_3(k) \end{bmatrix}, \ \bar{D}(k-s) \triangleq \begin{bmatrix} D_2(k-s) & 0 \end{bmatrix}, \\
\bar{\Lambda}_\gamma &\triangleq \mathrm{diag}\{\sqrt{\nu_1} I, \sqrt{\nu_2} I, \ldots, \sqrt{\nu_l} I\}.
\end{aligned}
$$

Proof *Consider the following Lyapunov functional candidate for system (2.9):*

$$
\begin{aligned}
V(k) &\triangleq V_1(k) + V_2(k) \\
&= \eta^T(k) P(k) \eta(k) + \sum_{j=1}^{l} \sum_{i=k-j}^{k-1} \eta^T(i) Q(i,j) \eta(i)
\end{aligned} \tag{2.14}
$$

where $P(k) > 0$ and $Q(i,j) > 0$ are symmetric positive definite matrices with appropriate dimensions. Calculate the difference of $V(k)$ along the solution of system (2.9) and take the mathematical expectation. Then, we have

$$
\begin{aligned}
& \mathbb{E}\{\Delta V_1(k)\} \\
= {} & \mathbb{E}\{V_1(k+1) - V_1(k)\} \\
= {} & \mathbb{E}\Bigg\{ \Bigg(\mathcal{Y}_l^T(k) P(k+1) \mathcal{Y}_l(k) + \bar{\alpha}(1-\bar{\alpha}) \bar{g}^T(k) S_1^T P(k+1) S_1 \bar{g}(k) + \eta^T(k) \\
& \times \Bigg(\sum_{i=1}^{r} \bar{A}_i^T(k) P(k+1) \bar{A}_i(k) \Bigg) \eta(k) + \sum_{s=0}^{l} \nu_s \Bigg(\bar{C}_2(k-s) \eta(k-s) \\
& + \bar{D}_2(k-s) \varpi(k-s) \Bigg)^T P(k+1) \Bigg(\bar{C}_2(k-s) \eta(k-s) \\
& + \bar{D}_2(k-s) \varpi(k-s) \Bigg) - \eta^T(k) P(k) \eta(k) \Bigg\}.
\end{aligned} \tag{2.15}
$$

Similarly, by noting the equation (2.14), one has

$$
\mathbb{E}\{\Delta V_2(k)\} = \mathbb{E}\Bigg\{ \sum_{j=1}^{l} \eta^T(k) Q(k,j) \eta(k) - \eta_l^T(k) \bar{Q}(k,l) \eta_l(k) \Bigg\} \tag{2.16}
$$

where $\eta_l(k) \triangleq \begin{bmatrix} \eta^T(k-1) & \eta^T(k-2) & \cdots & \eta^T(k-l) \end{bmatrix}^T$. *Therefore, by denoting*

$$
\varpi_l(k) \triangleq \begin{bmatrix} \varpi^T(k-1) & \cdots & \varpi^T(k-l) \end{bmatrix}^T,
$$

$$\tilde{\eta}(k) \triangleq \begin{bmatrix}\eta^T(k) & \eta_l^T(k) & \varpi^T(k) & \varpi_l^T(k) & \bar{g}^T(k) & \sigma^T(k)\end{bmatrix}^T$$

and combining (2.14)–(2.16), one immediately obtains

$$\mathbb{E}\{\Delta V(k)\} = \mathbb{E}\left\{\tilde{\eta}^T(k)\bar{\Gamma}(k)\tilde{\eta}(k)\right\}. \tag{2.17}$$

Moreover, it follows from the constraint (2.2) that

$$\left[\bar{g}(k) - (I \otimes \Phi(k))\eta(k)\right]^T \left[\bar{g}(k) - (I \otimes \Psi(k))\eta(k)\right] \leq 0. \tag{2.18}$$

Then, substituting (2.18) into (2.17) results in

$$\begin{aligned}\mathbb{E}\{\Delta V(k)\} \leq\ & \mathbb{E}\Big\{\tilde{\eta}^T(k)\bar{\Gamma}(k)\tilde{\eta}(k) - \lambda(k)\left[\bar{g}(k) - (I \otimes \Phi(k))\eta(k)\right]^T \\ & \times\left[\bar{g}(k) - (I \otimes \Psi(k))\eta(k)\right]\Big\}.\end{aligned} \tag{2.19}$$

Considering the event condition (2.6), we have

$$\begin{aligned}\mathbb{E}\{\Delta V(k)\} \leq\ & \mathbb{E}\Big\{\tilde{\eta}^T(k)\bar{\Gamma}(k)\tilde{\eta}(k) - \lambda(k)\left[\bar{g}(k) - (I \otimes \Phi(k))\eta(k)\right]^T\left[\bar{g}(k)\right. \\ & \left. -(I \otimes \Psi(k))\eta(k)\right] - \sigma^T(k)\Omega\sigma(k) + \delta y^T(k)\Omega y(k)\Big\}.\end{aligned} \tag{2.20}$$

Due to $\{\varpi(k)\}_{k\in[-l,\ -1]} = 0$*, adding the zero term*

$$\tilde{z}^T(k)\tilde{z}(k) - \gamma^2\varpi^T(k)U\varpi(k) - (\tilde{z}^T(k)\tilde{z}(k) - \gamma^2\varpi^T(k)U\varpi(k)) \tag{2.21}$$

to (2.20) *results in*

$$\begin{aligned}\mathbb{E}\{\Delta V(k)\} \leq\ & \mathbb{E}\Big\{\tilde{\eta}^T(k)\Gamma(k)\tilde{\eta}(k)\Big\} + \mathbb{E}\left\{\frac{\gamma^2}{l+1}\sum_{s=0}^{l}\|\varpi(k-s)\|_U^2\right. \\ & \left.-\gamma^2\|\varpi(k)\|_U^2\right\} - \mathbb{E}\left\{\|\tilde{z}(k)\|^2 - \gamma^2\|\varpi(k)\|_U^2\right\}.\end{aligned} \tag{2.22}$$

Summing up (2.22) *on both sides from* 0 *to* $N-1$ *with respect to* k*, we obtain*

$$\begin{aligned}& \sum_{k=0}^{N-1}\mathbb{E}\{\Delta V(k)\} \\ \leq\ & \mathbb{E}\Big\{\sum_{k=0}^{N-1}\tilde{\eta}^T(k)\Gamma(k)\tilde{\eta}(k)\Big\} + \mathbb{E}\left\{\frac{\gamma^2}{l+1}\sum_{s=0}^{l}\sum_{k=0}^{N-1}(\|\varpi(k-s)\|_U^2\right. \\ & \left.-\|\varpi(k)\|_U^2)\right\} - \mathbb{E}\left\{\sum_{k=0}^{N-1}(\|\tilde{z}(k)\|^2 - \gamma^2\|\varpi(k)\|_U^2)\right\}.\end{aligned} \tag{2.23}$$

It can be obtained from (2.12) and (2.13) that

$$\begin{aligned}
&\mathbb{E}\left\{\sum_{k=0}^{N-1}\left(\gamma^2\|\varpi(k)\|_U^2-\|\tilde{z}(k)\|^2\right)+\gamma^2\sum_{i=-l}^{0}\eta^T(i)V_i\eta(i)\right\}\\
>\ &\mathbb{E}\{V(N)\}+\mathbb{E}\left\{\gamma^2\sum_{k=-l}^{0}\eta^T(i)V_i\eta(i)-V(0)\right\}\geq 0 \qquad (2.24)
\end{aligned}$$

which is equivalent to (2.10), and the proof is now complete.

Based on the analysis results, we are now ready to solve the filter design problem for system (2.9) in the following theorem.

For convenience of later analysis, we denote

$$\begin{aligned}
\hat{\Gamma}_{11}(k) &\triangleq \begin{bmatrix}-P(k)+\sum_{j=1}^{l}Q(k,j)+\Gamma_{11}(k) & * & *\\ \delta\bar{\beta}_0(\Lambda_\beta\bar{\mathcal{C}}_l(k))^T\Omega\bar{C}(k) & -\bar{Q}(k,l)+\Gamma_{22}(k) & *\\ \Gamma_{31}(k) & \Gamma_{32}(k) & \Gamma_{33}(k)\end{bmatrix},\\
\hat{\Gamma}_{21}(k) &\triangleq \begin{bmatrix}\delta\bar{\beta}_0(\Lambda_\beta\bar{\mathcal{D}}_l(k))^T\Omega\bar{C}(k) & \Gamma_{42}(k) & \Gamma_{43}(k)\\ \lambda(k)\mathcal{U}_1(k) & 0 & 0\\ 0 & 0 & 0\end{bmatrix},\\
\hat{\Gamma}_{22}(k) &\triangleq \operatorname{diag}\{\Gamma_{44}(k),-\lambda(k)I,-\Omega I\},\\
\hat{\Gamma}_{32}(k) &\triangleq \begin{bmatrix}\Lambda_\beta\hat{K}(k)\bar{\mathcal{D}}_l(k) & \bar{\alpha}I & H_0K(k)\end{bmatrix},\\
\hat{\Gamma}_{31}(k) &\triangleq \big[\ \hat{A}_0(k)+\bar{\beta}_0H_0K(k)\hat{C}_0(k) \quad \Lambda_\beta\hat{K}(k)\hat{\mathcal{C}}_{0l}(k)\\
&\qquad \hat{D}_0(k)+H_0K(k)\hat{D}_3(k)\big],\\
\hat{\Gamma}_{41}(k) &\triangleq \begin{bmatrix}\sqrt{\nu_0}H_0K(k)\bar{C}(k) & 0 & \sqrt{\nu_0}H_0K(k)\bar{D}(k)\\ 0 & \Lambda_\beta\hat{K}(k)\hat{\mathcal{C}}_{0l}(k) & 0\\ 0 & \bar{\Lambda}_\gamma\hat{K}(k)\bar{\mathcal{C}}_l(k) & 0\end{bmatrix},\\
\hat{\Gamma}_{44}(k) &\triangleq \operatorname{diag}\{-R(k+1),-R(k+1),-\bar{R}(k+1)\},\\
\hat{\Gamma}_{51}(k) &\triangleq \begin{bmatrix}\hat{\Gamma}_{511}(k) & \hat{\Gamma}_{512}(k) & 0\end{bmatrix},\ \hat{K}(k)\triangleq I_l\otimes H_0K(k),\\
\bar{A}_r(k) &\triangleq \begin{bmatrix}\bar{A}_1^T(k) & \bar{A}_2^T(k) & \cdots & \bar{A}_r^T(k)\end{bmatrix}^T,\ \hat{\Gamma}_{511}(k)\triangleq\begin{bmatrix}0 & 0 & \bar{A}_r^T(k)\end{bmatrix}^T,\\
\hat{\Gamma}_{52}(k) &\triangleq \operatorname{diag}\{\bar{\Lambda}_\gamma\hat{K}(k)\bar{\mathcal{D}}_l(k),\sqrt{\bar{\alpha}(1-\bar{\alpha})}S_1,0\},\\
\hat{\Gamma}_{55}(k) &\triangleq \operatorname{diag}\{-\bar{R}(k+1),-R(k+1),-\hat{R}(k+1)\},\\
\hat{A}_0(k) &\triangleq I_2\otimes A(k),\ \bar{R}(k+1)\triangleq I_l\otimes R(k+1),\ \hat{R}(k+1)\triangleq I_r\otimes R(k+1),\\
\hat{C}_0(k) &\triangleq \begin{bmatrix}0 & C(k)\end{bmatrix},\ \hat{\Gamma}_{512}(k)\triangleq\begin{bmatrix}(\bar{\Lambda}_\gamma\hat{K}(k)\bar{\mathcal{C}}_l(k))^T & 0 & 0\end{bmatrix}^T,\\
\hat{D}_0(k) &\triangleq \mathbf{1}_2\otimes\begin{bmatrix}D_1(k) & 0\end{bmatrix},\ \hat{D}_3(k)\triangleq\begin{bmatrix}\bar{\beta}_0D_2(k) & D_3(k)\end{bmatrix},\\
\hat{\mathcal{C}}_{0l}(k) &\triangleq \operatorname{diag}\left\{\hat{C}_0(k-1),\hat{C}_0(k-2),\ldots,\hat{C}_0(k-l)\right\},\ H_0\triangleq\begin{bmatrix}0 & I\end{bmatrix}^T.
\end{aligned}$$

Theorem 2.2 *Consider the discrete time-varying nonlinear stochastic system*

(2.1) with the time-varying filter (2.8). For the given disturbance attenuation level $\gamma > 0$, the positive definite weighted matrices $U > 0$, $V_i > 0$ $(i = -l, -l+1, \ldots, 0)$, the event weighted matrix $\Omega > 0$ and the saclar $\delta \in [0,\ 1)$, the filtering error $\tilde{z}(k)$ satisfies the performance criterion (2.10) if there exist families of positive scalars $\{\lambda(k)\}_{k\in[0,N-1]}$, positive definite matrices $\{P(k)\}_{k\in[0,N]} > 0$, $\{Q(i,j)\}_{i\in[-l,N],j\in[1,l]} > 0$, $\{R(k)\}_{k\in[0,N]} > 0$ and real-valued matrices $K(k)_{k\in[0,N-1]}$ satisfying

$$\hat{\Gamma}(k) = \begin{bmatrix} \hat{\Gamma}_{11}(k) & * & * & * & * \\ \hat{\Gamma}_{21}(k) & \hat{\Gamma}_{22}(k) & * & * & * \\ \hat{\Gamma}_{31}(k) & \hat{\Gamma}_{32}(k) & -R(k+1) & * & * \\ \hat{\Gamma}_{41}(k) & 0 & 0 & \hat{\Gamma}_{44}(k) & * \\ \hat{\Gamma}_{51}(k) & \hat{\Gamma}_{52}(k) & 0 & 0 & \hat{\Gamma}_{55}(k) \end{bmatrix} < 0 \tag{2.25}$$

and the initial condition

$$\gamma^2 V_0 - P(0) > 0, \;\; \gamma^2 V_{-i} - \sum_{j=i}^{l} Q(-i,j) > 0 \;\; (i = 1, 2, \ldots, l) \tag{2.26}$$

with the parameters updated by $P(k+1) = R^{-1}(k+1)$.

Proof *In order to avoid partitioning the positive define matrices $\{P(k)\}_{k\in[0,N]}$, $\{Q(i,j)\}_{i\in[-l,N],j\in[1,l]}$ and $\{R(k)\}_{k\in[0,N]}$, we rewrite the parameters in Theorem 2.1 in the following form:*

$$\begin{aligned} &\bar{A}(k) = \hat{A}_0(k) + \bar{\beta}_0 H_0 K(k)\hat{C}_0(k), \;\; \bar{C}_1(k-s) = H_0 K(k)\hat{C}_0(k-s), \\ &\bar{\mathcal{C}}_{1l}(k) = \hat{K}(k)\hat{\mathcal{C}}_{0l}(k), \;\; \bar{C}_2(k-s) = H_0 K(k)\bar{C}(k-s), \\ &\bar{\mathcal{D}}_{2l}(k) = \hat{K}(k)\bar{\mathcal{D}}_l(k), \;\; \bar{D}_1(k) = \hat{D}_0(k) + H_0 K(k)\hat{D}_3(k), \\ &\bar{D}_2(k) = H_0 K(k)\bar{D}(k-s), \;\; \bar{K}(k) = H_0 K(k), \\ &\bar{\mathcal{C}}_{2l}(k) = \hat{K}(k)\bar{\mathcal{C}}_l(k), \;\; \bar{\mathcal{D}}_{2l}(k) = \hat{K}(k)\bar{\mathcal{D}}_l(k). \end{aligned} \tag{2.27}$$

Noticing (2.27) and using Lemma 2.1 (Schur Complement Lemma), (2.25) can be obtained by (2.12) after some straightforward algebraic manipulations. The proof of this theorem is now complete.

Remark 2.3 *Theorem 2.1 presents sufficient conditions for the existence of admissible filters. It is worth noting that the technique used for deriving these conditions is quite different from the previous results in the filtering area, e.g. [54, 146, 150]. In this subsection, to reduce the design conservatism, the positive definite matrices $\{P(k)\}_{k\in[0,N]}$, $\{Q(i,j)\}_{i\in[-l,N],j\in[1,l]}$ and $\{R(k)\}_{k\in[0,N]}$ remain in its original form. Therefore, the difficulty of dilating positive definite matrices does not occur in our result. Besides, it can be* *observed from Theorem 2.2 that the main results established contain all the information of the addressed general systems including the time-varying systems parameters, multiplicative noise, the threshold of event trigger, the occurrence probabilities of the random nonlinearity as well as the statistics characteristics of the channel coefficients.*

For implementation purpose, based on Theorem 2.2, we summarize the Finite-Horizon Filter Design (*FHFD*) algorithm as follows.

The Finite-Horizon Filter Design (*FHFD*) Algorithm:

Step 1.	Given the disturbance attenuation level γ, the positive definite weighted matrices $U > 0$, $V_i > 0$ $(i = -l, -l+1, \ldots, 0)$, the event weighted matrix $\Omega > 0$ and the saclar $\delta \in [0, 1)$.
Step 2.	Set $k = 0$. Solve the matrix inequalities (2.25) and the recursive matrix inequalities (2.13) to obtain the values of matrices $P(0)$, $\sum_{j=i}^{l} Q(-i, j)$ $(i = 1, 2, \ldots, l)$, $R(1)$ and the filter gain matrix $K(0)$.
Step 3.	Set $k = k + 1$, update the matrices $P(k+1) = R^{-1}(k+1)$ and then obtain the filter gain matrix $K(k)$ by solving the recursive matrix inequalities (2.25).
Step 4.	If $k < N$, then go to Step 3, else go to Step 5.
Step 5.	Stop.

2.2 Event-Triggered Variance-Constrained H_∞ Control

In this section, a general event-triggered framework is set up to deal with the variance-constrained H_∞ control problem for a class of discrete time-varying systems with randomly occurring saturations, stochastic nonlinearities and state-multip-icative noises. Based on the relative error with respect to the measurement signal, an event indicator variable is introduced and the corresponding event-triggered scheme is proposed in order to determine whether the measurement output is transmitted to the controller or not. The stochastic nonlinearities under consideration are characterized by statistical means which can cover several classes of well-studied nonlinearities. A set of unrelated random variables is exploited to govern the phenomena of randomly occurring saturations, stochastic nonlinearities and state-dependent noises.

2.2.1 Problem Formulation

Consider the following class of discrete time-varying stochastic systems

$$\begin{cases} x(k+1) = \big(A(k) + \sum_{i=1}^{r} A_i(k)w_i(k)\big)x(k) + B_1(k)u(k) + g(k,x(k)) \\ \qquad + D(k)v(k) \\ z(k) = L(k)x(k) + B_2(k)u(k) \end{cases} \tag{2.28}$$

and m sensor measurements with randomly occurring saturations

$$y_i(k) = \alpha_i(k)\sigma(C_i(k)x(k)) + (1-\alpha_i(k))C_i(k)x(k) + E_i(k)\varpi_i(k) \; (i = 1,2,\ldots,m) \tag{2.29}$$

where $x(k) \in \mathbb{R}^{n_x}$ represents the state vector; $y_i(k) \in \mathbb{R}$ is the measurement output measured by sensor i from the plant; $z(k) \in \mathbb{R}^{n_z}$ is the controlled output vector; $u(k) \in \mathbb{R}^{n_u}$ is the control input vector; $w_i(k) \in \mathbb{R}$ $(i = 1,2,\ldots,r)$, $v(k) \in \mathbb{R}^{n_v}$ and $\varpi_i(k) \in \mathbb{R}^{n_w}$ $(i = 1,2,\ldots,m)$ are, respectively, the multiplicative noise, the process noise and the measurement noise for sensor i. The noise sequences are mutually uncorrelated zero-mean Gaussian sequences with $\mathbb{E}\{w_i(k)w_i^T(k)\} = 1$, $\mathbb{E}\{v(k)v^T(k)\} = V(k)$ and $\mathbb{E}\{\varpi_i(k)\varpi_i^T(k)\} = \bar{W}_i(k)$. $A(k)$, $A_i(k)$, $B_1(k)$, $B_2(k)$, $D(k)$, $L(k)$, $C_i(k)$ and $E_i(k)$ are known, real, time-varying matrices with appropriate dimensions.

The nonlinear function $g(k,x(k))$ with $g(k,0) = 0$ is a stochastic nonlinear function having the following statistical characteristics:

$$\mathbb{E}\left\{g(k,x(k))\Big|x(k)\right\} = 0,$$

$$\mathbb{E}\left\{g(k,x(k))g^T(j,x(k))\big|x(k)\right\} = 0, \quad k \neq j$$

and

$$\begin{aligned}\mathbb{E}\left\{g(k,x(k))g^T(k,x(k))\big|x(k)\right\} &:= \textstyle\sum_{i=1}^{q} \pi_i\pi_i^T\mathbb{E}\left\{x^T(k)\Gamma_i(k)x(k)\right\} \\ &= \textstyle\sum_{i=1}^{q} \Theta_i(k)\mathbb{E}\left\{x^T(k)\Gamma_i(k)x(k)\right\}\end{aligned} \tag{2.30}$$

where q is a known nonnegative integer, $\Theta_i(k)$ and $\Gamma_i(k)$ $(i = 1,2,\ldots,q)$ are known matrices with appropriate dimensions.

The saturation function $\sigma(\cdot) : \mathbb{R} \mapsto \mathbb{R}$ is defined as

$$\sigma(\vartheta) \triangleq sign(\vartheta)\min\{1, |\vartheta|\}. \tag{2.31}$$

Here, $sign(\cdot)$ denotes the signum function. Without loss of generality, the saturation level is taken as unity.

The variable $\alpha_i(k)$ $(i = 1,2,\ldots,m)$ in (2.29), which governs the ROSS phenomenon, are Bernoulli distributed white sequences taking values on 0 or 1 with

$$\text{Prob}\{\alpha_i(k) = 1\} = \bar{\alpha}, \quad \text{Prob}\{\alpha_i(k) = 0\} = 1 - \bar{\alpha} \tag{2.32}$$

where $\bar{\alpha} \in [0,1]$ is a known constant. Throughout this section, the stochastic variables $\alpha_i(k)$ $(i = 1,\dots,m)$, $v(k)$, $\varpi_i(k)$ $(i = 1,2,\dots,m)$ and $w_i(k)$ $(i = 1,\dots,r)$ are mutually uncorrelated.

For notational brevity, we set

$$\begin{aligned} y(k) &\triangleq \begin{bmatrix} y_1^T(k) & y_2^T(k) & \cdots & y_m^T(k) \end{bmatrix}^T, \\ \Lambda_\alpha(k) &\triangleq \mathrm{diag}\{\alpha_1(k), \alpha_2(k), \dots, \alpha_m(k)\}, \\ E(k) &\triangleq \mathrm{diag}\{E_1(k), E_2(k), \dots, E_m(k)\}, \\ C(k) &\triangleq \begin{bmatrix} C_1^T(k) & C_2^T(k) & \cdots & C_m^T(k) \end{bmatrix}^T, \\ \varpi(k) &\triangleq \begin{bmatrix} \varpi_1^T(k) & \varpi_2^T(k) & \cdots & \varpi_m^T(k) \end{bmatrix}^T, \ \bar{\Lambda}_\alpha \triangleq I_m \otimes \bar{\alpha}. \end{aligned} \tag{2.33}$$

Then, the sensor model (2.29) can be expressed in the following compact form:

$$y(k) = \Lambda_\alpha(k)\sigma(C(k)x(k)) + (I - \Lambda_\alpha(k))C(k)x(k) + E(k)\varpi(k) \tag{2.34}$$

where

$$\sigma(C(k)x(k)) \triangleq \begin{bmatrix} \sigma^T(C_1(k)x(k)) & \sigma^T(C_2(k)x(k)) & \cdots & \sigma^T(C_m(k)x(k)) \end{bmatrix}^T.$$

In this section, the notation σ has been slightly abused to denote both the vector-valued and the scalar-valued saturation functions.

Definition 2.1 *[92] A nonlinearity $\Psi : \mathbb{R}^m \mapsto \mathbb{R}^m$ is said to satisfy a sector condition if*

$$(\Psi(\bar{v}) - H_1\bar{v})^T (\Psi(\bar{v}) - H_2\bar{v}) \le 0, \ \ \forall\, \bar{v} \in \mathbb{R}^r \tag{2.35}$$

for some real matrices $H_1, H_2 \in \mathbb{R}^{r\times r}$, where $H = H_2 - H_1$ is a positive-definite symmetric matrix. In this case, we say that Ψ belongs to the sector $[H_1\ H_2]$.

As in [194,201], for diagonal matrices K_1 and K_2 satisfying $0 \le K_1 < I \le K_2$, the saturation function in (2.34) can be decomposed into a linear and a nonlinear part as

$$\sigma(C(k)x(k)) = K_1C(k)x(k) + \Psi(C(k)x(k)) \tag{2.36}$$

where $\Psi(C(k)x(k))$ is a nonlinear vector-valued function satisfying the following sector condition:

$$\Psi^T(C(k)x(k))\,(\Psi(C(k)x(k)) - KC(k)x(k)) \le 0 \tag{2.37}$$

with $K \triangleq K_2 - K_1$.

For the purpose of reducing data communication frequency, the event generator is constructed which uses the previously measurement output to determine whether the newly measurement output will be sent out to the controller or not. As such, we define the event generator function $f(\cdot,\cdot)$ as follows:

$$f(\varphi(k),\delta) \triangleq \varphi^T(k)\varphi(k) - \delta y^T(k)y(k) \tag{2.38}$$

where $\varphi(k) \triangleq y(k_i) - y(k)$, $y(k_i)$ is the measurement at latest event time, $y(k)$ is the current measurement and $\delta \in [0,\ 1)$.

The execution is triggered as long as the condition

$$f(\varphi(k),\delta) > 0 \tag{2.39}$$

is satisfied. Therefore, the sequence of event-triggered instants $0 \le k_0 \le k_1 \le \cdots \le k_i \le \cdots$ is determined iteratively by

$$k_{i+1} = \inf\{k \in \mathbb{N} | k > k_i, f(\varphi(k),\delta) > 0\}. \tag{2.40}$$

Accordingly, any measurement data satisfying the event condition (2.39) will be transmitted to the controller.

Remark 2.4 *The event triggering mechanism is adopted here in order to effectively decrease the data communication frequency and network bandwidth usages. It should be noted that the mechanism in (2.38) is a relative error-based event-triggering criterion in the discrete-time setting, which can be regarded as the discretization of Tabuada's framework for the continuous-time system [162]. The importance of relative error-based event-triggering criterion has been widely recognized, but the corresponding results for discrete-time systems have been very few especially when the variance-constrained H_∞ control problem becomes a research focus. Generally speaking, the event triggering rule used here would help avoid frequent transmission in the case that the changing rate of the measurement signal is relatively small. Obviously, the set of event instants is only a subset of the time sequences, i.e., $\{k_0, k_1, k_2, \ldots\} \in \{0, 1, 2, \ldots\}$. Note that when $\delta = 0$, all the measurement sequences are transmitted and this reduces to the traditional time triggered scheme.*

For system (2.28), the following controller structure is adopted:

$$\begin{cases} x_c(k+1) = A_c(k)x_c(k) + B_c(k)y(k_i) \\ u(k) = C_c(k)x_c(k) \end{cases} \tag{2.41}$$

where $x_c(k) \in \mathbb{R}^{n_c}$ is the controller state, $A_c(k)$, $B_c(k)$ and $C_c(k)$ are the controller parameters to be designed.

Under the output feedback controller (2.41), the closed-loop system becomes

$$\begin{cases} \eta(k+1) = \mathcal{Y}_l(k) + H_0 g(k,x(k)) + \sum_{i=1}^{r} w_i(k)\bar{A}_i(k)\eta(k) \\ \qquad + \tilde{B}_c(k)\bar{C}(k)\eta(k) + \tilde{B}_c(k)\Psi(\hat{C}(k)\eta(k)) + \bar{D}(k)\xi(k) \\ z(k) = \bar{M}(k)\eta(k) \end{cases} \tag{2.42}$$

where

$$\eta(k) \triangleq \begin{bmatrix} x^T(k) & x_c^T(k) \end{bmatrix}^T, \ \tilde{\Lambda}_\alpha(k) \triangleq \Lambda_\alpha(k) - \bar{\Lambda}_\alpha, \ \xi(k) \triangleq \begin{bmatrix} v^T(k) & \varpi^T(k) \end{bmatrix}^T,$$
$$\mathcal{Y}_l(k) \triangleq (\bar{A}(k) + \bar{B}_c(k)\bar{\Lambda}_\alpha K_1\hat{C}(k))\eta(k) + \bar{B}_c(k)(\varphi(k) + \bar{\Lambda}_\alpha\Psi(\hat{C}(k)\eta(k))),$$
$$\bar{A}(k) \triangleq \begin{bmatrix} A(k) & B_1(k)C_c(k) \\ B_c(k)(I - \bar{\Lambda}_\alpha)C(k) & A_c(k) \end{bmatrix}, \ \bar{M}(k) \triangleq \begin{bmatrix} L(k) & B_2(k)C_c(k) \end{bmatrix},$$
$$\bar{B}_c(k) \triangleq \begin{bmatrix} 0 & B_c^T(k) \end{bmatrix}^T, \ \bar{A}_i(k) \triangleq \mathrm{diag}\{A_i(k), 0\}, \ \bar{C}(k) \triangleq \begin{bmatrix} (K_1 - I)C(k) & 0 \end{bmatrix},$$
$$\tilde{B}_c(k) \triangleq \begin{bmatrix} 0 & (B_c(k)\tilde{\Lambda}_\alpha(k))^T \end{bmatrix}^T, \ \hat{C}(k) \triangleq \begin{bmatrix} C(k) & 0 \end{bmatrix}, \ H_0 \triangleq \begin{bmatrix} I & 0 \end{bmatrix}^T,$$
$$\bar{D}(k) \triangleq \mathrm{diag}\{D(k), B_c(k)E(k)\}.$$

The state covariance matrix of the dynamical system (2.42) can be defined as

$$\mathbb{X}(k) \triangleq \mathbb{E}\left\{\eta(k)\eta^T(k)\right\} = \mathbb{E}\left\{\begin{bmatrix} x(k) \\ x_c(k) \end{bmatrix}\begin{bmatrix} x(k) \\ x_c(k) \end{bmatrix}^T\right\}. \tag{2.43}$$

Our objective of this section is to design a dynamic output feedback controller of the form (2.41) over the finite horizon $[0, N]$ such that the following two requirements are satisfied simultaneously:

- (Q1): For the given disturbance attenuation level $\gamma > 0$, the positive definite matrices U_1, U_2, S and the initial state $x(0)$, the controlled output $z(k)$ satisfies the following performance constraint:

$$J_1 \triangleq \mathbb{E}\left\{\sum_{k=0}^{N-1}\left(\|z(k)\|^2 - \gamma^2\|\xi(k)\|_U^2\right)\right\} - \gamma^2\mathbb{E}\left\{x^T(0)Sx(0)\right\} < 0$$
$$(\forall\{\xi(k)\} \neq 0) \tag{2.44}$$

  where $\|\xi(k)\|_U^2 \triangleq \xi^T(k)U\xi(k)$, $U \triangleq \mathrm{diag}\{U_1, U_2\}$.

- (Q2): The state covariances satisfy the following constraints:

$$J_2 \triangleq \mathbb{E}\left\{x(k)x^T(k)\right\} \leq \Upsilon(k) \tag{2.45}$$

  where $\Upsilon(k)$ $(0 \leq k < N)$ is a sequence of given matrices specifying the acceptable covariance upper bounds obtained from the engineering requirements.

Remark 2.5 *In the desired performance requirement (Q2), the estimation error variance at each sampling time point is required to be not more than an individual upper bound. Note that the specified error variance constraint may not be minimal but should meet engineering requirements, which gives rise to a practically acceptable 'window' with the hope to keep the estimated states within such a 'window'. On the other hand, since the variance constraint is relaxed from the minimum to the acceptable one, there would exist much freedom that can be used to attempt to directly achieve other desired performance requirements, such as the robustness and H_∞ disturbance rejection attenuation level as discussed in this section.*

2.2.2 Finite-Horizon Controller Design

In this subsection, let us investigate both the event-triggered controller performance analysis and controller design problem for system (2.42).

Before deriving the main results, we define the event indicator variable $\mu(k)$ as follows:

$$\mu(k) \triangleq \begin{cases} 1 & \textit{if the event generator condition is satisfied} \\ & \quad \textit{at the current instant } k \\ 0 & \textit{if no event is triggered} \end{cases} \tag{2.46}$$

A. Event-triggered H_∞ Performance

Firstly, we propose the following event-triggered H_∞ performance analysis results for a class of nonlinear time-varying systems with multiplicative noises and randomly occurring saturations. For presentation clearly, we denote

$$\begin{aligned}
\Xi_{11}(k) &\triangleq 2\big(\bar{A}(k)+\bar{B}_c(k)\bar{\Lambda}_\alpha K_1\hat{C}(k)\big)^T P(k+1)\big(\bar{A}(k)+\bar{B}_c(k)\bar{\Lambda}_\alpha K_1\hat{C}(k)\big) \\
&\quad +\bar{M}^T(k)\bar{M}(k)+\sum_{i=1}^{q}\bar{\Gamma}_i(k)\cdot\mathrm{tr}\Big[H_0^T P(k+1)H_0\Theta_i(k)\Big] \\
&\quad +\sum_{i=1}^{r}\bar{A}_i^T(k)P(k+1)\bar{A}_i(k)-P(k)+\bar{\alpha}(1-\bar{\alpha})\bar{C}^T(k)\hat{B}_{cp}^T(k) \\
&\quad \times P(k+1)\hat{B}_{cp}(k)\bar{C}(k)+\delta((\bar{\Lambda}_\alpha K_1+I-\bar{\Lambda}_\alpha)\hat{C}(k))^T \\
&\quad \times(\bar{\Lambda}_\alpha K_1+I-\bar{\Lambda}_\alpha)\hat{C}(k)+\delta\bar{\alpha}(1-\bar{\alpha})((I-K_1)\hat{C}(k))^T \\
&\quad \times(I-K_1)\hat{C}(k), \\
\Xi_{31}(k) &\triangleq 2(\bar{B}_c(k)\bar{\Lambda}_\alpha)^T P(k+1)\big(\bar{A}(k)+\bar{B}_c(k)\bar{\Lambda}_\alpha K_1\hat{C}(k)\big)+\bar{\alpha}(1-\bar{\alpha}) \\
&\quad \times\hat{B}_{cp}^T(k)P(k+1)\hat{B}_{cp}(k)\bar{C}(k)+\lambda_1(k)K\hat{C}(k)+\delta\bar{\Lambda}_\alpha^T(\bar{\Lambda}_\alpha K_1 \\
&\quad +I-\bar{\Lambda}_\alpha)\hat{C}(k)+\delta\bar{\alpha}(1-\bar{\alpha})(I-K_1)\hat{C}(k), \\
\Xi_{33}(k) &\triangleq 2(\bar{B}_c(k)\bar{\Lambda}_\alpha)^T P(k+1)\bar{B}_c(k)\bar{\Lambda}_\alpha+\bar{\alpha}(1-\bar{\alpha})\hat{B}_{cp}^T(k)P(k+1) \\
&\quad \times\hat{B}_{cp}(k)-\lambda_1(k)I+\delta\bar{\Lambda}_\alpha^T\bar{\Lambda}_\alpha\delta\bar{\alpha}(1-\bar{\alpha})I, \\
\Xi_{44}(k) &\triangleq 2\bar{D}^T(k)P(k+1)\bar{D}(k)+\delta\bar{E}^T(k)\bar{E}(k)-\gamma^2 U+\mu^2(k)\bar{E}^T(k) \\
&\quad \times\bar{B}_c^T(k)P(k+1)\bar{B}_c(k)\bar{E}(k), \\
\bar{\Gamma}_i(k) &\triangleq \mathrm{diag}\{\Gamma_i(k),0\},\ P(k+1)\triangleq\mathrm{diag}\{M(k+1),N(k+1)\}, \\
\hat{B}_{cp}(k) &\triangleq \begin{bmatrix}\breve{B}_c^T(k) & \hat{B}_c^T(k)\end{bmatrix}^T,\ \breve{B}_c(k)\triangleq\mathrm{diag}\{\breve{B}_{c_1}(k),\breve{B}_{c_2}(k),\ldots,\breve{B}_{c_m}(k)\}, \\
\bar{S} &\triangleq \mathrm{diag}\{S,0\},\ \hat{B}_c(k)\triangleq\mathrm{diag}\{\hat{B}_{c_1}(k),\hat{B}_{c_2}(k),\ldots,\hat{B}_{c_m}(k)\}, \\
\breve{B}_{c_i}(k) &\triangleq \begin{bmatrix}\bar{B}_{c1,i}^T(k) & \bar{B}_{c2,i}^T(k) & \cdots & \bar{B}_{c_{n_x},i}^T(k)\end{bmatrix}^T,\ \bar{E}(k)\triangleq\begin{bmatrix}0 & E(k)\end{bmatrix}, \\
\hat{B}_{c_i}(k) &\triangleq \begin{bmatrix}\bar{B}_{c_{n_x+1},i}^T(k) & \bar{B}_{c_{n_x+2},i}^T(k) & \cdots & \bar{B}_{c_{n_x+n_c},i}^T(k)\end{bmatrix}^T (i=1,2,\ldots,m).
\end{aligned}$$

Theorem 2.3 *Consider the discrete time-varying nonlinear stochastic system described by (2.28)–(2.29). Let the disturbance attenuation level $\gamma>0$, the* *positive definite weighted matrices $U_1 > 0$, $U_2 > 0$ and $S > 0$, the scalar $\delta \in [0,\ 1)$ and the controller parameters $A_c(k)$, $B_c(k)$ and $C_c(k)$ in (2.41) be given. The performance criterion defined in (2.44) is guaranteed for all nonzero $\xi(k)$ if, with the initial condition $P(0) = \mathrm{diag}\{M(0), N(0)\} \leq \gamma^2 \bar{S}$, there exist families of positive scalars $\{\lambda_1(k)\}_{k\in[0,\ N-1]}$ and a sequence of positive definite matrices $P(k) = \mathrm{diag}\{M(k), N(k)\}_{k\in[0,\ N]} > 0$ satisfying the following recursive matrix inequalities:*

$$
\Xi(k) = \begin{bmatrix} \Xi_{11}(k) & * & * & * \\ 0 & 2\bar{B}_c^T(k)P(k+1)\bar{B}_c(k) - I & * & * \\ \Xi_{31}(k) & 0 & \Xi_{33}(k) & * \\ \delta\bar{E}^T(k)(\bar{\Lambda}_\alpha K_1 + I - \bar{\Lambda}_\alpha)\hat{C}(k) & 0 & \delta\bar{E}^T(k)\bar{\Lambda}_\alpha & \Xi_{44}(k) \end{bmatrix} < 0 \tag{2.47}
$$

Proof *Define $J(k) \triangleq \eta^T(k+1)P(k+1)\eta(k+1) - \eta^T(k)P(k)\eta(k)$. Taking (2.42) into consideration, we have*

$$
\begin{aligned}
&\mathbb{E}\{J(k)\} = \mathbb{E}\left\{\eta^T(k+1)P(k+1)\eta(k+1) - \eta^T(k)P(k)\eta(k)\right\} \\
\leq\ & \mathbb{E}\Big\{2\varphi^T(k)\bar{B}_c^T(k)P(k+1)\bar{B}_c(k)\varphi(k) + \eta^T(k)\sum_{i=1}^{q}\bar{\Gamma}_i(k)\cdot \mathrm{tr}\big[H_0^T P(k+1) \\
&\times H_0\Theta_i(k)\big]\eta(k) + \eta^T(k)\sum_{i=1}^{r}\bar{A}_i^T(k)P(k+1)\bar{A}_i(k)\eta(k) + 2\mathcal{Y}_{ol}^T(k) \\
&\times P(k+1)\mathcal{Y}_{ol}(k) - \eta^T(k)P(k)\eta(k) + \bar{\alpha}(1-\bar{\alpha})\big(\bar{C}(k)\eta(k) \\
&+\Psi(\hat{C}(k)\eta(k))\big)^T\hat{B}_{cp}^T(k)P(k+1)\hat{B}_{cp}(k)\big(\bar{C}(k)\eta(k) \\
&+\Psi(\hat{C}(k)\eta(k))\big) + \xi^T(k)\big(2\bar{D}^T(k)P(k+1)\bar{D}(k) \\
&+\mu^2(k)\bar{E}^T(k)\bar{B}_c^T(k)P(k+1)\bar{B}_c(k)\bar{E}(k)\big)\xi(k)\Big\}.
\end{aligned}
$$

Adding the zero term $\|z(k)\|^2 - \gamma^2\|\xi(k)\|_U^2 - \|z(k)\|^2 + \gamma^2\|\xi(k)\|_U^2$ to $\mathbb{E}\{J(k)\}$, we obtain

$$
\mathbb{E}\{J(k)\} \leq \mathbb{E}\Big\{\bar{\eta}^T(k)\Xi(k)\bar{\eta}(k)\Big\} - \mathbb{E}\left\{\|z(k)\|^2 - \gamma^2\|\xi(k)\|_U^2\right\} \tag{2.48}
$$

where $\bar{\eta}(k) \triangleq \begin{bmatrix}\eta^T(k) & \varphi^T(k) & \Psi^T(\hat{C}(k)\eta(k)) & \xi^T(k)\end{bmatrix}^T$.

Summing up (2.48) on both sides from 0 *to* $N-1$ *with respect to* k*, we arrive at*

$$\begin{aligned}\sum_{k=0}^{N-1}\mathbb{E}\{J(k)\} &= \mathbb{E}\{\eta^T(N)P(N)\eta(N)-\eta^T(0)P(0)\eta(0)\}\\ &\leq \mathbb{E}\left\{\sum_{k=0}^{N-1}\bar{\eta}^T(k)\Xi(k)\bar{\eta}(k)\right\}\\ &\quad -\mathbb{E}\left\{\sum_{k=0}^{N-1}(\|z(k)\|^2-\gamma^2\|\xi(k)\|_U^2\right\}\end{aligned} \tag{2.49}$$

Hence, the performance index defined in (2.44) is given by

$$\begin{aligned}J_1 &= \mathbb{E}\left\{\sum_{k=0}^{N-1}\bar{\eta}^T(k)\Xi(k)\bar{\eta}(k)\right\}-\mathbb{E}\{\eta^T(N)P(N)\eta(N)\}\\ &\quad +\mathbb{E}\{\eta^T(0)(P(0)-\gamma^2\bar{S})\eta(0)\}.\end{aligned} \tag{2.50}$$

Noting that $\Xi(k)<0$, $P(N)>0$ *and the initial condition* $P(0)\leq\gamma^2\bar{S}$*, we know* $J_1<0$ *and the proof is now complete.*

B. Event-triggered Variance Analysis

Noting the saturation function in (2.36), we rearrange the closed-loop system (2.42) as follows:

$$\begin{cases}\eta(k+1)=\mathfrak{Y}_l(k)+H_0 g(k,x(k))+\sum_{i=1}^{r}w_i(k)\bar{A}_i(k)\eta(k)\\ \qquad +\tilde{B}_c(k)\sigma(\hat{C}(k)\eta(k))-\tilde{B}_c(k)\hat{C}(k)\eta(k)+\bar{D}(k)\xi(k)\\ z(k)=\bar{M}(k)\eta(k)\end{cases} \tag{2.51}$$

where

$$\mathfrak{Y}_l(k)\triangleq\bar{A}(k)\eta(k)+\bar{B}_c(k)\bar{\Lambda}_\alpha\sigma(\hat{C}(k)\eta(k))+\bar{B}_c(k)\varphi(k).$$

Let us now deal with the event-triggered variance analysis issue for the addressed nonlinear stochastic time-varying systems.

Theorem 2.4 *Consider the discrete time-varying nonlinear stochastic system described by (2.28)–(2.29). Let the scalar* $\delta\in[0,\ 1)$ *and the controller parameters* $A_c(k)$, $B_c(k)$ *and* $C_c(k)$ *in (2.41) be given. We have* $Q(k)\geqslant\mathbb{X}(k)$ $(\forall k\in\{1,2,\cdots,N+1\})$ *if, with initial condition* $Q(0)=\mathbb{X}(0)$*, there exists a sequence of positive definite matrices* $\{Q(k)\}_{1\leq k\leq N+1}$ *satisfying the following matrix inequality:*

$$Q(k+1)\geqslant\Phi(Q(k)) \tag{2.52}$$

where

$$\begin{aligned}
\Phi(Q(k)) &\triangleq 3\bar{A}(k)Q(k)\bar{A}^T(k) + 3\bar{B}_c(k)\Bigg(2\delta\Bigg(\mathrm{tr}\Big[m\big(\bar{\Lambda}_\alpha^T\bar{\Lambda}_\alpha + \bar{\alpha}(1-\bar{\alpha})I\big)\Big] \\
&\quad +\mathrm{tr}\Big[\hat{C}^T(k)\big((I-\bar{\Lambda}_\alpha)^T(I-\bar{\Lambda}_\alpha) + \bar{\alpha}(1-\bar{\alpha})I\big)\hat{C}(k)Q(k)\Big]\Bigg)I \\
&\quad +\delta\cdot\mathrm{tr}\Big[\bar{E}^T(k)\bar{E}(k)\bar{V}(k)\Big]I\Bigg)\bar{B}_c^T(k) + 3m\bar{B}_c(k)\bar{\Lambda}_\alpha\bar{\Lambda}_\alpha^T\bar{B}_c^T(k) \\
&\quad +\sum_{i=1}^{q} H_0\Theta_i(k)H_0^T\cdot\mathrm{tr}\left[\bar{\Gamma}_i(k)Q(k)\right] + \sum_{i=1}^{r}\bar{A}_i(k)Q(k)\bar{A}_i^T(k) \\
&\quad +2\bar{\alpha}(1-\bar{\alpha})\bar{B}_c(k)\hat{\mathcal{C}}(k)(I_m\otimes Q(k))\hat{\mathcal{C}}^T(k)\bar{B}_c^T(k) \\
&\quad +2\bar{\alpha}(1-\bar{\alpha})m\bar{B}_c(k)\bar{B}_c^T(k) + 2\bar{D}(k)\bar{V}(k)\bar{D}^T(k) \\
&\quad +\mu^2(k)\bar{B}_c(k)\bar{E}(k)\bar{V}(k)\bar{E}^T(k)\bar{B}_c^T(k), \\
\hat{C}_i(k) &\triangleq \Big[\hat{C}_{i,1}(k) \quad \hat{C}_{i,2}(k) \quad \cdots \quad \hat{C}_{i,n_x+n_c}(k)\Big] \ (i=1,2,\ldots,m), \\
\bar{V}(k) &\triangleq \mathrm{diag}\{V(k),\bar{W}(k)\},\ \hat{\mathcal{C}}(k)\triangleq\mathrm{diag}\{\hat{C}_1(k),\hat{C}_2(k),\ldots,\hat{C}_m(k)\}, \\
\bar{W}(k) &\triangleq \mathrm{diag}\{\bar{W}_1(k),\bar{W}_2(k),\ldots,\bar{W}_m(k)\}. \qquad (2.53)
\end{aligned}$$

Proof *From (2.51), we know that the Lyapunov-type equation governing the evolution of state covariance $\mathbb{X}(k)$ is given by*

$$\begin{aligned}
\mathbb{X}(k+1) &= \mathbb{E}\left\{\eta(k+1)\eta^T(k+1)\right\} \\
&= \mathbb{E}\Bigg\{\Bigg(\bar{\mathcal{Y}}_l(k) + H_0 g(k,x(k)) + \sum_{i=1}^{r} w_i(k)\bar{A}_i(k)\eta(k) + \tilde{B}_c(k) \\
&\quad \times\sigma(\hat{C}(k)\eta(k)) - \tilde{B}_c(k)\hat{C}(k)\eta(k) + \bar{D}(k)\xi(k)\Bigg)\Bigg(\bar{\mathcal{Y}}_l(k) \\
&\quad +H_0 g(k,x(k)) + \sum_{i=1}^{r} w_i(k)\bar{A}_i(k)\eta(k) \\
&\quad +\tilde{B}_c(k)\sigma(\hat{C}(k)\eta(k)) - \tilde{B}_c(k)\hat{C}(k)\eta(k) + \bar{D}(k)\xi(k)\Bigg)^T\Bigg\} \\
&\leq 3\bar{A}(k)\mathbb{X}(k)\bar{A}^T(k) + 3\bar{B}_c(k)\Bigg(2\delta\Bigg(\mathrm{tr}\Big[m\big(\bar{\Lambda}_\alpha^T\bar{\Lambda}_\alpha + \bar{\alpha}(1-\bar{\alpha})I\big)\Big] \\
&\quad +\mathrm{tr}\Big[\hat{C}^T(k)\big((I-\bar{\Lambda}_\alpha)^T(I-\bar{\Lambda}_\alpha) + \bar{\alpha}(1-\bar{\alpha})\big)\hat{C}(k)\mathbb{X}(k)\Big]\Bigg)I \\
&\quad +\delta\cdot\mathrm{tr}\Big[\bar{E}^T(k)\bar{E}(k)\bar{V}(k)\Big]I\Bigg)\bar{B}_c^T(k) + 3m\bar{B}_c(k)\bar{\Lambda}_\alpha\bar{\Lambda}_\alpha^T\bar{B}_c^T(k)
\end{aligned}$$

$$
\begin{aligned}
& +\sum_{i=1}^{q} H_0\Theta_i(k)H_0^T \cdot \mathrm{tr}\left[\bar{\Gamma}_i(k)\mathbb{X}(k)\right] + \sum_{i=1}^{r} \bar{A}_i(k)\mathbb{X}(k)\bar{A}_i^T(k) \\
& +2\bar{D}(k)\bar{V}(k)\bar{D}^T(k) + 2\bar{\alpha}(1-\bar{\alpha})\bar{B}_c(k)\hat{C}(k)(I_m \otimes \mathbb{X}(k))\hat{C}^T(k) \\
& \times \bar{B}_c^T(k) + 2\bar{\alpha}(1-\bar{\alpha})m\bar{B}_c(k)\bar{B}_c^T(k) \\
& +\mu^2(k)\bar{B}_c(k)\bar{E}(k)\bar{V}(k)\bar{E}^T(k)\bar{B}_c^T(k) \\
= \ & \Phi(\mathbb{X}(k)). \qquad (2.54)
\end{aligned}
$$

We now complete the proof by induction. Obviously, $Q(0) \geqslant \mathbb{X}(0)$*. Letting* $Q(k) \geqslant \mathbb{X}(k)$*, we arrive at*

$$
Q(k+1) \geqslant \Phi(Q(k)) \geqslant \Phi(\mathbb{X}(k)) \geqslant \mathbb{X}(k+1), \qquad (2.55)
$$

and therefore the proof is finished.

Remark 2.6 *In the calculation of the performance index* J_2*, namely, the state covariances, the triggering information at the time instant* k *is explicitly exploited in (33) in terms of the binary variable* $\mu(k)$*. This would definitely help reduce the conservatism and tighten the upper bound. Comparing to the traditional event-triggering mechanism where only the stability is the concern, the introduction of such a binary variable* $\mu(k)$ *would play an important role in estimating the state covariance performance.*

Next, in light of Theorem 2.4, we have the following corollary.

Corollary 2.1 *The inequality holds*

$$
\begin{aligned}
J_2 \ &:= \ \mathbb{E}\left\{x(k)x^T(k)\right\} = \begin{bmatrix} I & 0 \end{bmatrix} \mathbb{X}(k) \begin{bmatrix} I & 0 \end{bmatrix}^T = H_0^T\mathbb{X}(k)H_0 \\
&\leq \ H_0^T Q(k) H_0, \quad k \in [0, N-1].
\end{aligned}
$$

In what follows, we conclude the above analysis by presenting a theorem which aims to take both the H_∞ performance index in (2.44) and the covariance constraint in (2.45) into consideration in a unified framework.

For presentation simplicity, we denote

$$
\begin{aligned}
\bar{\Xi}_{11}(k) \ &\triangleq \ \mathrm{diag}\left\{ \sum_{i=1}^{q} \bar{\Gamma}_i(k)R_i(k) - P(k), -I \right\}, \\
\hat{A}_r(k) \ &\triangleq \ \begin{bmatrix} \bar{A}_1^T(k) & \bar{A}_2^T(k) & \cdots & \bar{A}_r^T(k) \end{bmatrix}^T, \\
\bar{\Xi}_{21}(k) \ &\triangleq \ \mathrm{diag}\left\{ \lambda_1(k)K\hat{C}(k), 0 \right\}, \ \bar{\Xi}_{22}(k) \triangleq \mathrm{diag}\left\{ -\lambda_1(k)I, -\gamma^2 U \right\}, \\
\bar{\Xi}_{31}(k) \ &\triangleq \ \begin{bmatrix} \bar{\Xi}_{311}(k) & 0 \end{bmatrix}, \ \bar{\Xi}_{422}(k) \triangleq \begin{bmatrix} 0 & \sqrt{2}\bar{D}^T(k) & \mu(k)\bar{E}^T(k)\bar{B}_c^T(k) \end{bmatrix}^T,
\end{aligned}
$$

$$
\begin{aligned}
\bar{\Xi}_{311}(k) &\triangleq \begin{bmatrix} \sqrt{2}\big(\bar{A}(k)+\bar{B}_c(k)\bar{\Lambda}_\alpha K_1\hat{C}(k)\big) \\ \hat{A}_r(k) \\ \sqrt{\bar{\alpha}(1-\bar{\alpha})}\hat{B}_{cp}(k)\bar{C}(k) \\ \sqrt{\delta}(\bar{\Lambda}_\alpha K_1+I-\bar{\Lambda}_\alpha)\hat{C}(k) \\ \sqrt{\delta}\sqrt{\bar{\alpha}(1-\bar{\alpha})}(I-K_1)\hat{C}(k) \\ \bar{M}(k) \end{bmatrix}, \\
\bar{\Xi}_{32}(k) &\triangleq \begin{bmatrix} \sqrt{2}\bar{B}_c(k)\bar{\Lambda}_\alpha & 0 \\ 0 & 0 \\ \sqrt{\bar{\alpha}(1-\bar{\alpha})}\hat{B}_{cp}(k) & 0 \\ \sqrt{\delta}\bar{\Lambda}_\alpha & \sqrt{\delta}\bar{E}(k) \\ \sqrt{\delta}\sqrt{\bar{\alpha}(1-\bar{\alpha})}I & 0 \\ 0 & 0 \end{bmatrix}, \\
\bar{\Xi}_{33}(k) &\triangleq \mathrm{diag}\Big\{-P^{-1}(k+1), -I_r\otimes P^{-1}(k+1), -I_m\otimes P^{-1}(k+1), \\
&\qquad -I,-I,-I\Big\},\ \hat{\Xi}_{22}(k)\triangleq \mathrm{diag}\left\{-Q^{-1}(k),-I,-I\right\}, \\
\bar{\Xi}_{41}(k) &\triangleq \begin{bmatrix}0 & \bar{\Xi}_{412}(k)\end{bmatrix},\ \bar{\Xi}_{42}(k)\triangleq\begin{bmatrix}0 & \bar{\Xi}_{422}(k)\end{bmatrix}, \\
\bar{\Xi}_{44}(k) &\triangleq -I_3\otimes P^{-1}(k+1),\ \bar{\Xi}_{412}(k)\triangleq\begin{bmatrix}\sqrt{2}\bar{B}_c^T(k) & 0 & 0\end{bmatrix}^T, \\
\hat{\Xi}_{21}(k) &\triangleq \begin{bmatrix}\sqrt{3}\bar{A}(k) & \tau(k)\bar{B}_c(k) & \sqrt{3m}\bar{B}_c(k)\bar{\Lambda}_\alpha\end{bmatrix}^T, \\
\hat{\Xi}_{41}(k) &\triangleq \begin{bmatrix}\breve{A}_r(k) & \sqrt{2\bar{\alpha}(1-\bar{\alpha})}\bar{B}_c(k)\hat{\mathcal{C}}(k) & \sqrt{2}\bar{D}(k) & \mu(k)\bar{B}_c(k)\bar{E}(k)\end{bmatrix}^T, \\
\bar{m}_1 &\triangleq \mathrm{tr}\Big[m\big(\bar{\Lambda}_\alpha^T\bar{\Lambda}_\alpha+\bar{\alpha}(1-\bar{\alpha})I\big)\Big], \\
\hat{\Xi}_{44}(k) &\triangleq \mathrm{diag}\Big\{-I_r\otimes Q^{-1}(k), -I_m\otimes Q^{-1}(k), -\bar{V}^{-1}(k), -\bar{V}^{-1}(k)\Big\}, \\
\breve{A}_r(k) &\triangleq \begin{bmatrix}\bar{A}_1(k) & \bar{A}_2(k) & \cdots & \bar{A}_r(k)\end{bmatrix}, \\
\tau(k) &\triangleq \sqrt{6\delta(\bar{m}_1+\rho(k))+3\delta\mathrm{tr}\Big[\bar{E}^T(k)\bar{E}(k)\bar{V}(k)\Big]+2\bar{\alpha}(1-\bar{\alpha})m}, \\
\Theta_{51} &\triangleq \begin{bmatrix}H_0\pi_1 & H_0\pi_2 & \cdots & H_0\pi_q\end{bmatrix}^T, \\
\Theta_{55} &\triangleq \mathrm{diag}\Big\{\chi_1 I, \chi_2 I, \ldots, \chi_q I\Big\}, \\
\chi_i &\triangleq \left(\mathrm{tr}\left[\bar{\Gamma}_i(k)Q(k)\right]\right)^{-1}\ \ (i=1,2,\ldots,q), \\
\rho(k) &\triangleq \mathrm{tr}\Big[\hat{C}^T(k)\big((I-\bar{\Lambda}_\alpha)^T(I-\bar{\Lambda}_\alpha)+\bar{\alpha}(1-\bar{\alpha})I\big)\hat{C}(k)Q(k)\Big].
\end{aligned} \tag{2.56}
$$

Theorem 2.5 *Consider the discrete time-varying nonlinear stochastic system described by (2.28)–(2.29). Let the disturbance attenuation level* $\gamma > 0$, *the positive definite weighted matrices* $U_1 > 0$, $U_2 > 0$ *and* $S > 0$, *the scalar* $\delta \in [0,\ 1)$ *and the controller parameters* $A_c(k)$, $B_c(k)$ *and*

$C_c(k)$ in (2.41) be given. Then, for the closed-loop system (2.42), we have $J_1 < 0$ and $J_2 < 0$ ($\forall k \in \{0,1,\ldots,N+1\}$) if there exist families of positive scalars $\{\lambda_1(k)\}_{k\in[0,\ N-1]}$, $\{R_i(k)\}_{0\leq k\leq N}$ ($i = 1,2,\cdots,q$) and families of positive definite matrices $\{M(k)\}_{1\leq k\leq N+1}$, $\{N(k)\}_{1\leq k\leq N+1}$ and $\{Q(k)\}_{1\leq k\leq N+1}$ satisfying the following recursive matrix inequalities:

$$\begin{bmatrix} -R_i(k) & * \\ H_0\pi_i & -P^{-1}(k+1) \end{bmatrix} < 0, (i = 1,2,\cdots,q) \quad (2.57)$$

$$\begin{bmatrix} \bar{\Xi}_{11}(k) & * & * & * \\ \bar{\Xi}_{21}(k) & \bar{\Xi}_{22}(k) & * & * \\ \bar{\Xi}_{31}(k) & \bar{\Xi}_{32}(k) & \bar{\Xi}_{33}(k) & * \\ \bar{\Xi}_{41}(k) & \bar{\Xi}_{42}(k) & 0 & \bar{\Xi}_{44}(k) \end{bmatrix} < 0, \quad (2.58)$$

$$\begin{bmatrix} -Q(k+1) & * & * & * \\ \hat{\Xi}_{21}(k) & \hat{\Xi}_{22}(k) & * & * \\ \Theta_{51} & 0 & -\Theta_{55} & * \\ \hat{\Xi}_{41}(k) & 0 & 0 & \hat{\Xi}_{44}(k) \end{bmatrix} \leq 0, \quad (2.59)$$

with the initial condition

$$\begin{cases} P(0) \leq \gamma^2\bar{S} \\ Q(0) = \mathbb{X}(0), \end{cases} \quad (2.60)$$

where the system data are defined in (2.56).

Proof *Based on some straightforward algebraic manipulations and under the initial conditions (2.60), we can see that the inequalities (2.57) and (2.58) imply (2.47), and the inequality (2.59) is equivalent to (2.52). Therefore, according to Theorem 2.3, Theorem 2.4 and Corollary 2.1, the H_∞ index defined in (2.44) satisfies $J_1 < 0$ and, at the same time, the system state covariances achieves $\mathbb{E}\left\{x(k)x^T(k)\right\} \leqslant \begin{bmatrix} I & 0 \end{bmatrix} Q(k) \begin{bmatrix} I & 0 \end{bmatrix}^T$, $\forall k \in \{0,1,\cdots,N+1\}$. The proof is complete.*

Up to now, the analysis problem has been dealt with for the event-triggered variance constrained control of stochastic systems with randomly occurring saturations.

Next, an algorithm would be developed to cope with the addressed event-triggered variance-constrained controller design problem for the discrete time-varying nonlinear stochastic system with randomly occurring saturations.

Theorem 2.6 *Let the disturbance attenuation level $\gamma > 0$, positive definite weighted matrices U_1, U_2 and S, a scalar $\delta \in [0,\ 1)$ and a sequence of prespecified variance upper bounds $\{\Upsilon(k)\}_{0\leqslant k\leqslant N+1}$ be given. The addressed event-triggered variance constrained controller design problem is solvable for the stochastic nonlinear system (2.42) if there exist families of positive* *definite matrices* $\{\digamma(k)\}_{1\leqslant k\leqslant N+1}$ ($\digamma = \mathcal{M},\mathcal{N},Q_1,Q_2$), *positive scalars* $\{\lambda_1(k)\}_{k\in[0,\ N-1]}$, $\{R_i(k)\}_{0\leq k\leq N}$ ($i=1,2,\cdots,q$) *and families of real-valued matrices* $\{Q_3(k)\}_{1\leqslant k\leqslant N+1}$, $\{A_c(k)\}_{0\leqslant k\leqslant N}$, $\{B_c(k)\}_{0\leqslant k\leqslant N}$ *and* $\{C_c(k)\}_{0\leqslant k\leqslant N}$ *satisfying the following recursive matrix inequalities:*

$$\begin{bmatrix} -R_i(k) & * & * \\ \pi_i & -\mathcal{M}(k+1) & * \\ 0 & 0 & -\mathcal{N}(k+1) \end{bmatrix} < 0,\ (i=1,2,\ldots,q) \tag{2.61}$$

$$\begin{bmatrix} \Phi_{11}(k) & * & * & * & * \\ \Phi_{21}(k) & \Phi_{22}(k) & * & * & * \\ \Phi_{31}(k) & \Phi_{32}(k) & \Phi_{33}(k) & * & * \\ \Phi_{41}(k) & 0 & 0 & \Phi_{44}(k) & * \\ 0 & \Phi_{52}(k) & 0 & 0 & \Phi_{55}(k) \end{bmatrix} < 0, \tag{2.62}$$

$$\begin{bmatrix} -Q(k+1) & * & * & * & * \\ \tau_1(k)\bar{\Phi}_{21}(k) & -Q(k) & * & * & * \\ \bar{\Phi}_{31}(k) & 0 & \bar{\Phi}_{33}(k) & * & * \\ \bar{\Phi}_{41}(k) & 0 & 0 & \bar{\Phi}_{44}(k) & * \\ \bar{\Phi}_{51}(k) & 0 & 0 & 0 & \bar{\Phi}_{55}(k) \end{bmatrix} < 0, \tag{2.63}$$

$$Q_1(k+1) - \Upsilon(k+1) \leqslant 0, \tag{2.64}$$

with the initial conditions

$$\begin{cases} \begin{bmatrix} M(0)-\gamma^2 S & 0 \\ 0 & N(0) \end{bmatrix} \leqslant 0 \\ \mathbb{E}\{x(0)x^T(0)\} = Q_1(0) \leq \Upsilon(0) \end{cases} \tag{2.65}$$

and the parameters updated by

$$M(k+1) = \mathcal{M}^{-1}(k+1), \quad N(k+1) = \mathcal{N}^{-1}(k+1) \tag{2.66}$$

where

$$\begin{aligned}
\bar{\mathcal{N}}(k+1) &\triangleq \text{diag}\{\mathcal{M}(k+1),\mathcal{N}(k+1)\},\ \phi_\alpha \triangleq \sqrt{\bar{\alpha}(1-\bar{\alpha})}, \\
C_k(k) &\triangleq (K_1 - I)C(k),\ Q_{13}(k) \triangleq \begin{bmatrix} Q_1^T(k) & Q_3^T(k) \end{bmatrix}^T, \\
\Phi_{11}(k) &\triangleq \text{diag}\Big\{\sum_{i=1}^{q}\Gamma_i(k)R_i(k) - M(k), -N(k), -I\Big\}, \\
\Phi_{21}(k) &\triangleq \text{diag}\{\lambda_1(k)KC(k),0,0\},\ \Phi_{31}(k) \triangleq \begin{bmatrix} \Omega_{311}(k) & \Omega_{312}(k) \\ \Omega_{321}(k) & 0 \end{bmatrix}, \\
\Phi_{22}(k) &\triangleq \text{diag}\{-\lambda_1(k)I, -\gamma^2 U_1, -\gamma^2 U_2\}, \\
\Omega_{312}(k) &\triangleq \begin{bmatrix} \sqrt{2}B_1(k)C_c(k) & 0 \\ \sqrt{2}A_c(k) & 0 \end{bmatrix},\ \Phi_{32}(k) \triangleq \begin{bmatrix} \Omega_{411}(k) & 0 \\ \Omega_{421}(k) & \Omega_{422}(k) \end{bmatrix},
\end{aligned}$$

$$
\begin{aligned}
\Omega_{411}(k) &\triangleq \begin{bmatrix} 0 \\ \sqrt{2}B_c(k)\bar{\Lambda}_\alpha \end{bmatrix},\ \Omega_{311}(k) \triangleq \begin{bmatrix} \sqrt{2}A(k) \\ \sqrt{2}B_c(k)(C(k)+\bar{\Lambda}_\alpha C_k(k)) \end{bmatrix},\\
\Omega_{422}(k) &\triangleq \begin{bmatrix} 0 & \Omega_{4222}(k) \end{bmatrix},\ \mathcal{D}(k) \triangleq \begin{bmatrix} 0 & \sqrt{2}D(k) \end{bmatrix},\\
\Omega_{321}(k) &\triangleq \Big[\ \tilde{A}_l^T(k) \quad \phi_\alpha(\hat{B}_{cp}(k)C_k(k))^T \\
&\qquad \sqrt{\delta}\big(\bar{\Lambda}_\alpha C_k(k)+C(k)\big)^T \quad -\phi_\alpha\sqrt{\delta}C_k^T(k)\ \Big]^T,\\
\Omega_{421}(k) &\triangleq \begin{bmatrix} 0 & \phi_\alpha\hat{B}_{cp}^T(k) & \sqrt{\delta}\bar{\Lambda}_\alpha & \phi_\alpha\sqrt{\delta}I \end{bmatrix}^T,\\
\Omega_{4222}(k) &\triangleq \begin{bmatrix} 0 & 0 & \sqrt{\delta}E^T(k) & 0 \end{bmatrix}^T,\ \Phi_{52}(k) \triangleq \mathrm{diag}\{\mathcal{D}(k), \mathcal{B}_c(k)\},\\
\Phi_{41}(k) &\triangleq \mathrm{diag}\Big\{ \begin{bmatrix} L(k) & B_2(k)C_c(k) \end{bmatrix}, \begin{bmatrix} 0 & \sqrt{2}B_c^T(k) \end{bmatrix}^T \Big\},\\
\Phi_{33}(k) &\triangleq \mathrm{diag}\big\{ -\mathcal{M}(k+1), -\mathcal{N}(k+1), -I_r\otimes\bar{\mathcal{N}}(k+1),\\
&\qquad -I_m\otimes\bar{\mathcal{N}}(k+1), -I, -I\big\},\\
\Phi_{44}(k) &\triangleq \mathrm{diag}\big\{ -I, -\mathcal{M}(k+1), -\mathcal{N}(k+1)\big\},\\
\mathcal{B}_c(k) &\triangleq \begin{bmatrix} \sqrt{2}E^T(k)B_c^T(k) & 0 & \mu(k)E^T(k)B_c^T(k) \end{bmatrix}^T,\\
\tilde{A}_l(k) &\triangleq \begin{bmatrix} A_{l1}^T(k) & A_{l2}^T(k) & \cdots & A_{lr}^T(k) \end{bmatrix}^T,\ \Phi_{55}(k) \triangleq -I_2\otimes\bar{\mathcal{N}}(k+1),\\
A_{li}(k) &\triangleq \begin{bmatrix} A_i^T(k) & 0 \end{bmatrix}^T\ (i=1,2,\ldots,r),\ \bar{\Phi}_{33}(k) \triangleq \mathrm{diag}\{-I,-I\},\\
\bar{\Phi}_{21}(k) &\triangleq \Big[\ \begin{matrix} (A(k)Q_1(k)+B_1(k)C_c(k)Q_3(k))^T \\ (A(k)Q_3^T(k)+B_1(k)C_c(k)Q_2(k))^T \end{matrix}\\
&\qquad \begin{matrix} (B_c(k)(I-\bar{\Lambda}_\alpha)C(k)Q_1(k)+A_c(k)Q_3(k))^T \\ (B_c(k)(I-\bar{\Lambda}_\alpha)C(k)Q_3^T(k)+A_c(k)Q_2(k))^T \end{matrix}\ \Big],\\
\bar{\Omega}_{31}(k) &\triangleq \begin{bmatrix} \hat{\tau}(k)B_c(k) & \sqrt{3m}B_c(k)\bar{\Lambda}_\alpha \end{bmatrix}^T, \bar{\Phi}_{31}(k) \triangleq \begin{bmatrix} 0 & \bar{\Omega}_{31}(k) \end{bmatrix},\\
\bar{\Phi}_{41}(k) &\triangleq \begin{bmatrix} \hat{\Theta}_{51}^T & \hat{A}_l^T(k) & 0 & \sqrt{2}D(k) \\ 0 & 0 & \sqrt{2}\phi_\alpha\hat{C}_l^T(k) & 0 \end{bmatrix}^T,\\
\hat{\chi}_i &\triangleq \Big(\mathrm{tr}\Big[\breve{\Gamma}_i(k)\Big]\Big)^{-1}\ (i=1,2,\ldots,q),\ \bar{\Phi}_{51}(k) \triangleq \begin{bmatrix} 0 & \bar{\Omega}_{51}(k) \end{bmatrix},\\
\bar{\Phi}_{44}(k) &\triangleq \mathrm{diag}\{-\hat{\Theta}_{55}, -I_r\otimes Q(k), -I_m\otimes Q(k), -V^{-1}(k)\},\\
\bar{\Phi}_{55}(k) &\triangleq \mathrm{diag}\Big\{ -\bar{W}^{-1}(k), -V^{-1}(k), -\bar{W}^{-1}(k)\Big\},\\
\hat{A}_l(k) &\triangleq \begin{bmatrix} \Pi_1^T(k) & \Pi_2^T(k) & \cdots & \Pi_r^T(k) \end{bmatrix}^T,\ \hat{\Theta}_{51} \triangleq \begin{bmatrix} \pi_1 & \pi_2 & \cdots & \pi_q \end{bmatrix}^T,\\
\hat{\rho}(k) &\triangleq \mathrm{tr}\Big(\begin{bmatrix} C(k) & 0 \end{bmatrix}^T ((I-\bar{\Lambda}_\alpha)^T(I-\bar{\Lambda}_\alpha)+\phi_a^2 I)\\
&\qquad \times \begin{bmatrix} C(k)Q_1(k) & C(k)Q_3^T(k) \end{bmatrix}\Big),
\end{aligned}
$$

$$
\begin{aligned}
\hat{\tau}(k) &\triangleq \sqrt{6\delta(\bar{m}_1+\hat{\rho}(k))+3\delta\mathrm{tr}\Big[\bar{E}^T(k)\bar{E}(k)\bar{V}(k)\Big]+2\bar{\alpha}(1-\bar{\alpha})m},\\
\hat{\Theta}_{55} &\triangleq \mathrm{diag}\Big\{\hat{\chi}_1 I,\hat{\chi}_2 I,\ldots,\hat{\chi}_q I\Big\},\\
\hat{C}_l(k) &\triangleq \begin{bmatrix}\hat{\Pi}_1^T(k) & \hat{\Pi}_2^T(k) & \cdots & \hat{\Pi}_m^T(k)\end{bmatrix}^T,\\
\Pi_i(k) &\triangleq Q_{13}(k)A_i^T(k)\ (i=1,2,\ldots,r),\\
\bar{\Omega}_{51}(k) &\triangleq \begin{bmatrix}\sqrt{2}B_c(k)E(k) & 0 & \mu(k)B_c(k)E(k)\end{bmatrix}^T,\\
\hat{\Pi}_i(k) &\triangleq Q_{13}(k)C_i^T(k)\big[B_{c_{1,i}}^T(k)\ B_{c_{2,i}}^T(k)\ \cdots\ B_{c_{n_x,i}}^T(k)\big](i=1,2,\ldots,m).
\end{aligned}
$$

Proof *Decompose the variable* $Q(k)$ *as* $Q(k) = \begin{bmatrix}Q_1(k) & * \\ Q_3(k) & Q_2(k)\end{bmatrix}$. *Meanwhile, noticing that* $P(k) = \mathrm{diag}\{M(k), N(k)\}$, *we define* $P^{-1}(k) \triangleq \mathrm{diag}\{\mathcal{M}(k),\mathcal{N}(k)\}$. *It is easy to see that the inequalities (2.57)–(2.59) are equivalent to (2.61)–(2.63), respectively. Therefore, according to Theorem 2.5, we have* $J_1 < 0$ *and* $\mathbb{E}\{x(k)x^T(k)\} \leqslant \begin{bmatrix} I & 0\end{bmatrix} Q(k) \begin{bmatrix} I & 0\end{bmatrix}^T$, $\forall k \in \{0,1,\cdots,N+1\}$. *From (2.64), it is obvious that*

$$
\mathbb{E}\{x(k)x^T(k)\} \leqslant \begin{bmatrix} I & 0\end{bmatrix} Q(k) \begin{bmatrix} I & 0\end{bmatrix}^T \leq \Upsilon_k, \quad \forall k \in \{0,1,\cdots,N\}.
$$

It can now be concluded that the requirements $(Q1)$ *and* $(Q2)$ *are simultaneously satisfied. The proof is complete.*

By means of Theorem 2.6, we can summarize the Event-triggered Variance Constrained Controller Design (*EVCCD*) algorithm as follows:

The *EVCCD* algorithm:	
Step 1.	Given the disturbance attenuation level γ, the positive definite weighted matrices $U_1 > 0$, $U_2 > 0$, $S > 0$ and the scalar $\delta \in [0,\ 1)$.
Step 2.	Set $k = 0$ and solve the matrix inequalities (2.65) to obtain the matrices $M(0)$, $N(0)$ and $Q_1(0)$ $(i = 1,2,\ldots,l)$.
Step 3.	Obtain the matrices $\{\mathcal{M}(k+1),\mathcal{N}(k+1),Q_1(k+1), Q_2(k+1),Q_3(k+1)\}$ and the desired controller parameters $\{A_c(k),B_c(k),C_c(k)\}$ by solving the matrix inequalities (2.61)–(2.64).
Step 4.	Set $k = k+1$ and obtain $\{M(k+1)\}$ and $\{N(k+1)\}$ by the parameter update formula (2.66).
Step 5.	If $k < N$, then go to Step 3, else go to Step 6.
Step 6.	Stop.

Remark 2.7 *It can be observed from Algorithm EVCCD that the main results established contain all the information of the addressed general systems including the time-varying systems parameters, multiplicative noise, the threshold of event trigger, the occurrence probabilities of the random nonlinearity and the sensor saturation. On the other hand, we point out that both the current system measurement and previous system states are employed to control the current system state. Such a recursive control process is particularly useful for real-time implementation such as online process control.*

2.3 Illustrative Examples

In this section, two simulation examples are presented to demonstrate the effectiveness and applicability of the theory presented in this chapter.

2.3.1 Example 1

In this example, we aim to demonstrate the effectiveness and applicability of the proposed method in Section 2.1. The system model is concerned with one of the test runs of an aircraft which is powered by energy from two F-404 engines. Both engines are mounted close together in the aft fuselage. We are interested in tracking such an aircraft through wireless communications subject to fading channels and multiplicative noises. In this simulation, the nominal system matrix A and the measurement output matrix C are taken from the linearized model of an F-404 aircraft engine system in [49]:

$$A(k) = \begin{bmatrix} -1.4600 & 0 & 2.4280 \\ 0.1643 & -0.4000 & -0.3788 \\ 0.3107 & 0 & -2.2300 \end{bmatrix}, \; C(k) = \begin{bmatrix} 1 & 0 & 0 \\ 0 & 1 & 0 \end{bmatrix}.$$

Setting the sampling time $T = 0.5s$, we obtain the following discretized nominal system matrices

$$A(k) = \begin{bmatrix} 0.5227 & 0 & 0.5009 \\ 0.0458 & 0.8187 & -0.0783 \\ 0.0641 & 0 & 0.3638 \end{bmatrix}, \; C(k) = \begin{bmatrix} 0.6487 & 0 & 0 \\ 0 & 0.6487 & 0 \end{bmatrix}.$$

As discussed in [185], virtually all aircraft engine systems are in some way disturbed by uncontrolled external forces. The disturbances may assume a myriad of forms such as wind gusts, gravity gradients, structural vibrations, or sensor and actuator noise, and may enter the systems in many different ways. These perturbations generally degrade the performance of the system and, in some cases, may even jeopardize the outcome of the engineering task.

For example, the random vibration of an aircraft engine system would have a major impact on the accurate fatigue analysis as well as the design of engine control systems [87]. As in [63], we suppose that the motion of the F-404 aircraft engine can be determined by the system of stochastic differential equations derived from the basic aerodynamics, and the stochastic part of the motion is due to the changing wind.

In the F-404 aircraft engine model, $x_1(k)$ and $x_2(k)$ represent the horizontal position and $x_3(k)$ is the altitude of the aircraft. Our purpose is to design a time-varying filter in the form of (2.8) in a network environment. The movement of the aircraft is affected by the wind that acts as stochastic disturbances. In fact, when modelling the aircraft engine system, there exist modelling errors (state-multiplicative noises) and linearization errors (nonlinear disturbances). Moreover, in the scenario of tracking the aircraft through wireless communications, both fading channels and multiplicative noises are often unavoidable. To this end, the corresponding parameters are given as follows:

$$A_1(k) = \begin{bmatrix} 0.05 & -0.1 & 0 \\ 0 & 0.02\sin(k) & 0.1 \\ 0.01 & 0 & 0.2 \end{bmatrix}, \quad A_2(k) = \begin{bmatrix} 0.05\sin(k) & 0 & 0 \\ 0 & 0.02 & 0 \\ 0.1 & 0.01 & 0.02 \end{bmatrix},$$

$$D_1(k) = \begin{bmatrix} 0.2 & -0.05 & 0.01 \end{bmatrix}^T, \ L(k) = \begin{bmatrix} 1 & 1 & 1 \end{bmatrix},$$
$$D_2(k) = \begin{bmatrix} 0.3 & -0.05 \end{bmatrix}^T, \ D_3(k) = \begin{bmatrix} 0 & 0.1 \end{bmatrix}^T.$$

To track the state of the F-404 aircraft engine system, the RONs should be taken into account due to the unpredictable changes of the environmental circumstances. In practice, the probability $\alpha(k)$ can be determined beforehand thorough statistical tests. In this illustrative example, the probability of randomly occurring nonlinearities is taken as $\bar{\alpha} = 0.7$ and the nonlinear vector-valued function $g(k, x(k))$ is chosen as

$$g(k, x(k)) = \begin{bmatrix} -0.5x_1(k) + 0.4x_2(k) + 0.1x_3(k) \\ 0.1x_1(k) + \frac{\sin x_1(k)}{\sqrt{x_1^2(k)+x_2^2(k)+10}} \\ 0.5x_2(k) \end{bmatrix}$$

where $x_i(k)$ $(i = 1, 2, 3)$ denotes the i-th element of the system state $x(k)$. It is easy to see that the constraint (2.2) is met with

$$\Phi(k) = \begin{bmatrix} -0.2 & 0.4 & 0.1 \\ 0.05 & 0 & 0 \\ 0 & 0.2 & 0 \end{bmatrix}, \quad \Psi(k) = \begin{bmatrix} -0.8 & 0.4 & 0.1 \\ 0.15 & 0 & 0 \\ 0 & 0.8 & 0 \end{bmatrix}.$$

The order of the fading model is $l = 1$ and the probability density functions of channel coefficients are

$$\begin{cases} \varrho(\beta_0(k)) = 0.0005(e^{9.89\beta_0(k)} - 1), & 0 \le \beta_0(k) \le 1, \\ \varrho(\beta_1(k)) = 8.5017e^{-8.5\beta_1(k)}, & 0 \le \beta_1(k) \le 1. \end{cases}$$

TABLE 2.1
The Filter Parameters $K(k)$

k	0	1	2
$K(k)$	$\begin{bmatrix} 0.3376 & 0.4775 \\ 0.4476 & 0.4285 \\ 0.4575 & 0.4726 \end{bmatrix}$	$\begin{bmatrix} 0.3302 & 0.4091 \\ 0.4149 & 0.2967 \\ 0.4657 & 0.4363 \end{bmatrix}$	$\begin{bmatrix} 0.1377 & 0.0046 \\ 0.2056 & -0.0236 \\ 0.3276 & 0.0040 \end{bmatrix}$
k	...	...	50
$K(k)$	...	...	$\begin{bmatrix} 0.1270 & -0.1241 \\ 0.3614 & -0.2708 \\ 0.1956 & -0.1346 \end{bmatrix}$

The mathematical expectation $\bar{\beta}_s$ and variance ν_s $(s = 0, 1)$ can be obtained as 0.8991, 0.1174, 0.0133 and 0.01364, respectively.

The H_∞ performance level γ, the positive definite weighted matrices U, V_i $(i = -1, 0)$ are chosen as $\gamma = 1$, $U = I$, $V_{-1} = V_0 = 5I$, respectively. Choose event weighted matrix $\Omega = I$ and the threshold $\delta = 0.6$. As long as it goes beyond the established threshold, updates are triggered such that the value $\|\sigma(k)\|$ is reset to zero again. By applying Algorithm *FHFD*, the desired filter parameters are obtained and listed in Table 2.1.

In the simulation, the initial value of the state is $x(0) = [\ -0.55 \quad -0.16 \quad 0]^T$ and the exogenous disturbance inputs are selected as

$$\xi(k) = 0.5e^{-2k}\sin(4k), \quad v(k) = \frac{4}{k+20}\sin(k). \tag{2.67}$$

Fig. 2.1 plots the measurement $y(k)$ and the measurement $y(k_i)$ for event-triggered instants, and the outputs $z(k)$ and the filtering errors $\tilde{z}(k)$ are depicted in Fig. 2.2 and Fig. 2.3, respectively.

For $\delta = 0$, that is, no event triggering happens, Fig. 2.4 plots the measurement $y(k)$ and the measurement $y(k_i)$ for event-triggered instants. The corresponding outputs $z(k)$ and the filtering errors $\tilde{z}(k)$ are depicted in Fig. 2.5 and Fig. 2.6, respectively. It can be seen from the simulation results that the larger δ the worse the filtering performance, which is in agreement with the fact that event triggering is based on the relative error with respect to the output signal. Clearly, the bandwidth utilization cannot be reduced too much in order to guarantee certain filtering performance. All the simulation results confirm that the approach addressed in Section 2.1 provides a satisfactory filtering performance.

2.3.2 Example 2

In this example, we aim to demonstrate the effectiveness and applicability of the proposed method in Section 2.2.

Following [55], we consider the networked control problem for an industrial continuous-stirred tank reactor system, where chemical species A react to form

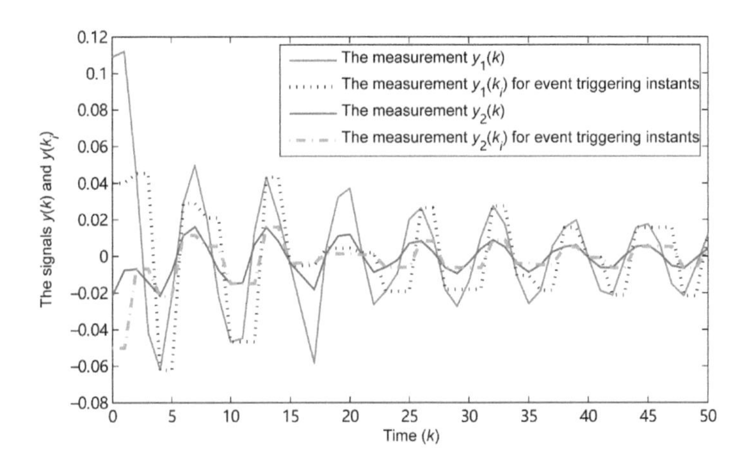

FIGURE 2.1
The measurement $y(k)$ and the measurement $y(k_i)$ for event-triggered instants when $\delta = 0.6$.

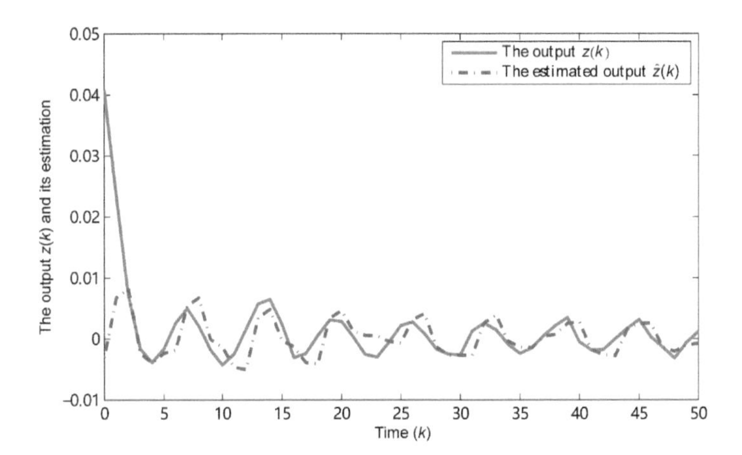

FIGURE 2.2
The output $z(k)$ and its estimation when $\delta = 0.6$.

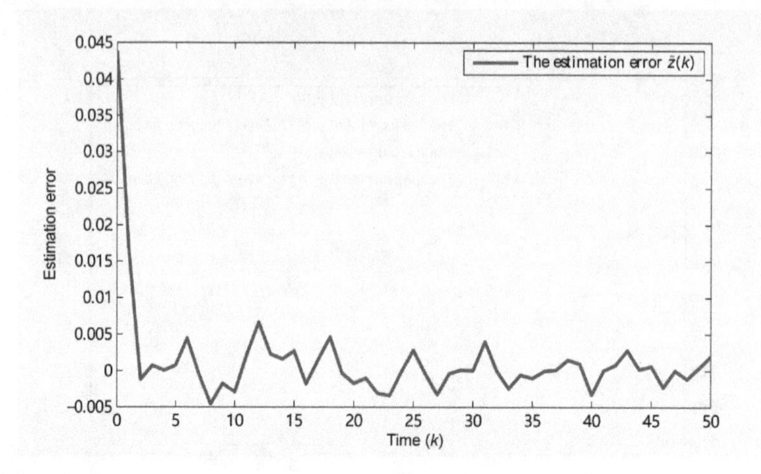

FIGURE 2.3
The estimation error $\tilde{z}(k)$ when $\delta = 0.6$.

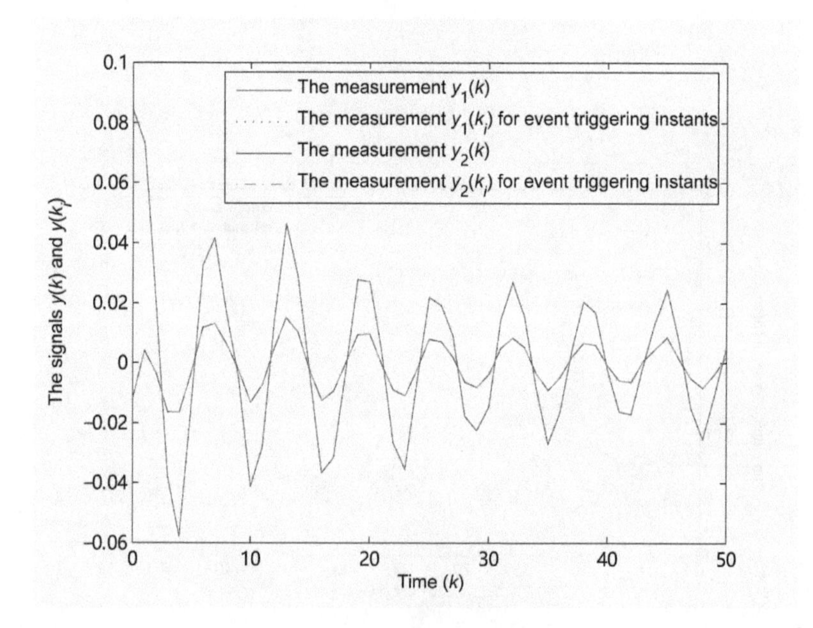

FIGURE 2.4
The measurement $y(k)$ and the measurement $y(k_i)$ for event-triggered instants when $\delta = 0$.

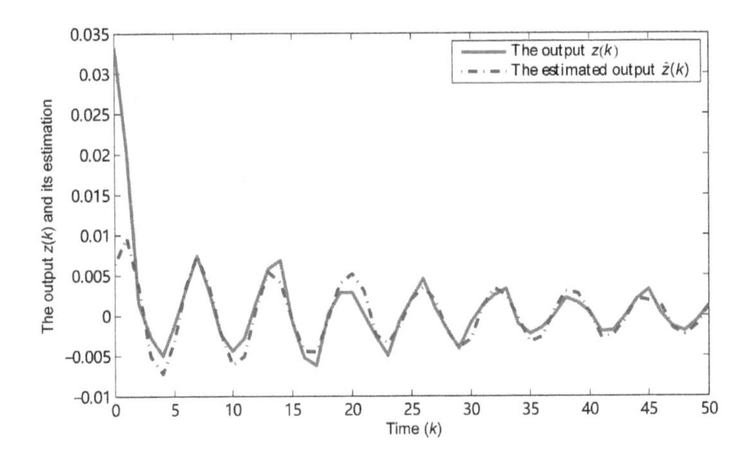

FIGURE 2.5
The output $z(k)$ and its estimation when $\delta = 0$.

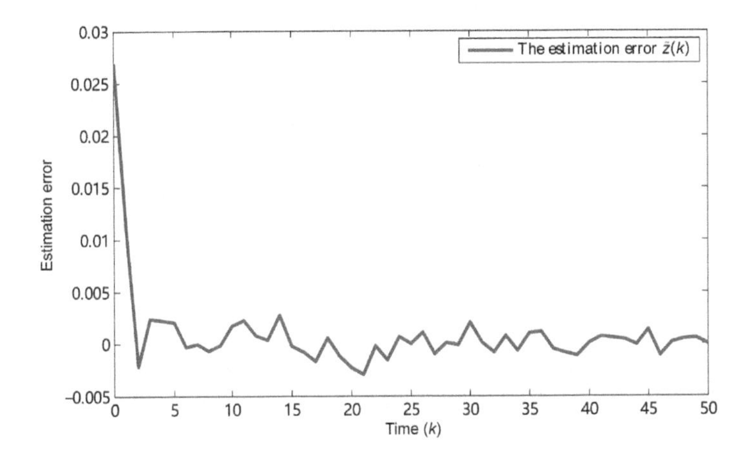

FIGURE 2.6
The estimation error $\tilde{z}(k)$ when $\delta = 0$.

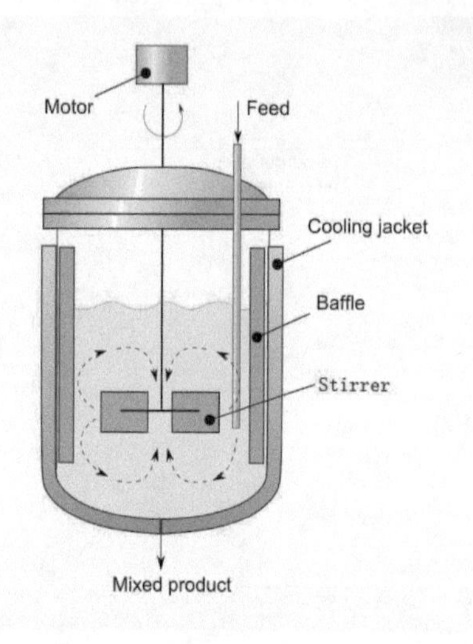

FIGURE 2.7
Cross-sectional diagram of continuous flow stirred-tank reactor.

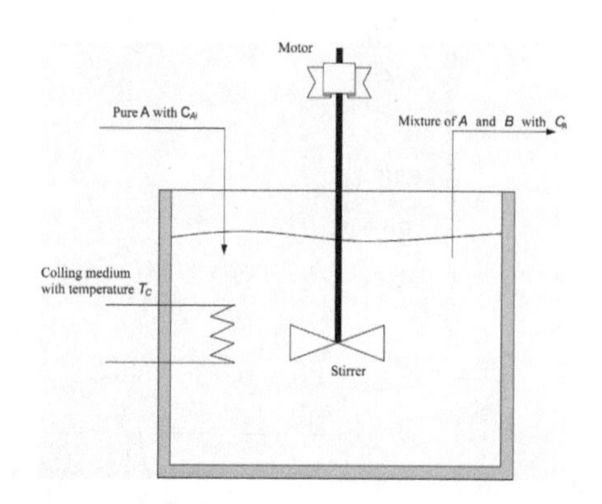

FIGURE 2.8
A continuous-stirred tank reactor model.

species B. Fig. 2.7 shows the cross-sectional diagram of continuous flow stirred-tank reactor and Fig. 2.8 illustrates the physical structure of the system, where C_{Ai}, C_A, T, T_C are, respectively, the input concentration of a key reactant A, the output concentration of chemical species A, the reaction temperature and the cooling medium temperature.

When modelling the industrial continuous-stirred tank reactor system, there exist modelling errors (state-multiplicative noises) and linearization errors (nonlinear disturbances). Moreover, since the system is in a network environment, the sensor saturations may occur in a probabilistic way and are randomly changeable in terms of their types and/or levels due to the random occurrence of networked-induced phenomena such as random sensor failures, sensor aging, or sudden environment changes.

By selecting the state and input variables as $x = \begin{bmatrix} C_A^T & T^T \end{bmatrix}^T$, $u = \begin{bmatrix} T_C^T & C_{Ai}^T \end{bmatrix}^T$. A discrete-space model is obtained as the form of (2.28)–(2.29), where system matrix $A(k)$ and the control matrix $B_1(k)$ are taken from the linearized model of an industrial continuous-stirred tank reactor system in [55]:

$$A(k) = \begin{bmatrix} 0.9719 & 0.0013 \\ 0.0340 & 0.8628 \end{bmatrix}, \quad B_1(k) = \begin{bmatrix} 0.0839 & 0.0232 \\ 0.0761 & 0.4144 \end{bmatrix}.$$

Our purpose is to design a time-varying controller in the form of (2.41) in order to control the cooling medium temperature T_C and the input concentration C_{Ai} of a key reactant A in a network environment. To this end, other parameters are set as follows:

$$A_1(k) = \begin{bmatrix} 0.05\cos(k) & -0.1 \\ 0.1 & 0.02\sin(k) \end{bmatrix}, \ A_2(k) = \begin{bmatrix} 0.05\sin(k) & 0.1 \\ \sin(k) & 0.02 \end{bmatrix},$$
$$D(k) = \begin{bmatrix} 0.2 & -0.05 \end{bmatrix}^T, \ L(k) = \begin{bmatrix} 0.1\sin(k) & -0.3 \end{bmatrix},$$
$$B_2(k) = \begin{bmatrix} 0.3\sin(k) & 0.1 \end{bmatrix}, \ C_1(k) = \begin{bmatrix} -0.1\sin(k) & 0.05 \end{bmatrix},$$
$$C_2(k) = \begin{bmatrix} -0.2 & 0.05 \end{bmatrix}, \ C_3(k) = \begin{bmatrix} 0.1 & -0.3\sin(k) \end{bmatrix},$$
$$E_1(k) = E_2(k) = E_3(k) = 0.1.$$

The disturbances $w_1(k)$, $w_2(k)$, $v(k)$ and $\varpi(k)$ are mutually independent Gaussian distributed sequences with unity variances. The probability is taken as $\bar{\alpha} = 0.8$. The saturation level is set to be 1, and other parameters are chosen as $K_1 = 0.4$, $K = 0.6$, $V(k) = 0.5$, $\bar{W}_1(k) = 0.8$, $\bar{W}_2(k) = 0.8$ and $\bar{W}_3(k) = 1$.

The nonlinear function $g(k, x(k))$ is chosen as follows:

$$g(k, x(k)) = \begin{bmatrix} 0.1 \\ 0.3 \end{bmatrix} \times \big(0.2x_1(k)\xi_1(k) + 0.3x_2(k)\xi_2(k)\big)$$

where $x_i(k)$ $(i = 1, 2)$ is the ith element of $x(k)$, and $\xi_i(k)$ $(i = 1, 2)$ are zero mean, uncorrelated Gaussian white noise processes with unity variances that are also uncorrelated with $w_1(k)$, $w_2(k)$, $v(k)$ and $\varpi(k)$. The positive definite weighted matrices U_1, U_2, S and variance upper bounds $\{\Upsilon(k)\}_{0 \le k \le N+1}$ are chosen as $U_1 = U_2 = I$, $S = 2I$ and $\{\Upsilon(k)\}_{0 \le k \le N+1} = \text{diag}\{1.5, 2.5\}$, respectively. The event-triggered transmission threshold is chosen as $\delta = 0.04$. As long as it goes beyond the established threshold, updates are triggered such that the value $\|\varphi(k)\|$ is reset to zero again. By applying Algorithm *EVCCD*, the desired controller parameters are obtained and listed in Table 2.2 with the optimized performance index $\gamma^* = 0.0882$.

TABLE 2.2
The Controller Parameters $A_c(k)$, $B_c(k)$ and $C_c(k)$

k	0	1
$A_c(k)$	$\begin{bmatrix} -0.2929 & 1.2502 \\ 1.7750 & 1.2936 \end{bmatrix}$	$\begin{bmatrix} 1.0060 & -0.2968 \\ 2.8505 & -0.7368 \end{bmatrix}$
$B_c(k)$	$\begin{bmatrix} 0.0576 & 0.0576 & 0.0572 \\ 0.0599 & 0.0599 & 0.0593 \end{bmatrix}$	$\begin{bmatrix} 0.0605 & 0.0604 & 0.0558 \\ 0.0641 & 0.0640 & 0.0564 \end{bmatrix}$
$C_c(k)$	$\begin{bmatrix} 1.0172 & 2.2310 \\ 1.6678 & 2.7745 \end{bmatrix}$	$\begin{bmatrix} 3.8643 & -1.5779 \\ 4.8421 & -1.6803 \end{bmatrix}$
k	...	30
$A_c(k)$	...	$\begin{bmatrix} 1.1233 & -0.3616 \\ 3.4398 & -0.9951 \end{bmatrix}$
$B_c(k)$	...	$\begin{bmatrix} 0.0592 & 0.0592 & 0.0558 \\ 0.0648 & 0.0648 & 0.0562 \end{bmatrix}$
$C_c(k)$	...	$\begin{bmatrix} 3.9808 & -1.6727 \\ 5.5280 & -1.9993 \end{bmatrix}$

TABLE 2.3
The Performance Comparison with Different Threshold δ

The event -triggered transmission threshold δ	The optimized performance index γ^*	The range of variance upper bound $\mathrm{Var}_{\max}$ and actual range of variance Var for $x_1(k)$	$\mathrm{Var}_{\max}$ and Var for $x_2(k)$
$\delta = 0$	$\gamma^* = 0.0812$	$\mathrm{Var}_{\max}(x_1)$: 0.5397 − 1.1822 $\mathrm{Var}(x_1)$: 0.4899 − 0.0236	$\mathrm{Var}_{\max}(x_2)$: 0.36 − 2.0223 $\mathrm{Var}(x_2)$: 0.2037 − 0.0186
$\delta = 0.08$	$\gamma^* = 0.0935$	$\mathrm{Var}_{\max}(x_1)$: 0.5397 − 1.6470 $\mathrm{Var}(x_1)$: 0.4899 − 0.0236	$\mathrm{Var}_{\max}(x_2)$: 0.36 − 3.5291 $\mathrm{Var}(x_2)$: 0.2037 − 0.0186
$\delta = 0.2$	$\gamma^* = 0.1039$	$\mathrm{Var}_{\max}(x_1)$: 0.5397 − 2.3572 $\mathrm{Var}(x_1)$: 0.4899 − 0.0232	$\mathrm{Var}_{\max}(x_2)$: 0.36 − 5.7347 $\mathrm{Var}(x_2)$: 0.2037 − 0.0181
$\delta = 0.6$	$\gamma^* = 0.1151$	$\mathrm{Var}_{\max}(x_1)$: 0.5397 − 4.4912 $\mathrm{Var}(x_1)$: 0.4899 − 0.0178	$\mathrm{Var}_{\max}(x_2)$: 0.36 − 19.5683 $\mathrm{Var}(x_2)$: 0.2037 − 0.0181
$\delta = 0.8$	$\gamma^* = 0.1975$	$\mathrm{Var}_{\max}(x_1)$: 0.5397 − 6.8032 $\mathrm{Var}(x_1)$: 0.4899 − 0.0232	$\mathrm{Var}_{\max}(x_2)$: 0.36 − 49.1282 $\mathrm{Var}(x_2)$: 0.2037 − 0.0179

TABLE 2.4
The Average Event-Triggered Ratio with Different Threshold δ

The event-triggered threshold δ	$\delta = 0$	$\delta = 0.08$	$\delta = 0.2$	$\delta = 0.6$	$\delta = 0.8$
The event-triggered law	100%	68.2%	51.6%	46.7%	43.3%

In the simulation, the initial value of the state is $x(0) = \begin{bmatrix} -0.7 & 0.6 \end{bmatrix}^T$. The simulation results are shown in Figs. 2.9–2.12, where Fig. 2.9 depicts the state evolution $x(k)$ and Fig. 2.10 plots the input control signal $u(k)$. Fig. 2.11 shows the measurement $y(k)$ and the measurement $y(k_i)$ for event-triggered instants, and the variance upper bound and actual variance are given in Fig. 2.12.

Now, let us show how the event triggering threshold δ in the equation (2.38) affect the H_∞ performance and the variance. In Table 2.3, the threshold δ versus the optimal disturbance attenuation level γ^*, the range of variance upper bound and actual range of variance are provided. It can be seen clearly that the larger δ the worse the H_∞ performance and the larger range of variance upper bound, which is in agreement with the fact that event triggering is based on the relative error with respect to the output signal. Clearly, the bandwidth utilization cannot be reduced too much in order

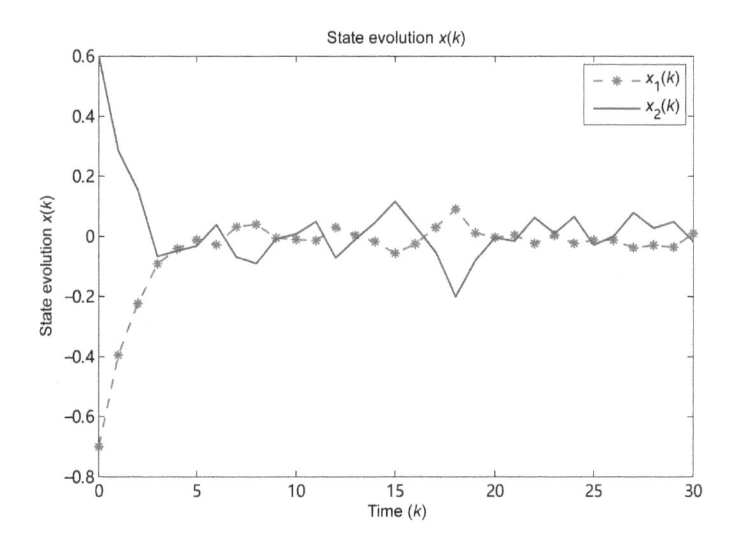

FIGURE 2.9
The state evolution $x(k)$.

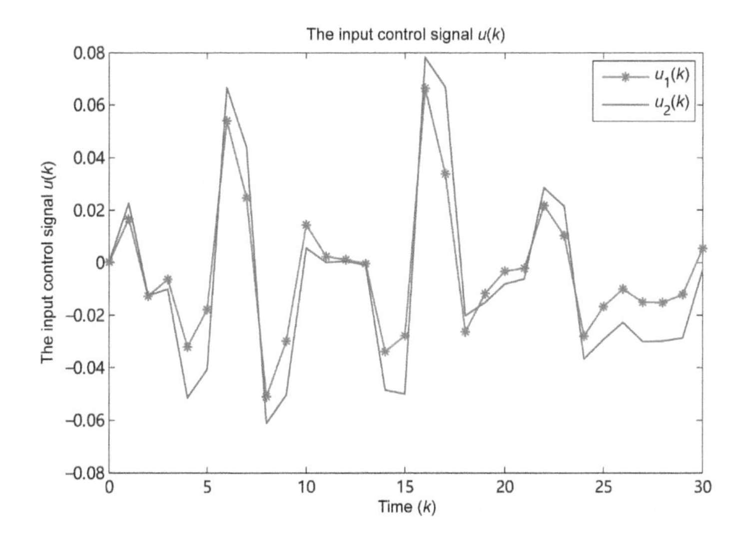

FIGURE 2.10
The input control signal $u(k)$.

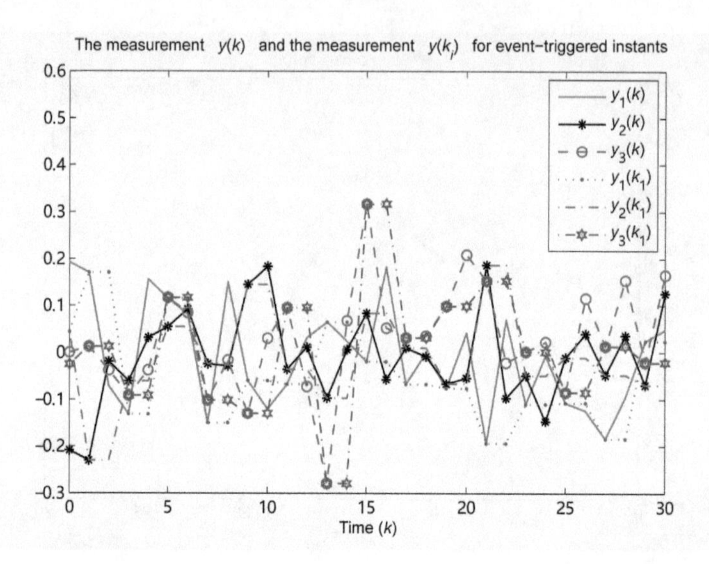

FIGURE 2.11
The measurement $y(k)$ and the measurement $y(k_i)$ for event-triggered instants.

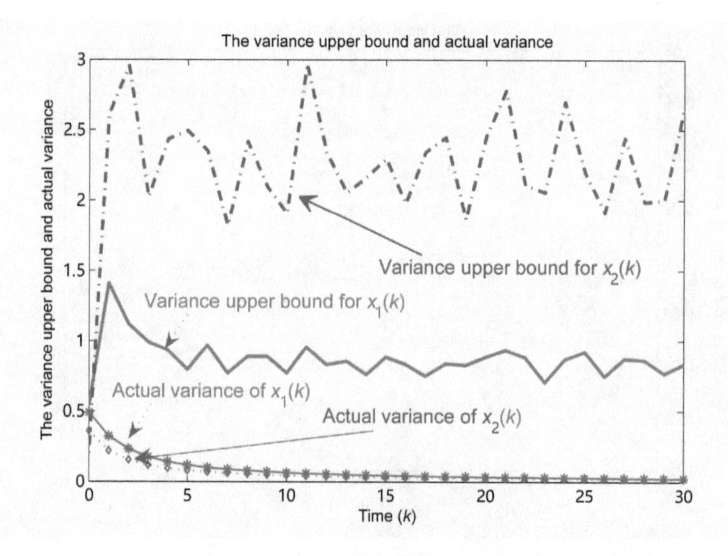

FIGURE 2.12
The variance upper bound and actual variance.

to guarantee certain control performance. Furthermore, by conducting 200 independent simulation trials, the average event-triggered ratio (number of event-triggered updates over total number of time points) is displayed in Table 2.4, which shows that the number of event-triggered updates is quite small and the communication burden is effectively reduced. All the simulation results confirm that the approach addressed in Section 2.2 provides a desired finite-horizon performance and the proposed *EVCCD* algorithm is indeed effective.

2.4 Summary

In this chapter, we have dealt with the event-based filtering problem for time-varying systems with multiplicative noise. Firstly, an event indicator variable has been created and the corresponding event-triggered scheme has been proposed. Some uncorrelated random variables have been introduced, respectively, to govern the phenomena of state-multiplicative noises and stochastic nonlinearities. Sufficient conditions have been provided to ensure the addressed problem and the desired filter gain matrices have been derived in terms of solving recursive matrix inequalities. Finally, the effectiveness and applicability of the developed algorithms have been demonstrated by two simulation examples.

3

Finite-Horizon Reliable Control Subject to Output Quantization

In this chapter, we consider the finite-horizon reliable H_∞ output feedback control problem for a class of discrete time-varying systems with ROUs, RONs as well as measurement quantizations. Both the deterministic actuator failures and probabilistic sensor failures are considered in order to reflect the reality. The actuator failure is quantified by a deterministic variable varying in a given interval and the sensor failure is governed by an individual random variable taking value on $[0, 1]$. Both the nonlinearities and the uncertainties enter into the system in random ways according to Bernoulli distributed white sequences with known conditional probabilities. The main purpose of the problem addressed is to design a time-varying output feedback controller over a given finite horizon such that, in the simultaneous presence of ROUs, RONs, actuator and sensor failures as well as measurement quantizations, the closed-loop system achieves a prescribed performance level in terms of the H_∞-norm. Sufficient conditions are first established for the robust H_∞ performance through intensive stochastic analysis, and then a recursive linear matrix inequality approach is employed to design the desired output feedback controller achieving the prescribed H_∞ disturbance rejection level. Finally, a numerical example is given to demonstrate the effectiveness of the proposed design scheme.

3.1 Problem Formulation

In this chapter, we consider the following uncertain discrete time-varying nonlinear stochastic system defined on $k \in [0, N]$:

$$\begin{cases} x(k+1) = (A(k) + \alpha(k)\Delta A(k))x(k) + B_1(k)u(k) \\ \qquad\qquad + \beta(k)g(k, x(k)) + D(k)w(k) \\ z(k) = M(k)x(k) + B_2(k)u(k) \\ x(0) = \varphi_0 \end{cases} \tag{3.1}$$

where $x(k) \in \mathbb{R}^{n_x}$ represents the state vector; $u(k) \in \mathbb{R}^{n_u}$ is the control input; $z(k) \in \mathbb{R}^{n_z}$ is the controlled output; $w(k) \in \mathbb{R}^{n_w}$ is the disturbance

DOI: 10.1201/9781003189497-3

input which belongs to $l[0,N]$; and φ_0 is a given real initial value. $A(k)$, $B_1(k)$, $B_2(k)$, $D(k)$ and $M(k)$ are known, real, time-varying matrices with appropriate dimensions.

The nonlinear function $g(k,x(k))$ satisfies the following condition:

$$\|g(k,x(k))\|^2 \leq \varepsilon(k)\|E(k)x(k)\|^2 \tag{3.2}$$

where $\varepsilon(k)>0$ is a known positive scalar and $E(k)$ is a known time-varying matrix.

The real-valued matrix $\Delta A(k)$ represents the norm-bounded parameter uncertainties of the following structure:

$$\Delta A(k) = H_a(k)F(k)N(k), \tag{3.3}$$

where $H_a(k)$ and $N(k)$ are known time-varying matrices, while $F(k)$ is an unknown time-varying matrix function satisfying the following condition

$$F^T(k)F(k) \leq I. \tag{3.4}$$

The stochastic variables $\alpha(k)\in\mathbb{R}$ and $\beta(k)\in\mathbb{R}$ are Bernoulli distributed white sequences taking values on either 0 or 1 with

$$\begin{aligned}\text{Prob}\{\alpha(k)=1\} &= \bar{\alpha}, \quad \text{Prob}\{\alpha(k)=0\}=1-\bar{\alpha},\\ \text{Prob}\{\beta(k)=1\} &= \bar{\beta}, \quad \text{Prob}\{\beta(k)=0\}=1-\bar{\beta},\end{aligned} \tag{3.5}$$

where $\bar{\alpha}\in[0,1]$ and $\bar{\beta}\in[0,1]$ are known constants.

Remark 3.1 *The random variables $\alpha(k)$ and $\beta(k)$ are introduced to characterize the phenomena of the ROUs and the RONs, respectively. The uncertainty ΔA and the nonlinearity g enter into the system in random ways according to individual Bernoulli distributions. Such a statistics description is more suitable for reflecting parameter/nonlinear variations that are unpredictable but appear in a random way with known probability laws.*

In this chapter, the measurement $\bar{y}(k)$ with probabilistic sensor failures is described by

$$\begin{aligned}\bar{y}(k) &= \Xi(k)C(k)x(k)+\bar{E}(k)v(k)\\ &= \sum_{j=1}^{n_y}\rho_j(k)C_j(k)x(k)+\bar{E}(k)v(k)\end{aligned} \tag{3.6}$$

where $v(k)\in\mathbb{R}^{n_v}$ is the measure noises belonging to $l[0,N]$. $\Xi(k) := \text{diag}\{\rho_1(k),\ldots,\rho_{n_y}(k)\}$ where $\rho_j(k)$ has the probability density function $\zeta_j(s)$ $(j=1,\ldots,n_y)$ on the interval $[0,1]$ with mathematical expectation $\bar{\mu}_j$ and variance χ_j^2. $C_j(k)$ is defined by

$$C_j(k) \triangleq \text{diag}\{\underbrace{0,\cdots,0}_{j-1},1,\underbrace{0,\cdots,0}_{n_y-j}\}C(k).$$

In the sequel, we denote $\bar{\Xi} \triangleq \mathbb{E}\{\Xi(k)\} = \mathrm{diag}\{\bar{\mu}_1, \bar{\mu}_2, \ldots, \bar{\mu}_{n_y}\}$. Throughout the chapter, we assume that $\alpha(k)$, $\beta(k)$ and $\rho_j(k)$ $(j = 1, ..., n_y)$ are unrelated random variables.

In a networked environment, it is often the case that $\bar{y}(k)$ is quantized before transmitted to the controller. Let us denote the quantizer as $q(\cdot) \triangleq \begin{bmatrix} q_1(\cdot) & q_2(\cdot) & \cdots & q_{n_y}(\cdot) \end{bmatrix}^T$ which is symmetric, i.e., $q_j(-v) = -q_j(v)$ $(j = 1, \ldots, n_y)$. The map of the quantization process is

$$y(k) = q(\bar{y}(k)) \triangleq \begin{bmatrix} q_1(\bar{y}^{(1)}(k)) & q_2(\bar{y}^{(2)}(k)) & \cdots & q_{n_y}(\bar{y}^{(n_y)}(k)) \end{bmatrix}^T$$

where $y(k) \in \mathbb{R}^{n_y}$ and $\bar{y}^{(i)}(k)$ $(i = 1, \ldots, n_y)$ denotes the ith element of $\bar{y}(k)$. In this chapter, we are interested in the logarithmic static and time-invariant quantizer. For each $q_j(\cdot)$ $(1 \le j \le m)$, the set of quantization levels is described by

$$\mathscr{U}_j \triangleq \{\pm\hat{\mu}_i^{(j)}, \hat{\mu}_i^{(j)} = \bar{\chi}_j^i \hat{\mu}_0^{(j)}, i = 0, \pm 1, \pm 2, \cdots\} \cup \{0\}, \;\; 0 < \bar{\chi}_j < 1, \; \hat{\mu}_0^{(j)} > 0,$$

and each of the quantization level corresponds to a segment such that the quantizer maps the whole segment to this quantization level.

According to [53], the logarithmic quantizer is given by

$$q_j(\bar{y}^{(j)}(k)) = \begin{cases} \hat{\mu}_i^{(j)}, & \frac{1}{1+\delta_j}\hat{\mu}_i^{(j)} \le \bar{y}^{(j)}(k) \le \frac{1}{1-\delta_j}\hat{\mu}_i^{(j)} \\ 0, & \bar{y}^{(j)}(k) = 0 \\ -h_j(-\bar{y}^{(j)}(k)), & \bar{y}^{(j)}(k) < 0 \end{cases}$$

where $\delta_j \triangleq (1 - \bar{\chi}_j)/(1 + \bar{\chi}_j)$. It can be easily seen from the above definition that $q_j(\bar{y}^{(j)}(k)) = (1 + \Delta_k^{(j)})\bar{y}^{(j)}(k)$ with $|\Delta_k^{(j)}| \le \delta_j$. According to the transformation discussed above, the quantizing effect can be transformed into the sector-bounded uncertainties.

Defining $\Delta_k \triangleq \mathrm{diag}\{\Delta_k^{(1)}, \ldots, \Delta_k^{(n_y)}\}$, the measurements with quantization effect and sensor failures can be expressed as

$$y(k) \;\; = \;\; (I + \Delta_k)\bar{y}(k). \tag{3.7}$$

Furthermore, by defining $\bar{\Delta} \triangleq \mathrm{diag}\{\delta_1, \cdots, \delta_{n_y}\}$ and $\bar{F}(k) \triangleq \Delta_k \bar{\Delta}^{-1}$, we obtain an unknown real-valued time-varying matrix satisfying $\bar{F}(k)\bar{F}^T(k) \le I$.

When the actuators experience failures, we use $u^F(k)$ to describe the control signal sent from actuators and the control input of actuator fault can therefore be described as follows:

$$u^F(k) = G(k)u(k) \tag{3.8}$$

where $G(k)$ is the actuator fault matrix with

$$\begin{cases} G(k) = \mathrm{diag}\{g_1(k), \ldots, g_{n_u}(k)\} \\ 0 \le \mathrm{diag}\{\underline{g}_1(k), \ldots, \underline{g}_{n_u}(k)\} = \underline{G}(k) \le G(k) \le \bar{G}(k) \\ \bar{G}(k) = \mathrm{diag}\{\bar{g}_1(k), \ldots, \bar{g}_{n_u}(k)\} \le I \end{cases} \tag{3.9}$$

and the variables $g_i(k)$ $(i = 1, 2, \ldots, n_u)$ quantify the failures of the actuators. $\underline{g}_i(k)$ and $\bar{g}_i(k)$ serve as the lower and the upper bounds on $g_i(k)$, respectively.

Defining

$$\begin{aligned}
G_0(k) =&\text{diag}\{\hat{g}_1(k), \ldots, \hat{g}_{n_u}\}\\
&\triangleq \frac{\underline{G}(k) + \bar{G}(k)}{2} = \text{diag}\left\{\frac{\underline{g}_1(k) + \bar{g}_1(k)}{2}, \ldots, \frac{\underline{g}_{n_u}(k) + \bar{g}_{n_u}(k)}{2}\right\},\\
\tilde{G}(k) =&\text{diag}\{\tilde{g}_1(k), \ldots, \tilde{g}_{n_u}(k)\}\\
&\triangleq \frac{\bar{G}(k) - \underline{G}(k)}{2} = \text{diag}\left\{\frac{\bar{g}_1(k) - \underline{g}_1(k)}{2}, \ldots, \frac{\bar{g}_{n_u}(k) - \underline{g}_{n_u}(k)}{2}\right\},
\end{aligned}$$

the matrix $G(k)$ can be rewritten as

$$\begin{aligned}
G(k) \quad = \quad & G_0(k) + \Delta_G(k) = G_0(k) + \text{diag}\{\tau_1(k), \ldots, \tau_{n_u}(k)\},\\
& |\tau_i(k)| \leq \tilde{g}_i(k) \quad (i = 1, 2, \ldots, n_u).
\end{aligned} \tag{3.10}$$

In this chapter, we adopt the following time-varying output-feedback controller for system (3.1):

$$\begin{cases} x_c(k+1) = A_c(k)x_c(k) + B_c(k)y(k)\\ u(k) = C_c(k)x_c(k)\end{cases} \tag{3.11}$$

where $x_c(k) \in \mathbb{R}^{n_c}$ is the controller state, $A_c(k)$, $B_c(k)$ and $C_c(k)$ are the controller parameters to be designed.

Letting $\eta(k) \triangleq \begin{bmatrix} x^T(k) & x_c^T(k)\end{bmatrix}^T$ and $\varpi(k) \triangleq \begin{bmatrix} w^T(k) & v^T(k)\end{bmatrix}^T$, we have the following closed-loop system to be investigated:

$$\begin{cases} \eta(k+1) = (\mathcal{A}(k) + \tilde{\alpha}(k)\Delta\mathcal{A}(k) + \tilde{\mathcal{B}}_c(k))\eta(k)\\ \qquad\qquad + (\bar{\beta} + \tilde{\beta}(k))\mathcal{G}(k, x(k)) + \mathcal{D}(k)\varpi(k)\\ z(k) = \mathcal{M}(k)\eta(k)\end{cases} \tag{3.12}$$

where

$$\begin{aligned}
\mathcal{A}(k) &\triangleq \bar{\mathcal{A}}(k) + \Delta\bar{\mathcal{A}}(k) + \bar{\alpha}\Delta\mathcal{A}(k),\ \Delta\bar{\mathcal{A}}(k) \triangleq \bar{\mathcal{H}}_a(k)\bar{F}(k)\bar{E}_c(k),\\
\Delta\mathcal{A}(k) &\triangleq \mathcal{H}_a(k)F(k)\mathcal{N}(k),\ \bar{\mathcal{A}}(k) \triangleq \begin{bmatrix} A(k) & B_1(k)G(k)C_c(k)\\ B_c(k)\bar{\Xi}C(k) & A_c(k)\end{bmatrix},\\
\bar{E}_c(k) &\triangleq \begin{bmatrix}\bar{\Lambda}\bar{\Xi}C(k) & 0\end{bmatrix},\ \bar{\mathcal{H}}_a(k) \triangleq \begin{bmatrix} 0 & B_c^T(k)\end{bmatrix}^T,\\
\tilde{\mathcal{B}}_c(k) &\triangleq \tilde{\mathcal{B}}_{c0}(k) + \Delta\tilde{\mathcal{B}}_{c0}(k),\ \Delta\tilde{\mathcal{B}}_{c0}(k) \triangleq \bar{\mathcal{H}}_a(k)\bar{F}(k)\tilde{E}_c(k),\\
\tilde{\mathcal{B}}_{c0}(k) &\triangleq \begin{bmatrix} 0 & 0\\ B_c(k)\tilde{\Xi}(k)C(k) & 0\end{bmatrix},\ \tilde{E}_c(k) \triangleq \begin{bmatrix}\bar{\Lambda}\tilde{\Xi}(k)C(k) & 0\end{bmatrix},\\
\mathcal{D}(k) &\triangleq \bar{\mathcal{D}}(k) + \Delta\bar{\mathcal{D}}(k),\ \bar{\mathcal{D}}(k) \triangleq \text{diag}\left\{D(k), B_c(k)\bar{E}(k)\right\},
\end{aligned}$$

$$\begin{aligned}
\Delta\bar{\mathcal{D}}(k) &\triangleq \bar{\mathcal{H}}_a(k)\bar{F}(k)\bar{E}_E(k),\ \bar{E}_E(k) \triangleq \begin{bmatrix} 0 & \bar{\Delta}\bar{E}(k) \end{bmatrix}, \\
\mathcal{M}(k) &\triangleq \begin{bmatrix} M(k) & B_2(k)G(k)C_c(k) \end{bmatrix},\ \mathcal{G}(k, x(k)) \triangleq \begin{bmatrix} g^T(k, x(k)) & 0 \end{bmatrix}^T, \\
\mathcal{N}(k) &\triangleq \begin{bmatrix} N(k) & 0 \end{bmatrix},\ \tilde{\alpha}(k) \triangleq \alpha(k) - \bar{\alpha},\ \tilde{\beta}(k) \triangleq \beta(k) - \bar{\beta}, \\
\tilde{\Xi}(k) &\triangleq \Xi(k) - \bar{\Xi},\ \mathcal{H}_a(k) \triangleq \begin{bmatrix} H_a^T(k) & 0 \end{bmatrix}^T.
\end{aligned}$$

The objective of this chapter is to find a sequence of reliable controller parameters $A_c(k)$, $B_c(k)$ and $C_c(k)$ such that the closed-loop system (3.12) satisfies the following performance requirement:

$$J \triangleq \mathbb{E}\left\{ \sum_{k=0}^{N-1} \left(\|z(k)\|^2 - \gamma^2 \|\varpi(k)\|_R^2 \right) - \gamma^2 \eta^T(0) S \eta(0) \right\} < 0 \tag{3.13}$$

where $\|\varpi(k)\|_R^2 \triangleq \varpi^T(k) R \varpi(k)$, and R and S are known positive definite weighted matrices.

3.2 Reliable Controller Design

In this section, we investigate both the controller analysis and controller design problem for the discrete time-varying nonlinear stochastic system (3.1) with randomly occurring uncertainties, nonlinearities, actuator and sensor failures subject to output quantization. The following lemmas will be needed for the derivation of our main results.

Lemma 3.1 *[14] (S-procedure) Let $L = L^T$ and M_a and N_a be real matrices of appropriate dimensions with F_a satisfying $F_a F_a^T \leq I$, then $L + M_a F_a N_a + N_a^T F_a^T M_a^T < 0$, if and only if there exists a positive scalar $\varphi > 0$ such that $L + \varphi^{-1} M_a M_a^T + \varphi N_a^T N_a < 0$ or*

$$\begin{bmatrix} L & M_a & \varphi N_a^T \\ M_a^T & -\varphi I & 0 \\ \varphi N_a & 0 & -\varphi I \end{bmatrix} < 0. \tag{3.14}$$

Lemma 3.2 *[195] Let $x \in \mathbb{R}^n$ and $y \in \mathbb{R}^n$. Then, for any scalar $\mu > 0$, we have*

$$x^T y + y^T x \leq \mu x^T x + \mu^{-1} y^T y \tag{3.15}$$

In the following theorem, we present the analysis results with known parameter matrix describing the actuator failures.

Theorem 3.1 *Consider the closed-loop system (3.12) with known actuator* *failure parameter matrix* $G(k)$. *Let the disturbance attenuation level* $\gamma > 0$, *families of positive scalars* $\{\varepsilon(k)\}_{0\leq k\leq N-1} > 0$, *the positive definite matrices* $R > 0$, $S > 0$ *and the controller feedback gain matrices* $\{A_c(k)\}_{0\leq k\leq N-1}$, $\{B_c(k)\}_{0\leq k\leq N-1}$ *and* $\{C_c(k)\}_{0\leq k\leq N-1}$ *be given. The* $\mathcal{H}_\infty$ *performance requirement defined in (3.13) is achieved for all nonzero* $\varpi(k)$ *if, with the initial condition* $P(0) \leq \gamma^2 S$, *there exist families of positive definite matrices* $\{P(k)\}_{0\leq k\leq N} > 0$, $\{Q(k)\}_{0\leq k\leq N} > 0$ *and families of positive scalars* $\{\lambda(k) > 0\}_{0\leq k\leq N-1}$ *satisfying the following recursive matrix inequalities*

$$\Omega(k) = \begin{bmatrix} \Omega_{11}(k) & * & * \\ 0 & \Omega_{22}(k) & * \\ \Omega_{31}(k) & \Omega_{32}(k) & \Omega_{33}(k) \end{bmatrix} < 0 \tag{3.16}$$

with the parameters updated by

$$P(k+1) = Q^{-1}(k+1) \tag{3.17}$$

where

$$\begin{aligned}
\Omega_{11}(k) &\triangleq -P(k) + \lambda(k)\varepsilon(k)H^T E^T(k)E(k)H, \ H \triangleq \begin{bmatrix} I & 0 \end{bmatrix}, \\
\Omega_{22}(k) &\triangleq \operatorname{diag}\{-\lambda(k)H_I^T H_I, -\gamma^2 R\}, \ \Delta\bar{\mathcal{B}}_{ci}(k) \triangleq \bar{\mathcal{H}}_a(k)\bar{F}(k)\bar{E}_{ci}(k), \\
\Omega_{31}(k) &\triangleq \Big[\ \mathcal{A}^T(k) \quad (\hat{\mathcal{B}}_{ci}(k) + \Delta\hat{\mathcal{B}}_{ci}(k))^T \quad \mathcal{M}^T(k) \\
&\qquad \sqrt{\bar{\alpha}(1-\bar{\alpha})}\Delta\mathcal{A}^T(k) \quad 0 \ \Big]^T, \\
\Omega_{32}(k) &\triangleq \begin{bmatrix} \Omega_{321}(k) & \Omega_{322}(k) \end{bmatrix}, \\
\Omega_{321}(k) &\triangleq \begin{bmatrix} \bar{\beta} I & 0 & 0 & 0 & \sqrt{\bar{\beta}(1-\bar{\beta})} I \end{bmatrix}^T, \\
\Omega_{322}(k) &\triangleq \begin{bmatrix} \mathcal{D}^T(k) & 0 & 0 & 0 & 0 \end{bmatrix}^T, \ Q(k+1) \triangleq P^{-1}(k+1), \\
\hat{\mathcal{B}}_{ci}(k) &\triangleq \begin{bmatrix} \chi_1 \bar{\mathcal{B}}_{c_1}^T(k) & \chi_2 \bar{\mathcal{B}}_{c_2}^T(k) & \cdots \chi_{n_y} \bar{\mathcal{B}}_{c_{n_y}}^T(k) \end{bmatrix}^T, \\
\Omega_{33}(k) &\triangleq \operatorname{diag}\Big\{ -Q(k+1), -\hat{Q}(k+1), -I, -Q(k+1), -Q(k+1) \Big\}, \\
\bar{\mathcal{B}}_{ci}(k) &\triangleq \begin{bmatrix} 0 & 0 \\ B_c(k)C_i(k) & 0 \end{bmatrix}, \ \hat{Q}(k+1) \triangleq I_{n_y} \otimes Q(k+1), \\
\Delta\hat{\mathcal{B}}_{ci}(k) &\triangleq \begin{bmatrix} \chi_1 \Delta\bar{\mathcal{B}}_{c_1}^T(k) & \chi_2 \Delta\bar{\mathcal{B}}_{c_2}^T(k) & \cdots & \chi_{n_y} \Delta\bar{\mathcal{B}}_{c_{n_y}}^T(k) \end{bmatrix}^T, \\
H_I &\triangleq \begin{bmatrix} I & I \end{bmatrix}, \ \bar{E}_{ci}(k) \triangleq \begin{bmatrix} \bar{\Delta} C_i(k) & 0 \end{bmatrix}.
\end{aligned} \tag{3.18}$$

Proof *Define*

$$J(k) \triangleq \eta^T(k+1)P(k+1)\eta(k+1) - \eta^T(k)P(k)\eta(k). \tag{3.19}$$

Taking (3.6) into consideration, we have

$$\mathbb{E}\{(\rho_i(k) - \bar{\mu}_i)(\rho_l(k) - \bar{\mu}_l)\} = \begin{cases} \chi_i^2 & i = l \\ 0 & i \neq l, \end{cases} \qquad (i, l = 1, 2, \ldots, n_y)$$

and therefore

$$
\begin{aligned}
& \mathbb{E}\{\tilde{\mathcal{B}}_c^T(k)P(k+1)\tilde{\mathcal{B}}_c(k)\} \\
= & \sum_{i=1}^{n_y}\sum_{l=1}^{n_y}\mathbb{E}(\rho_i(k)-\bar{\mu}_i)(\rho_l(k)-\bar{\mu}_l)\begin{bmatrix} 0 & 0 \\ B_c(k)(I+\Delta_k)C_i(k) & 0 \end{bmatrix}^T \\
& \times P(k+1)\begin{bmatrix} 0 & 0 \\ B_c(k)(I+\Delta_k)C_l(k) & 0 \end{bmatrix} \\
= & \mathbb{E}\Big\{\sum_{i=1}^{n_y}\chi_i^2\begin{bmatrix} 0 & 0 \\ B_c(k)(I+\Delta_k)C_i(k) & 0 \end{bmatrix}^T P(k+1) \\
& \times\begin{bmatrix} 0 & 0 \\ B_c(k)(I+\Delta_k)C_i(k) & 0 \end{bmatrix}\Big\} \\
= & \mathbb{E}\left\{\left(\hat{\mathcal{B}}_{ci}(k)+\Delta\hat{\mathcal{B}}_{ci}(k)\right)^T(I_{n_y}\otimes P(k+1))\left(\hat{\mathcal{B}}_{ci}(k)+\Delta\hat{\mathcal{B}}_{ci}(k)\right)\right\}.
\end{aligned}
$$

Adding the zero term $z^T(k)z(k)-\gamma^2\varpi^T(k)R\varpi(k)-z^T(k)z(k)+\gamma^2\varpi^T(k)R\varpi(k)$ *to* $\mathbb{E}\{J(k)\}$ *results in*

$$
\begin{aligned}
& \mathbb{E}\{J(k)\} \\
= & \mathbb{E}\Bigg\{\begin{bmatrix} \eta^T(k) & \mathcal{G}^T(k,x(k)) & \varpi^T(k) \end{bmatrix}\bar{\Omega}(k)\begin{bmatrix} \eta(k) \\ \mathcal{G}(k,x(k)) \\ \varpi(k) \end{bmatrix} \\
& -z^T(k)z(k)+\gamma^2\varpi^T(k)R\varpi(k)\Bigg\} \\
= & \mathbb{E}\{\xi^T(k)\bar{\Omega}(k)\xi(k)-z^T(k)z(k)+\gamma^2\varpi^T(k)R\varpi(k)\} \qquad (3.20)
\end{aligned}
$$

where

$$
\begin{aligned}
\bar{\Omega}(k) & \triangleq \begin{bmatrix} \bar{\Omega}_{11}(k) & * & * \\ \bar{\beta}P(k+1)\mathcal{A}(k) & \bar{\Omega}_{22}(k) & * \\ \mathcal{D}^T(k)P(k+1)\mathcal{A}(k) & \bar{\beta}\mathcal{D}^T(k)P(k+1) & \mathcal{D}^T(k)P(k+1)\mathcal{D}(k)-\gamma^2R \end{bmatrix}, \\
\xi(k) & \triangleq \begin{bmatrix} \eta^T(k) & \mathcal{G}^T(k,x(k)) & \varpi^T(k) \end{bmatrix}^T, \\
\bar{\Omega}_{22}(k) & \triangleq \bar{\beta}^2P(k+1)+\bar{\beta}(1-\bar{\beta})P(k+1), \\
\bar{\Omega}_{11}(k) & \triangleq \mathcal{A}^T(k)P(k+1)\mathcal{A}(k)+\bar{\alpha}(1-\bar{\alpha})\Delta\mathcal{A}^T(k)P(k+1)\Delta\mathcal{A}(k) \\
& +(\hat{\mathcal{B}}_{ci}(k)+\Delta\hat{\mathcal{B}}_{ci}(k))^T(I_{n_y}\otimes P(k+1))\left(\hat{\mathcal{B}}_{ci}(k)+\Delta\hat{\mathcal{B}}_{ci}(k)\right) \\
& +\mathcal{M}^T(k)\mathcal{M}(k)-P(k) \qquad (3.21)
\end{aligned}
$$

It follows from the constraint (3.2) *that*

$$\|H_I\mathcal{G}(k,x(k))\|^2 \leq \varepsilon(k)\|E(k)H\eta(k)\|^2 \tag{3.22}$$

and therefore we have

$$\begin{aligned}\mathbb{E}\{J(k)\} &\leq \mathbb{E}\{\xi^T(k)\bar{\Omega}(k)\xi(k) - \lambda(k)(\|H_I\mathcal{G}(k,x(k))\|^2 \\ &\quad -\varepsilon(k)\|E(k)H\eta(k)\|^2)\} - \mathbb{E}\left\{z^T(k)z(k) - \gamma^2\varpi^T(k)R\varpi(k)\right\} \\ &= \mathbb{E}\left\{\xi^T(k)\tilde{\Omega}(k)\xi(k)\right\} \\ &\quad -\mathbb{E}\left\{z^T(k)z(k) - \gamma^2\varpi^T(k)R\varpi(k)\right\}\end{aligned} \tag{3.23}$$

where

$$\tilde{\Omega}(k) \triangleq \bar{\Omega}(k) + \mathrm{diag}\{\lambda(k)\varepsilon(k)H^TE^T(k)E(k)H, -\lambda(k)H_I^TH_I, 0\}. \tag{3.24}$$

It follows from Lemma 2.1 (Schur complement Lemma) and (3.16) that $\tilde{\Omega}(k) < 0$. *Summing up (3.23) on both sides from* 0 *to* $N-1$ *with respect to* k, *we obtain*

$$\begin{aligned}\sum_{k=0}^{N-1}\mathbb{E}\{J(k)\} &= \mathbb{E}\left\{\eta^T(N)P(N)\eta(N)\right\} - \eta^T(0)P(0)\eta(0) \\ &\leq \mathbb{E}\left\{\sum_{k=0}^{N-1}\xi^T(k)\tilde{\Omega}(k)\xi(k)\right\} \\ &\quad - \mathbb{E}\left\{\sum_{k=0}^{N-1}\left(z^T(k)z(k) - \gamma^2\varpi^T(k)R\varpi(k)\right)\right\}.\end{aligned} \tag{3.25}$$

Hence, the H_∞ *performance index defined in (3.13) is given by*

$$\begin{aligned}J &= \mathbb{E}\left\{\sum_{k=0}^{N-1}\left(\|z(k)\|^2 - \gamma^2\|\varpi(k)\|_R^2\right) - \gamma^2\eta^T(0)S\eta(0)\right\} \\ &\leq \mathbb{E}\left\{\sum_{k=0}^{N-1}\xi^T(k)\tilde{\Omega}(k)\xi(k)\right\} \\ &\quad -\mathbb{E}\left\{\eta^T(N)P(N)\eta(N) - \eta^T(0)(P(0)-\gamma^2S)\eta(0)\right\}.\end{aligned} \tag{3.26}$$

Noting that $P(N) > 0$, $\tilde{\Omega}(k) < 0$ *and the initial condition* $P(0) \leq \gamma^2S$, *it can be obtained that* $J < 0$ *and the proof is now complete.*

Based on the analysis results with the desired output feedback controllers, we are now ready to solve the controller design problem for system (3.1) in the following theorem. For convenience of later analysis, we denote

$$\Gamma_{11}(k) \triangleq \mathrm{diag}\{-P(k) + \lambda(k)\varepsilon(k)H^TE^T(k)E(k)H, -\lambda(k)H_I^TH_I, -\gamma^2R\},$$

$$
\begin{aligned}
\Gamma_{21}(k) &\triangleq \begin{bmatrix} \bar{\mathcal{A}}_0(k) + E_I K(k) R_1(k) + R_2(k) G(k) \bar{C}_c(k) & \bar{\beta} I \\ (I_{n_y} \otimes E_I K(k)) R_5(k) & 0 \end{bmatrix. \\
&\qquad \left. \begin{matrix} E_I K(k) R_3(k) + \bar{\mathcal{D}}_0(k) \\ 0 \end{matrix} \right], \\
\Gamma_{22}(k) &\triangleq \operatorname{diag}\left\{-Q(k+1), -\hat{Q}(k+1)\right\}, \; \bar{\mathcal{D}}_0(k) \triangleq \operatorname{diag}\{D(k), 0\}, \\
\Gamma_{31}(k) &\triangleq \operatorname{diag}\left\{B_2(k) G(k) \bar{C}_c(k) + \mathcal{M}_0(k), \sqrt{\bar{\beta}(1-\bar{\beta})} I, 0\right\}, \\
\Gamma_{33}(k) &\triangleq \operatorname{diag}\{-I, -Q(k+1), -Q(k+1)\}, \; R_2(k) \triangleq \begin{bmatrix} B_1(k) \\ 0 \end{bmatrix}, \\
\Gamma_{42}(k) &\triangleq \operatorname{diag}\left\{(E_I K(k) E_I)^T, (I_{n_y} \otimes E_I K(k) E_I)^T\right\}, \\
\Gamma_{44}(k) &\triangleq \operatorname{diag}\{-\psi_1(k) I, -\psi_1(k) I\}, \; \Gamma_{51}(k) \triangleq \begin{bmatrix} \hat{E}_a(k) & 0 & \hat{E}_b(k) \end{bmatrix}, \\
\Gamma_{52}(k) &\triangleq \begin{bmatrix} \hat{E}_c(k) & 0 \end{bmatrix}, \; \hat{E}_b(k) \triangleq \begin{bmatrix} \psi_1(k) \bar{E}_E^T(k) & 0 & 0 & 0 \end{bmatrix}^T, \\
\hat{E}_a(k) &\triangleq \begin{bmatrix} \psi_1(k) \bar{E}_c^T(k) & \psi_1(k) \tilde{E}_{ci}^T(k) & 0 & \psi_2(k) \mathcal{N}^T(k) \end{bmatrix}^T, \\
\hat{E}_c(k) &\triangleq \begin{bmatrix} 0 & 0 & \bar{\alpha} \mathcal{H}_a(k) & 0 \end{bmatrix}^T, \\
\Gamma_{55}(k) &\triangleq \operatorname{diag}\{I_2 \otimes (-\psi_1(k) I), I_2 \otimes (-\psi_2(k) I)\}, \\
E_I &\triangleq \begin{bmatrix} 0 & I \end{bmatrix}^T, \; \Gamma_{53}(k) \triangleq \begin{bmatrix} 0 & 0 & \hat{E}_d(k) \end{bmatrix}, \\
\hat{E}_d(k) &\triangleq \begin{bmatrix} 0 & 0 & \sqrt{\bar{\alpha}(1-\bar{\alpha})} \mathcal{H}_a(k) & 0 \end{bmatrix}^T, \\
\bar{\mathcal{A}}_0(k) &\triangleq \operatorname{diag}\{A(k), 0\}, \; R_1(k) \triangleq \begin{bmatrix} 0 & I \\ \bar{\Xi} C(k) & 0 \end{bmatrix}, \; R_{4i}(k) \triangleq \begin{bmatrix} 0 & 0 \\ C_i(k) & 0 \end{bmatrix}, \\
\tilde{E}_{ci}(k) &\triangleq \begin{bmatrix} \chi_1 \bar{E}_{c_1}^T(k) & \chi_2 \bar{E}_{c_2}^T(k) & \cdots & \chi_{n_y} \bar{E}_{c_{n_y}}^T(k) \end{bmatrix}^T, \\
\mathcal{M}_0(k) &\triangleq \begin{bmatrix} M(k) & 0 \end{bmatrix}, \; R_3(k) \triangleq \operatorname{diag}\{0, \bar{E}(k)\}, \\
R_5(k) &\triangleq \begin{bmatrix} \chi_1 R_{41}^T(k) & \chi_2 R_{42}^T(k) & \cdots \chi_{n_y} R_{4n_y}^T(k) \end{bmatrix}^T. \qquad (3.27)
\end{aligned}
$$

Theorem 3.2 *Consider the closed-loop system (3.12) with known actuator failure parameter matrix $G(k)$. Let the disturbance attenuation level $\gamma > 0$, families of positive scalars $\{\varepsilon(k)\}_{0 \le k \le N-1} > 0$, the positive definite matrices $R > 0$ and $S > 0$ be given. The H_∞ performance requirement defined in (3.13) is achieved for all nonzero $\varpi(k)$ if, with the initial condition $P(0) \le \gamma^2 S$, there exist families of positive definite matrices $\{P(k)\}_{0 \le k \le N} > 0$, $\{Q(k)\}_{0 \le k \le N} > 0$, families of real-valued matrices $\{K(k)\}_{0 \le k \le N-1}$, $\{\bar{C}_c(k)\}_{0 \le k \le N-1}$, families of positive scalars $\{\lambda(k) > 0\}_{0 \le k \le N-1}$, $\{\psi_1(k) > 0\}_{0 \le k \le N-1}$ and*

$\{\psi_2(k) > 0\}_{0\le k\le N-1}$ *satisfying the following recursive matrix inequalities*

$$\Gamma(k) = \begin{bmatrix} \Gamma_{11}(k) & * & * & * & * \\ \Gamma_{21}(k) & \Gamma_{22}(k) & * & * & * \\ \Gamma_{31}(k) & 0 & \Gamma_{33}(k) & * & * \\ 0 & \Gamma_{42}(k) & 0 & \Gamma_{44}(k) & * \\ \Gamma_{51}(k) & \Gamma_{52}(k) & \Gamma_{53}(k) & 0 & \Gamma_{55}(k) \end{bmatrix} < 0 \tag{3.28}$$

with the parameters updated by

$$P(k+1) = Q^{-1}(k+1) \tag{3.29}$$

where the other parameters are defined in (3.27) and Theorem 3.1. Furthermore, if $(P(k), Q(k+1), K(k), \bar{C}_c(k))$ is the feasible solution of (3.28), then the output feedback controller parameters in the form of (3.11) are given as follows:

$$\begin{bmatrix} A_c(k) & B_c(k) \end{bmatrix} = K(k), \quad C_c(k) = \bar{C}_c(k)E_I. \tag{3.30}$$

Proof *To deal with the parameter uncertainties in (3.12), we rewrite (3.16) in the following form:*

$$\begin{aligned} &\hat{\Omega}(k) + \tilde{\mathcal{M}}_1(k)\tilde{\mathcal{F}}(k)\tilde{\mathcal{N}}_1(k) + \tilde{\mathcal{N}}_1^T(k)\tilde{\mathcal{F}}^T(k)\tilde{\mathcal{M}}_1^T(k) + \tilde{\mathcal{M}}_2(k)F(k)\tilde{\mathcal{N}}_2(k) \\ &+\tilde{\mathcal{N}}_2^T(k)F^T(k)\tilde{\mathcal{M}}_2^T(k) < 0 \end{aligned} \tag{3.31}$$

where

$$\hat{\Omega}(k) \triangleq \begin{bmatrix} \Omega_{11}(k) & * & * \\ 0 & \Omega_{22}(k) & * \\ \hat{\Omega}_{31}(k) & \hat{\Omega}_{32}(k) & \Omega_{33}(k) \end{bmatrix}, \ \hat{\Omega}_{32}(k) \triangleq \begin{bmatrix} \Omega_{321}(k) & \hat{\Omega}_{322}(k) \end{bmatrix},$$

$$\begin{aligned}
\tilde{\mathcal{M}}_1(k) &\triangleq \begin{bmatrix} 0 & 0 & 0 & \bar{\mathcal{H}}_a^T(k) & 0 & 0 & 0 & 0 \\ 0 & 0 & 0 & 0 & (I_{n_y} \otimes \bar{\mathcal{H}}_a(k))^T & 0 & 0 & 0 \end{bmatrix}^T, \\
\hat{\Omega}_{322}(k) &\triangleq \begin{bmatrix} \bar{\mathcal{D}}^T(k) & 0 & 0 & 0 & 0 \end{bmatrix}^T, \ \tilde{\mathcal{F}}(k) \triangleq \mathrm{diag}\{\bar{F}(k), I_{n_y} \otimes \bar{F}(k)\}, \\
\hat{\Omega}_{31}(k) &\triangleq \begin{bmatrix} \bar{\mathcal{A}}^T(k) & \hat{\mathcal{B}}_{ci}^T(k) & \mathcal{M}^T(k) & 0 & 0 \end{bmatrix}^T, \\
\tilde{\mathcal{N}}_1(k) &\triangleq \begin{bmatrix} \bar{E}_c(k) & 0 & \bar{E}_E(k) & 0 & 0 & 0 & 0 & 0 \\ \tilde{E}_{ci}(k) & 0 & 0 & 0 & 0 & 0 & 0 & 0 \end{bmatrix}, \\
\tilde{\mathcal{N}}_2(k) &\triangleq \begin{bmatrix} \mathcal{N}(k) & 0 & 0 & 0 & 0 & 0 & 0 & 0 \end{bmatrix}, \\
\tilde{\mathcal{M}}_2(k) &\triangleq \begin{bmatrix} 0 & 0 & 0 & \bar{\alpha}\mathcal{H}_a^T(k) & 0 & 0 & \sqrt{\bar{\alpha}(1-\bar{\alpha})}\mathcal{H}_a^T(k) & 0 \end{bmatrix}^T
\end{aligned} \tag{3.32}$$

According to Lemma 3.1, the following inequality holds:

$$\begin{bmatrix} \hat{\Omega}(k) & * & * & * & * \\ \tilde{\mathcal{M}}_1^T(k) & -\psi_1(k)I & * & * & * \\ \psi_1(k)\tilde{\mathcal{N}}_1(k) & 0 & -\psi_1(k)I & * & * \\ \tilde{\mathcal{M}}_2^T(k) & 0 & 0 & -\psi_2(k)I & 0 \\ \psi_2(k)\tilde{\mathcal{N}}_2(k) & 0 & 0 & 0 & -\psi_2(k)I \end{bmatrix} < 0 \quad (3.33)$$

In order to avoid partitioning the positive define matrices $P(k)$ and $Q(k)$, the parameters in (3.33) are expressed as follows:

$$\begin{aligned} \bar{\mathcal{A}}(k) &= \bar{\mathcal{A}}_0(k) + E_I K(k) R_1(k) + R_2(k) G(k) \bar{C}_c(k), \\ \bar{\mathcal{H}}_a(k) &= E_I K(k) E_I, \ \hat{\mathcal{B}}_{ci}(k) = (I_{n_y} \otimes E_I K(k)) R_5(k), \\ \mathcal{M}(k) &= \mathcal{M}_0(k) + B_2(k) G(k) \bar{C}_c(k), \ \bar{\mathcal{B}}_{ci}(k) = E_I K(k) R_{4i}(k) \\ \bar{\mathcal{D}}(k) &= \bar{\mathcal{D}}_0(k) + E_I K(k) R_3(k) \end{aligned} \quad (3.34)$$

where

$$K(k) \triangleq \begin{bmatrix} A_c(k) & B_c(k) \end{bmatrix}, \ \bar{C}_c(k) \triangleq \begin{bmatrix} 0 & C_c(k) \end{bmatrix}. \quad (3.35)$$

Noticing (3.34) and (3.35), (3.28) is obtained by (3.33) after applying Lemma 2.1 and some straightforward algebraic manipulations, and the proof of this theorem is then complete.

In Theorem 3.2, with known actuator failure parameter, the H_∞ performance requirement defined in (3.13) is obtained for the closed-loop system (3.12) and the output feedback controller is designed based on the recursive matrix inequalities approach. In the following Theorem, a design procedure for the desired controller parameters is given in the case that the failure parameter matrix $G(k)$ is unknown but satisfies the constraints (3.9)–(3.10).

Theorem 3.3 *Consider the closed-loop system (3.12). Let the disturbance attenuation level $\gamma > 0$, families of positive scalars $\{\varepsilon(k)\}_{0\le k\le N-1} > 0$, $\{\mu(k)\}_{0\le k\le N-1} > 0$, the positive definite matrices $R > 0$ and $S > 0$ be given. The H_∞ performance requirement defined in (3.13) is achieved for all nonzero $\varpi(k)$ if, with the initial condition $P(0) \le \gamma^2 S$, there exist families of positive definite matrices $\{P(k)\}_{0\le k\le N} > 0$, $\{Q(k)\}_{0\le k\le N} > 0$, families of real-valued matrices $\{K(k)\}_{0\le k\le N-1}$, $\{\bar{C}_c(k)\}_{0\le k\le N-1}$, families of positive scalars $\{\lambda(k) > 0\}_{0\le k\le N-1}$, $\{\psi_1(k)(k) > 0\}_{0\le k\le N-1}$ and $\{\psi_2(k) > 0\}_{0\le k\le N-1}$ satisfying the following recursive matrix inequalities*

$$\tilde{\Gamma}(k) = \begin{bmatrix} \hat{\Gamma}(k) & * & * \\ \mu(k)\hat{\mathcal{B}}(k) & -\mu(k)I & * \\ \hat{C}_c(k) & 0 & -\mu(k)(\tilde{G}^T(k)\tilde{G}(k))^{-1} \end{bmatrix} < 0 \quad (3.36)$$

with the parameters updated by

$$P(k+1) = Q^{-1}(k+1) \tag{3.37}$$

where

$$\begin{aligned}
\hat{\Gamma}(k) &\triangleq \begin{bmatrix} \Gamma_{11}(k) & * & * & * & * \\ \hat{\Gamma}_{21}(k) & \Gamma_{22}(k) & * & * & * \\ \hat{\Gamma}_{31}(k) & 0 & \Gamma_{33}(k) & * & * \\ 0 & \Gamma_{42}(k) & 0 & \Gamma_{44}(k) & * \\ \Gamma_{51}(k) & \Gamma_{52}(k) & \Gamma_{53}(k) & 0 & \Gamma_{55}(k) \end{bmatrix}, \\
\hat{\Gamma}_{21}(k) &\triangleq \left[\begin{matrix} \bar{\mathcal{A}}_0(k) + E_I K(k) R_1(k) + R_2(k) G_0(k) \bar{C}_c(k) & \bar{\beta} I \\ (I_{n_y} \otimes E_I K(k)) R_5(k) & 0 \end{matrix} \right. \\
&\qquad \left. \begin{matrix} E_I K(k) R_3(k) + \bar{\mathcal{D}}_0(k) \\ 0 \end{matrix} \right], \\
\hat{\Gamma}_{31}(k) &\triangleq \operatorname{diag}\left\{ B_2(k) G_0(k) \bar{C}_c(k) + \mathcal{M}_0(k), \sqrt{\bar{\beta}(1-\bar{\beta})} I, 0 \right\}, \\
\hat{\mathcal{B}}(k) &\triangleq \begin{bmatrix} 0 & 0 & 0 & R_2^T(k) & 0 & B_2^T(k) & 0 & 0 & 0 & 0 & 0 & 0 & 0 & 0 \end{bmatrix}, \\
\hat{\mathcal{C}}_c(k) &\triangleq \begin{bmatrix} \bar{C}_c(k) & 0 & 0 & 0 & 0 & 0 & 0 & 0 & 0 & 0 & 0 & 0 & 0 & 0 \end{bmatrix}.
\end{aligned} \tag{3.38}$$

Furthermore, if $(P(k), Q(k+1), K(k), \bar{C}_c(k))$ is the feasible solution of (3.36), then the output feedback controller parameters in the form of (3.11) are given as follows:

$$[A_c(k) \quad B_c(k)] = K(k), \quad C_c(k) = \bar{C}_c(k) E_I. \tag{3.39}$$

Proof *From (3.10), we know that $\Gamma(k)$ in Theorem 3.2 can be rewritten as*

$$\Gamma(k) = \hat{\Gamma}(k) + \hat{\mathcal{B}}^T(k) \Delta_G(k) \hat{\mathcal{C}}_c(k) + \hat{\mathcal{C}}_c^T(k) \Delta_G(k) \hat{\mathcal{B}}(k). \tag{3.40}$$

Noticing Lemma 3.2 and inequality (3.10), it is obtained that

$$\Gamma(k) \le \hat{\Gamma}(k) + \mu(k) \hat{\mathcal{B}}^T(k) \hat{\mathcal{B}}(k) + \mu^{-1}(k) \hat{\mathcal{C}}_c^T(k) \tilde{G}^T(k) \tilde{G}(k) \hat{\mathcal{C}}_c(k) = \tilde{\Psi}(k). \tag{3.41}$$

From Lemma 2.1, (3.36) in Theorem 3.3 implies that $\Gamma(k) \le \tilde{\Psi}(k) < 0$. This completes the proof.

By means of Theorem 3.3, the algorithm for designing the reliable robust controller can be outlined as follows.

Remark 3.2 *In the system model under investigation in this chapter, there are mainly six factors that constitute the complexity and complicate the design of reliable controller, which are ROUs, RONs, actuator failure, sensor failures,*

The Reliable Controller Design Algorithm:	
Step 1.	Give the $\mathcal{H}_\infty$ performance index γ, the positive definite matrices R, S and the state initial condition $\eta(0)$. Select the initial value for matrix $P(0)$ which satisfy the condition $P(0) \leq \gamma^2 S$ and set $k = 0$.
Step 2.	For the sampling instant k, solving the recursive matrix inequality (3.36) to obtain the values of matrix $Q(k+1)$ as well as the desired controller parameters $\{A_c(k), B_c(k), C_c(k)\}$.
Step 3.	Set $k = k + 1$ and obtain $P(k + 1)$ by the parameter update formula (3.37).
Step 4.	If $k < N$, where N is the maximum number of iterations allowed, then go to Step 2, else go to Step 5.
Step 5.	Stop.

quantization and time-varying parameters. It can be seen that all these six factors are explicitly reflected in our main results. In Theorem 3.3, the finite-horizon reliable controller is designed by solving a series of recursive linear matrix inequalities under which both the current system measurement and previous system states are employed to control the current system state. Such a recursive control process is particularly suitable for real-time implementation such as online process control.

3.3 An Illustrative Example

In this section, we present a simulation example to illustrate the effectiveness of the proposed reliable controller design scheme for discrete time-varying stochastic systems with randomly occurring uncertainties, nonlinearities, actuator and sensor failures subject to output quantization. The system data are given as follows:

$$
\begin{aligned}
A(k) &= \begin{bmatrix} -0.6 & 0.2 \\ 1.1\sin(5k) & 0.5 \end{bmatrix}, \ H_a(k) = \begin{bmatrix} 0.1 & 0.3 \end{bmatrix}^T, \ N(k) = \begin{bmatrix} 0.2 & 0 \end{bmatrix}, \\
F(k) &= \sin(k), \ C(k) = \begin{bmatrix} -2 + 0.3\sin(5k) & 0.5 \\ 0 & 1 \end{bmatrix}, \\
B_1(k) &= \begin{bmatrix} -2 \\ 3\sin(5k) \end{bmatrix}, \ D(k) = \begin{bmatrix} 0.1\sin(3k) \\ -0.3 \end{bmatrix}, \\
M(k) &= \begin{bmatrix} -0.4 & 0.5\sin(5k) \end{bmatrix}, \ B_2(k) = 0.2, \ \bar{E}(k) = \begin{bmatrix} 0.1\sin(3k) \\ 0.2 \end{bmatrix}.
\end{aligned}
$$

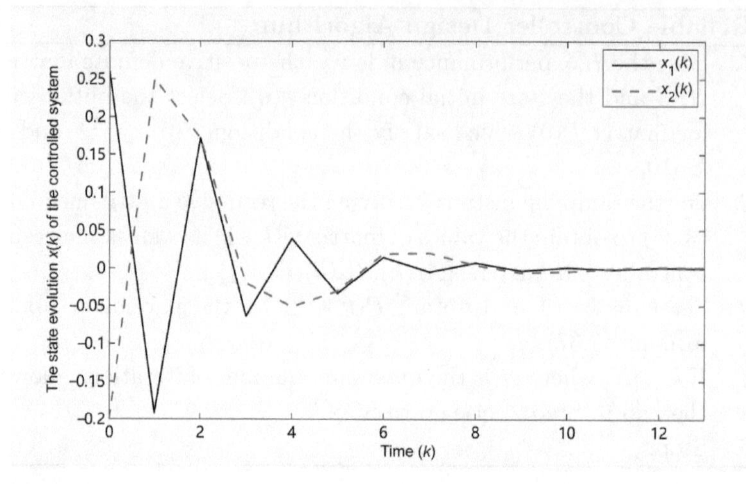

FIGURE 3.1
The state responses of the controlled system.

The nonlinear function $g(k, x(k))$ is selected as $g(k, x(k)) = 0.5x_1(k) \times \sin(x_2(k))$. It is easy to see that the constraint (3.2) is met with $\varepsilon(k) = 1$ and $E(k) = \text{diag}\{0.2, 0.15\}$. The parameters of the logarithmic quantizer are chosen as $\hat{\mu}_0 = 2$, $\bar{\chi}_1 = 0.8$ and $\bar{\chi}_2 = 0.5$.

Let $\bar{\alpha} = 0.9$ and $\bar{\beta} = 0.8$. Assume that the probability density functions of $\rho_1(k)$ and $\rho_2(k)$ in $[0, 1]$ are described by

$$\zeta_1(s_1) = \begin{cases} 0 & s_1 = 0 \\ 0.1 & s_1 = 0.5 \\ 0.9 & s_1 = 1 \end{cases}, \qquad \zeta_2(s_2) = \begin{cases} 0 & s_2 = 0 \\ 0.2 & s_2 = 0.5 \\ 0.8 & s_2 = 1 \end{cases}, \tag{3.42}$$

from which the expectations and variances can be easily calculated as $\bar{\mu}_1 = 0.95$, $\bar{\mu}_2 = 0.9$, $\chi_1^2 = 0.15^2$ and $\chi_2^2 = 0.04$.

The actuator fault matrix $G(k)$ is assumed to satisfy $0.85 \leq G(k) \leq 0.9$. Then, we can obtain that $G_0(k) = 0.875$ and $\tilde{G}(k) = 0.025$. In the simulation, let $\mu(k) = 1$. The H_∞ performance level is chosen as $\gamma = 1$, the initial values of the states are $x_0 = \begin{bmatrix} 0.26 & -0.2 \end{bmatrix}^T$, $\hat{x}_0 = \begin{bmatrix} 0.2 & -0.16 \end{bmatrix}^T$ and the positive definite matrices $R = \text{diag}\{2, 2\}$ and $S = \text{diag}\{2, 2, 2, 2\}$. The exogenous disturbance input is selected as $w(k) = 0.5\sin(4k)$ and $v(k) = 0.2\cos(4k)$. According to reliable controller design algorithm, the desired controller parameters in Theorem 3.3 can be solved recursively subject to given initial conditions and prespecified performance index. Table 3.1 lists the desired parameters of reliable controller $A_c(k)$, $B_c(k)$ and $C_c(k)$ from the time $k = 0$ to $k = 3$.

The simulation results are shown in Figs. 3.1–3.2, where Fig. 3.1 plots

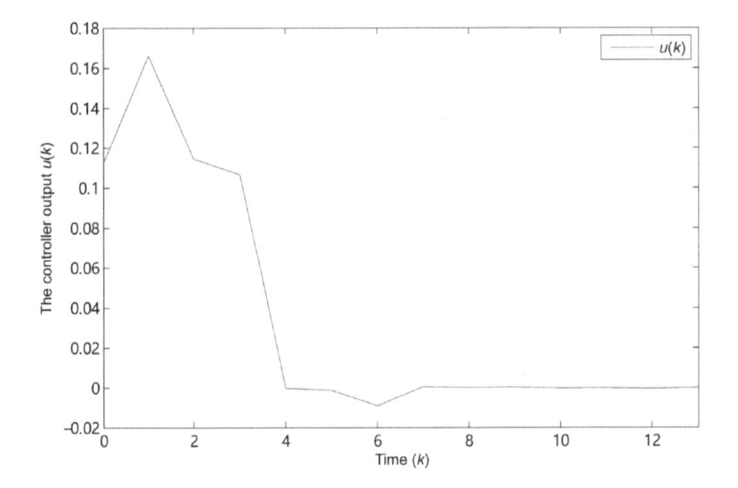

FIGURE 3.2
The controller output $u(k)$.

the state simulation results of the closed-loop system (3.12) and Fig. 3.2 depicts the controller output. The simulation results confirm that the desired finite-horizon performance is well achieved and the proposed reliable controller design algorithm is indeed effective.

3.4 Summary

In this chapter, the finite-horizon reliable H_∞ output-feedback control problem has been investigated for a class of discrete time-varying systems with ROUs, RONs as well as measurement quantizations. The actuator failures have been quantified by a variable varying in a given interval and the sensor failures have been governed by an individual random variable satisfying a certain probabilistic distribution in the interval $[0, 1]$. Both the RONs and the ROUs have been modelled by the Bernoulli distributed white sequences with known conditional probabilities. In the presence of output quantization, a time-varying output feedback controller has been designed to preserve a guaranteed H_∞ performance. A simulation example has been utilized to demonstrate the effectiveness of the finite-horizon reliable control techniques presented in this chapter. Other possible future research directions include real-time applications of the proposed reliable control theory in telecommunications, and further extensions of the present results to more

TABLE 3.1
Recursive Process

k	**0**	**1**	**2**
$A_c(k)$	$\begin{bmatrix} 0.5314 & 0.5314 \\ 0.5314 & 0.5314 \end{bmatrix}$	$\begin{bmatrix} 0.0141 & 0.0141 \\ 0.0141 & 0.0141 \end{bmatrix}$	$\begin{bmatrix} 0.0791 & 0.0110 \\ 0.0241 & 0.0698 \end{bmatrix}$
$B_c(k)$	$\begin{bmatrix} -2.0606 & 10.4288 \\ -2.0606 & 10.4288 \end{bmatrix}$	$\begin{bmatrix} -0.0785 & 1.0544 \\ -0.0785 & 1.0544 \end{bmatrix}$	$\begin{bmatrix} -0.0414 & 0.5060 \\ -0.0414 & 0.5060 \end{bmatrix}$
$C_c(k)$	$\begin{bmatrix} 1.8772 & 1.8772 \end{bmatrix}$	$\begin{bmatrix} 0.5637 & 0.5637 \end{bmatrix}$	$\begin{bmatrix} 2.2341 & 2.2341 \end{bmatrix}$
k	**3**	...	
$A_c(k)$	$\begin{bmatrix} 0.3136 & 0.3136 \\ 0.3136 & 0.3136 \end{bmatrix}$	...	
$B_c(k)$	$\begin{bmatrix} -0.3578 & 6.4767 \\ -0.3578 & 6.4767 \end{bmatrix}$	...	
$C_c(k)$	$\begin{bmatrix} 5.2543 & 5.2543 \end{bmatrix}$	...	

complex systems with unreliable communication links, such as sampled data systems, bilinear systems and a more general class of nonlinear stochastic systems.

4

Finite-Horizon Estimation of Randomly Occurring Faults

In this chapter, the finite-horizon H_∞ fault estimation problem is first studied for a class of nonlinear stochastic time-varying systems with both randomly occurring faults and fading channels. The system model (dynamical plant) is subject to Lipschitz-like nonlinearities and the faults occur in a random way governed by a set of Bernoulli distributed white sequences. The system measurements are transmitted through fading channels described by a modified stochastic Rice fading model. The purpose of the addressed problem is to design a time-varying fault estimator such that, in the presence of channel fading and randomly occurring faults, the influence from the exogenous disturbances onto the estimation errors is attenuated at the given level quantified by a H_∞-norm in the mean square sense. By utilizing the stochastic analysis techniques, sufficient conditions are established to ensure that the dynamic system under consideration satisfies the prespecified performance constraint on the fault estimation, and then a recursive linear matrix inequality approach is employed to design the desired fault estimator gains. Second, the finite-horizon estimation problem of ROFs is dealt with for a class of nonlinear systems whose parameters are all time-varying. The faults are assumed to occur in a random way governed by two sets of Bernoulli distributed white sequences. The stochastic nonlinearities entering the system are described by statistical means that can cover several classes of well-studied nonlinearities. The aim of the problem is to estimate the random faults, over a finite horizon, such that the influence from the exogenous disturbances onto the estimation errors is attenuated at the given level quantified by a H_∞-norm in the mean square sense. Finally, two simulation examples are given to illustrate the effectiveness of the proposed schemes.

4.1 On H_∞ Estimation of ROFs with Fading Channels

In this section, we concerned the finite-horizon H_∞ fault estimation problem for a class of nonlinear stochastic time-varying systems with both randomly occurring faults and fading channels.

DOI: 10.1201/9781003189497-4

4.1.1 Problem Formulation

Consider the following class of discrete time-varying nonlinear stochastic systems defined on $k \in [0, N]$:

$$\begin{cases} x(k+1) = g(k, x(k)) + \alpha(k) D_f(k) f(k) + E_1(k) w(k) \\ \tilde{y}(k) = C(k) x(k) + E_2(k) v(k) \\ x(0) = \varphi_0 \end{cases} \tag{4.1}$$

where $x(k) \in \mathbb{R}^{n_x}$ represents the state vector; $\tilde{y}(k) \in \mathbb{R}^{n_y}$ is the process output; $w(k) \in \mathbb{R}^{n_w}$, $v(k) \in \mathbb{R}^{n_v}$ and $f(k) \in \mathbb{R}^{n_l}$ are, respectively, the disturbance input, the measurement noises and the fault signal, all of which belong to $l_2[0, N]$; and φ_0 is a given initial value. $D_f(k)$, $E_1(k)$, $C(k)$ and $E_2(k)$ are known, real, time-varying matrices with appropriate dimensions.

The nonlinear function $g(\cdot, \cdot)$ is assumed to satisfy $g(k, 0) = 0$ and the following condition:

$$\|g(k, x(k) + \sigma(k)) - g(k, x(k)) - A(k)\sigma(k)\| \leq b(k) \|\sigma(k)\| \tag{4.2}$$

where $A(k)$ is a known matrix, $\sigma(k) \in \mathbb{R}^{n_x}$ is any vector and $b(k)$ is a known positive scalar.

Remark 4.1 *The nonlinear description (4.2) with the system parameter $A(k)$ reflects the distance between the originally nonlinear model (4.1) and the nominal linear model. In fact, such a nonlinear description resembles the Lipschitz conditions on the nonlinear functions. In applications, the linearization technique is utilized to quantify the maximum possible deviation from the nominal model.*

The dynamic characteristics of the fault vector $f(k)$ can be described as follows:

$$f(k+1) = A_f(k) f(k) \tag{4.3}$$

where $A_f(k)$ is a known matrix with appropriate dimensions.

The variable $\alpha(k)$ in (4.1), which accounts for the randomly occurring fault phenomena, is a Bernoulli distributed white sequences taking values on 0 or 1 with

$$\text{Prob}\{\alpha(k) = 1\} = \bar{\alpha}, \quad \text{Prob}\{\alpha(k) = 0\} = 1 - \bar{\alpha} \tag{4.4}$$

where $\bar{\alpha} \in [0, 1]$ is a known constant.

Remark 4.2 *The time-varying system (4.1) provides a way of accounting for the ROF phenomenon by resorting to the random variable $\alpha(k)$. At the kth time point, if $\alpha(k) = 1$, the fault occurs; and if $\alpha(k) = 0$, the system works normally. The fault obeying (4.3) may occur in a probabilistic way based* *on an individual probability distribution that can be specified* a prior *through statistical tests. Such a ROF concept could better reflect the probabilistically intermittent faults for the finite-horizon fault estimation problems, which render more practical significance for the time-varying system (4.1).*

In this section, consider the case when an unreliable wireless network medium is utilized for the signal transmission. In this case, the phenomenon of fading channels becomes an issue that constitutes another focus of our present research. The measurement signal $y(k)$ with probabilistic fading channels is described by

$$y(k) = \sum_{s=0}^{l_k} \beta_s(k)\tilde{y}(k-s) + E_3(k)\xi(k) \tag{4.5}$$

with $l_k = \min\{l, k\}$. Here, l is the given number of paths. $\beta_s(k)$ $(s = 0, 1, \cdots, l_k)$ are channel coefficients which are mutually independent random variables taking values over $[0, 1]$ with mathematical expectations $\bar{\beta}_s$ and variances ν_s. $\xi(k) \in l_2[0, N]$ is also an external disturbance. For simplicity, we set $\{\tilde{y}(k)\}_{k\in[-l,-1]} = 0$, i.e., $\{x(k)\}_{k\in[-l,-1]} = 0$ and $\{v(k)\}_{k\in[-l,-1]} = 0$.

Throughout the section, we assume that $\alpha(k)$ and $\beta_s(k)$ $(s = 0, \ldots, l_k)$ are uncorrelated random variables. The probabilistic fading measurement (4.5) is actually a weighted sum of the signals from channels of different delays where the weights are random variables taking values on the interval $[0, 1]$. Such fading measurement includes the traditional packet dropouts and random communication delays as special cases. For example, $l = 0$ corresponds to the case of probabilistically degraded measurements (without time-delays) and $l = 1$ corresponds to the case that degraded measurement and one-step communication delay could occur simultaneously.

Letting $x_f(k) \triangleq \begin{bmatrix} x^T(k) & f^T(k) \end{bmatrix}^T$ and $z(k) \triangleq f(k)$, we have from (4.1), (4.3) and (4.5) that

$$\begin{cases} x_f(k+1) = (\bar{A}_f(k) + \tilde{\alpha}(k)\bar{D}_f(k))x_f(k) + \bar{E}_1(k)w(k) \\ \qquad\qquad + Hg(k, H^T x_f(k)) \\ y(k) = \displaystyle\sum_{s=0}^{l_k} \beta_s(k)[\bar{C}(k-s)x_f(k-s) + E_2(k-s)v(k-s)] \\ \qquad\qquad + E_3(k)\xi(k) \\ z(k) = Lx_f(k) \end{cases} \tag{4.6}$$

where

$$\begin{aligned} \bar{A}_f(k) &\triangleq \begin{bmatrix} 0 & \bar{\alpha}D_f(k) \\ 0 & A_f(k) \end{bmatrix}, \ \bar{D}_f(k) \triangleq \begin{bmatrix} 0 & D_f(k) \\ 0 & 0 \end{bmatrix}, \ H \triangleq \begin{bmatrix} I \\ 0 \end{bmatrix}, \ L \triangleq \begin{bmatrix} 0 \\ I \end{bmatrix}^T, \\ \bar{C}(k) &\triangleq \begin{bmatrix} C(k) & 0 \end{bmatrix}, \ \bar{E}_1(k) \triangleq \begin{bmatrix} E_1^T(k) & 0 \end{bmatrix}^T, \ \tilde{\alpha}(k) \triangleq \alpha(k) - \bar{\alpha}. \end{aligned}$$

For the purpose of simplicity, for $-l \leq i \leq -1$, we assume that $C(i) = 0$, $\tilde{y}(i) = 0$ and $\begin{bmatrix} v^T(i) & \xi^T(i) \end{bmatrix} = 0$. Based on the actually received signal $y(k)$, the following time-varying fault estimator is constructed for system (4.6):

$$\begin{cases} \hat{x}_f(k+1) = \bar{A}_f(k)\hat{x}_f(k) + Hg(k, H^T\hat{x}_f(k)) - K(k)\Big(y(k) \\ \qquad\qquad - \sum_{s=0}^{l} \bar{\beta}_s \bar{C}(k-s)\hat{x}_f(k-s)\Big) \\ \hat{z}(k) = L\hat{x}_f(k) \end{cases} \tag{4.7}$$

where $\hat{x}_f(k) \in \mathbb{R}^{n_x+n_l}$ is the estimate of the state $x_f(k)$, $\hat{z}(k) \in \mathbb{R}^{n_l}$ represents the estimate of the fault $f(k)$ and $K(k)$ is the gain matrix of the fault estimator to be designed.

Remark 4.3 *It is worth pointing out that the constructed fault estimator (4.7) can be regarded as a Luenberger-type observer. In comparison with other kinds of estimators, the computational complexity with respect to (4.7) is relatively light as one parameter $K(k)$ needs to be designed, where $A(k)$ and $b(k)$ are involved. In addition, for the fault estimation purpose, the designed estimator should be physically implementable in practical engineering, and therefore the unknown (but bounded) disturbance inputs $w(k)$, $v(k)$ and $\xi(k)$ are excluded in (4.7).*

For notational simplicity, we denote

$$m(k) \triangleq g(k, H^T x_f(k)) - g(k, H^T\hat{x}_f(k)) - A(k)H^T(x_f(k) - \hat{x}_f(k)). \tag{4.8}$$

Letting $e(k) \triangleq x_f(k) - \hat{x}_f(k)$, $\tilde{\beta}_s(k) \triangleq \beta_s(k) - \bar{\beta}_s$, $\eta(k) \triangleq \begin{bmatrix} x_f^T(k) & e^T(k) \end{bmatrix}^T$, $\tilde{z}(k) \triangleq z(k) - \hat{z}(k)$ and $\varpi(k) \triangleq \begin{bmatrix} w^T(k) & \xi^T(k) \end{bmatrix}^T$, we have the following dynamic system to be investigated:

$$\begin{cases} \eta(k+1) = \mathcal{Y}_l(k) + \tilde{\alpha}(k)\mathcal{D}_f(k)\eta(k) + \sum_{s=0}^{l_k} \tilde{\beta}_s(k)\mathcal{C}(k-s)\eta(k-s) \\ \qquad\qquad + \sum_{s=0}^{l_k} \tilde{\beta}_s(k)\mathcal{E}_2(k-s)v(k-s) \\ \tilde{z}(k) = \mathcal{L}(k)\eta(k) \end{cases} \tag{4.9}$$

where

$$\begin{aligned} \mathcal{Y}_l(k) \triangleq\ & \mathcal{A}_f(k)\eta(k) + \mathcal{H}\mathcal{F}(\eta(k)) + \sum_{s=1}^{l_k} \bar{\beta}_s \bar{\mathcal{C}}(k-s)\eta(k-s) \\ & + \sum_{s=0}^{l_k} \bar{\beta}_s \mathcal{E}_2(k-s)v(k-s) + \mathcal{E}_1(k)\varpi(k), \end{aligned}$$

$$\begin{aligned}
\mathcal{F}(\eta(k)) &\triangleq \left[\left(g(k,H^Tx_f(k))-A(k)H^Tx_f(k)\right)^T \quad m^T(k)\right]^T,\\
\mathcal{A}_f(k) &\triangleq \mathrm{diag}\{A_H, A_H+\bar{\beta}_0K(k)\bar{C}(k)\},\ \mathcal{H}\triangleq \mathrm{diag}\{H,H\},\\
A_H &\triangleq \bar{A}_f(k)+HA(k)H^T,\ \mathcal{D}_f(k)\triangleq \mathbf{1}_2\otimes\left[\bar{D}_f(k)\quad 0\right],\\
\mathcal{C}(k-s) &\triangleq \begin{bmatrix} 0 & 0\\ A_K & 0\end{bmatrix}, \mathcal{E}_1(k)\triangleq \begin{bmatrix}\bar{E}_1(k) & 0\\ \bar{E}_1(k) & K(k)E_3(k)\end{bmatrix},\\
\bar{\mathcal{C}}(k-s) &\triangleq \mathrm{diag}\{0,A_K\},\ \mathcal{L}(k)\triangleq\left[0\quad L\right],\\
A_K &\triangleq K(k)\bar{C}(k-s)\ \mathcal{E}_2(k-s)\triangleq\left[0^T\quad (K(k)E_2(k-s))^T\right]^T.
\end{aligned}$$

Our objective of this section is to find a fault estimate $\hat{z}(k)$ $(0\le k\le N-1)$ such that, for the given positive scalar γ, the dynamic system (4.9) satisfies the following fault estimation performance requirement:

$$\begin{aligned}
J\triangleq&\mathbb{E}\left\{\sum_{k=0}^{N-1}\left(\|\tilde{z}(k)\|^2-\gamma^2\|\varpi(k)\|^2_{P_a}-\gamma^2\|v(k)\|^2_{P_b}\right)\right\}\\
&-\gamma^2\sum_{i=-l}^{0}\mathbb{E}\left\{\eta^T(i)P_{ci}\eta(i)\right\}<0,\forall(\{\varpi(k)\},\{v(k)\},\eta(0))\neq 0 \qquad (4.10)
\end{aligned}$$

where P_a, P_b and P_{ci} are known positive definite weighted matrices, $\|\varpi(k)\|^2_{P_a}\triangleq\varpi^T(k)P_a\varpi(k)$ and $\|v(k)\|^2_{P_b}\triangleq v^T(k)P_bv(k)$.

Remark 4.4 *The fault estimation performance requirement (4.10) is adopted from the classical gain-based $\mathcal{H}_\infty$ control theory, which means that the influence from disturbances $\varpi(k)$, $v(k)$ and initial states $\eta(i)$ $(i=-l,-l+1,\cdots,0)$ onto the fault estimation error $\tilde{z}(k)$ over the given finite-horizon should be constrained by means of the given disturbance attenuation level γ.*

4.1.2 Main Results

In this subsection, let us investigate both the fault estimator performance analysis and design problems for system (4.9). First, we propose the following finite-horizon fault estimation performance analysis results for a class of nonlinear time-varying systems with ROFs and fading channels.

For convenience of later analysis, we denote

$$\begin{aligned}
&\bar{\Gamma}(k)\triangleq\left[\Gamma_{ij}(k)\right]_{\{i=1,2,\ldots,5;j=1,2,\ldots,5\}},\quad \bar{P}(k+1)\triangleq I_l\otimes P(k+1),\\
&\bar{Q}(k,l)\triangleq\mathrm{diag}\{Q(k-1,1),Q(k-2,2),\cdots,Q(k-l,l)\},\\
&\Gamma_{11}(k)\triangleq\mathcal{A}_f^T(k)P(k+1)\mathcal{A}_f(k)+\bar{\alpha}(1-\bar{\alpha})\mathcal{D}_f^T(k)P(k+1)\mathcal{D}_f(k)-P(k)\\
&\qquad+\nu_0\mathcal{C}^T(k)P(k+1)\mathcal{C}(k)+\sum_{j=1}^{l}Q(k,j),
\end{aligned}$$

$$\begin{aligned}
\Gamma_{21}(k) \triangleq& \mathcal{H}^T P(k+1)\mathcal{A}_f(k),\ \Gamma_{31}(k) \triangleq (\Lambda_\beta \bar{\mathcal{C}}_l(k))^T P(k+1)\mathcal{A}_f(k),\\
\Gamma_{22}(k) \triangleq& \mathcal{H}^T P(k+1)\mathcal{H},\ \Gamma_{32}(k) \triangleq (\Lambda_\beta \bar{\mathcal{C}}_l(k))^T P(k+1)\mathcal{H},\\
\Gamma_{33}(k) \triangleq& (\Lambda_\beta \bar{\mathcal{C}}_l(k))^T P(k+1)\Lambda_\beta \bar{\mathcal{C}}_l(k) - \bar{Q}(k,l) + (\bar{\Lambda}_\gamma \mathcal{C}_l(k))^T \bar{P}(k+1)\bar{\Lambda}_\gamma \mathcal{C}_l(k),\\
\Gamma_{42}(k) \triangleq& (\bar{\Lambda}_\beta \bar{\mathcal{E}}_{2l}(k))^T P(k+1)\mathcal{H},\\
\Gamma_{41}(k) \triangleq& (\bar{\Lambda}_\beta \bar{\mathcal{E}}_{2l}(k))^T P(k+1)\mathcal{A}_f(k) + \nu_0 \mathcal{E}_2^T(k)P(k+1)\mathcal{C}(k)\bar{H}_3,\\
\Gamma_{43}(k) \triangleq& (\bar{\Lambda}_\beta \bar{\mathcal{E}}_{2l}(k))^T P(k+1)\Lambda_\beta \bar{\mathcal{C}}_l(k) + \bar{H}_0(\bar{\Lambda}_\gamma \hat{\mathcal{E}}_{2l}(k))^T \bar{P}(k+1)\bar{\Lambda}_\gamma \mathcal{C}_l(k),\\
\Gamma_{44}(k) \triangleq& (\bar{\Lambda}_\beta \bar{\mathcal{E}}_{2l}(k))^T P(k+1)\bar{\Lambda}_\beta \bar{\mathcal{E}}_{2l}(k) + (\hat{\Lambda}_\gamma \bar{\mathcal{E}}_{2l}(k))^T \hat{P}(k+1)\hat{\Lambda}_\gamma \bar{\mathcal{E}}_{2l}(k),\\
\Gamma_{51}(k) \triangleq& \mathcal{E}_1^T(k)P(k+1)\mathcal{A}_f(k),\ \hat{P}(k+1) \triangleq I_{l+1}\otimes P(k+1),\\
\Gamma_{52}(k) \triangleq& \mathcal{E}_1^T(k)P(k+1)\mathcal{H},\ \Gamma_{53}(k) \triangleq \mathcal{E}_1^T(k)P(k+1)\Lambda_\beta \bar{\mathcal{C}}_l(k),\\
\Gamma_{54}(k) \triangleq& \mathcal{E}_1^T(k)P(k+1)\bar{\Lambda}_\beta \bar{\mathcal{E}}_{2l}(k),\ \Gamma_{55}(k) \triangleq \mathcal{E}_1^T(k)P(k+1)\mathcal{E}_1(k),\\
\phi(k) \triangleq& \rho(k)b^2(k)(\bar{H}_1^T H H^T \bar{H}_1 + \bar{H}_2^T H H^T \bar{H}_2), \Lambda_\beta \triangleq \begin{bmatrix} \bar{\beta}_1 I & \bar{\beta}_2 I & \cdots & \bar{\beta}_l I \end{bmatrix},\\
\bar{\mathcal{C}}_l(k) \triangleq& \mathrm{diag}\{\bar{\mathcal{C}}(k-1), \bar{\mathcal{C}}(k-2), \ldots, \bar{\mathcal{C}}(k-l)\}, \bar{H}_0 \triangleq \begin{bmatrix} 0_{n_v, l\cdot n_v} & I_{l\cdot n_v, l\cdot n_v} \end{bmatrix}^T,\\
\bar{\mathcal{E}}_{2l}(k) \triangleq& \mathrm{diag}\{\mathcal{E}_2(k), \mathcal{E}_2(k-1), \ldots, \mathcal{E}_2(k-l)\}, \bar{\Lambda}_\beta \triangleq \begin{bmatrix} \bar{\beta}_0 I & \bar{\beta}_1 I & \cdots & \bar{\beta}_l I \end{bmatrix},\\
\mathcal{C}_l(k) \triangleq& \mathrm{diag}\{\mathcal{C}(k-1), \mathcal{C}(k-2), \ldots, \mathcal{C}(k-l)\}, \bar{H}_1 \triangleq \begin{bmatrix} I_{n_x+n_l} & 0_{n_x+n_l} \end{bmatrix},\\
\hat{\mathcal{E}}_{2l}(k) \triangleq& \mathrm{diag}\{\mathcal{E}_2(k-1), \mathcal{E}_2(k-2), \ldots, \mathcal{E}_2(k-l)\}, \bar{H}_2 \triangleq \begin{bmatrix} 0_{n_x+n_l} & I_{n_x+n_l} \end{bmatrix},\\
\bar{\Lambda}_\gamma \triangleq& \mathrm{diag}\{\sqrt{\nu_1}I, \sqrt{\nu_2}I, \ldots, \sqrt{\nu_l}I\}, \hat{\Lambda}_\gamma \triangleq \mathrm{diag}\{\sqrt{\nu_0}I, \sqrt{\nu_1}I, \ldots, \sqrt{\nu_l}I\},\\
\bar{P}_b \triangleq& \frac{\gamma^2}{l+1} I_{l+1}\otimes P_b, \bar{H}_3 \triangleq \begin{bmatrix} I_{n_v, 2(n_x+n_l)} & 0_{l\cdot n_v, 2(n_x+n_l)} \end{bmatrix}^T.
\end{aligned}$$

Theorem 4.1 *Consider the discrete time-varying nonlinear stochastic system described by (4.1)–(4.5). Let the disturbance attenuation level* $\gamma > 0$*, the positive definite matrices* $P_a > 0$, $P_b > 0$, $P_{ci} > 0$ $(i = -l, -l+1, \ldots, 0)$ *and the gain matrices of the fault estimator* $\{K(k)\}_{k\in[0,N-1]}$ *in (4.7) be given. The fault estimator* $\hat{z}(k)$ $(k = 0, 1, \ldots, N-1)$ *satisfies the performance criterion (4.10) if there exist families of positive scalars* $\{\rho(k)\}_{k\in[0,N-1]}$*, positive definite matrices* $\{P(k)\}_{k\in[0,N]} > 0$ *and* $\{Q(i,j)\}_{i\in[-l,N], j\in[1,l]} > 0$ *satisfying*

$$\begin{aligned}
\Gamma(k) &= \bar{\Gamma}(k) + \mathrm{diag}\{\mathcal{L}^T(k)\mathcal{L}(k) + \phi(k), -\rho(k)I, 0, -\bar{P}_b, -\gamma^2 P_a\}\\
&< 0 \qquad (4.11)
\end{aligned}$$

and the initial condition

$$\gamma^2 P_{c0} - P(0) > 0, \gamma^2 P_{-ci} - \sum_{j=i}^{l} Q(-i,j) > 0 \ (i = 1, 2, \ldots, l). \qquad (4.12)$$

Proof *Consider the following Lyapunov-like functional candidate for* *system (4.9):*

$$V(k) \triangleq V_1(k) + V_2(k) = \eta^T(k)P(k)\eta(k) + \sum_{j=1}^{l}\sum_{i=k-j}^{k-1} \eta^T(i)Q(i,j)\eta(i) \quad (4.13)$$

where $P(k) > 0$ and $Q(i,j) > 0$ are symmetric positive definite matrices with appropriate dimensions. Calculating the difference of $V(k)$ along the solution of system (4.9) and taking the mathematical expectation, we have

$$\begin{aligned}
&\mathbb{E}\{\Delta V_1(k)\} = \mathbb{E}\{V_1(k+1) - V_1(k)\}\\
=&\mathbb{E}\Big\{\Big(\mathcal{Y}_l^T(k)P(k+1)\mathcal{Y}_l(k) - \eta^T(k)P(k)\eta(k) + \bar{\alpha}(1-\bar{\alpha})\eta^T(k)\mathcal{D}_f^T(k)\\
&\times P(k+1)\mathcal{D}_f(k)\eta(k) + \sum_{s=0}^{l}\nu_s(\mathcal{C}(k-s)\eta(k-s) + \mathcal{E}_2(k-s)v(k-s))^T\\
&\times P(k+1)(\mathcal{C}(k-s)\eta(k-s) + \mathcal{E}_2(k-s)v(k-s))\Big\}. \quad (4.14)
\end{aligned}$$

Similarly, by noting the equation (4.13), one has

$$\mathbb{E}\{\Delta V_2(k)\}=\mathbb{E}\Big\{\sum_{j=1}^{l}\eta^T(k)Q(k,j)\eta(k) - \eta_l^T(k)\bar{Q}(k,l)\eta_l(k)\Big\} \quad (4.15)$$

where $\eta_l(k) \triangleq \begin{bmatrix}\eta^T(k-1) & \eta^T(k-2) & \cdots & \eta^T(k-l)\end{bmatrix}^T$. Therefore, by denoting

$$\begin{aligned}
v_l(k) &\triangleq \begin{bmatrix}v^T(k) & v^T(k-1) & \cdots & v^T(k-l)\end{bmatrix}^T,\\
\tilde{\eta}(k) &\triangleq \begin{bmatrix}\eta^T(k) & \mathcal{F}^T(\eta(k)) & \eta_l^T(k) & v_l^T(k) & \varpi^T(k)\end{bmatrix}^T
\end{aligned}$$

and combining (4.13)–(4.15), one immediately obtains

$$\mathbb{E}\{\Delta V(k)\} = \mathbb{E}\{\Delta V_1(k) + \Delta V_2(k)\} = \mathbb{E}\{\tilde{\eta}^T(k)\bar{\Gamma}(k)\tilde{\eta}(k)\}. \quad (4.16)$$

Moreover, it follows from the constraint (4.2) that

$$\|\mathcal{F}(\eta(k))\|^2 \le b^2(k)\eta^T(k)(\bar{H}_1^T H H^T \bar{H}_1 + \bar{H}_2^T H H^T \bar{H}_2)\eta(k). \quad (4.17)$$

Hence we have

$$\begin{aligned}
\mathbb{E}\{\Delta V(k)\} \le&\mathbb{E}\Big\{\tilde{\eta}^T(k)\bar{\Gamma}(k)\tilde{\eta}(k) - \rho(k)(\|\mathcal{F}(\eta(k))\|^2\\
&- b^2(k)\eta^T(k)(\bar{H}_1^T H H^T \bar{H}_1 + \bar{H}_2^T H H^T \bar{H}_2)\eta(k))\Big\}. \quad (4.18)
\end{aligned}$$

Summing up (4.18) on both sides from 0 to $N-1$ with respect to k, we obtain

$$\begin{aligned}
&\sum_{k=0}^{N-1} \mathbb{E}\{\Delta V(k)\} = \mathbb{E}\{V(N)\} - \mathbb{E}\{V(0)\} \\
\leq\ & \mathbb{E}\Big\{\sum_{k=0}^{N-1} \tilde{\eta}^T(k)\Gamma(k)\tilde{\eta}(k)\Big\} + \mathbb{E}\left\{\frac{\gamma^2}{l+1}\sum_{s=0}^{l}\sum_{k=0}^{N-1}(\|v(k-s)\|^2_{P_b} - \|v(k)\|^2_{P_b})\right\} \\
&-\mathbb{E}\left\{\sum_{k=0}^{N-1}(\|\tilde{z}(k)\|^2 - \gamma^2\|\varpi(k)\|^2_{P_a} - \gamma^2\|v(k)\|^2_{P_b})\right\}. \qquad (4.19)
\end{aligned}$$

It can be obtained from (4.11) and (4.12) that

$$\begin{aligned}
&\mathbb{E}\left\{\sum_{k=0}^{N-1}\Big(\gamma^2\|\varpi(k)\|^2_{P_a} + \gamma^2\|v(k)\|^2_{P_b} - \|\tilde{z}(k)\|^2\Big) + \gamma^2\sum_{i=-l}^{0}\eta^T(i)P_{ci}\eta(i)\right\} \\
>\ & \mathbb{E}\{V(N)\} + \mathbb{E}\left\{\gamma^2\sum_{k=-l}^{0}\eta^T(i)P_{Ci}\eta(i) - V(0)\right\} \geq 0 \qquad (4.20)
\end{aligned}$$

which is equivalent to (4.10), and the proof is now complete.

Remark 4.5 *White noise disturbances are frequently encountered in practice where Kalman filter (KF) or extended Kalman filter (EKF) approaches can be used to deal with the state estimation problem. In H_∞ estimation, the noise sources are arbitrary deterministic signals with bounded energy or average power, and an H_∞ estimator is sought which ensures a prescribed upper-bound on the L_2-induced gain from the noise signals to the estimation error. Such a H_∞ estimation approach is particularly appropriate to applications where the statistics of the noise signals are not exactly known. In fact, the H_∞ estimator has been widely adopted in practical engineering due to its capability of providing a bound for the worst-case estimation error. It should be pointed out that the problem addressed in this section is equipped with the following features: (1) the considered external disturbances are unknown but bounded and therefore do not possess known statistics; (2) the nonlinearities satisfy the given bounded conditions only; and (3) the plant under consideration is quite comprehensive that covers fading measurements, ROFs, nonlinearity and time-varying parameters. Unfortunately, the above features prevent the existing methods (such as KF, EKF) from being applied to the H_∞ state estimation problem for the underlying system in this section, and the proposed scheme in this section is particularly suitable for handling the addressed networked complex systems.*

Based on the analysis results, we are now ready to solve the fault estimator design problem for system (4.9) in the following theorem.

For convenience of later analysis, we denote

$$
\begin{aligned}
\hat{\Gamma}_{11}(k) &\triangleq \mathrm{diag}\left\{-P(k)+\sum_{j=1}^{l}Q(k,j)+\mathcal{L}^T(k)\mathcal{L}(k)+\phi(k),-\rho(k)I\right\},\\
H_0 &\triangleq \begin{bmatrix}0 & I\end{bmatrix}^T,\ \Lambda_{\bar{\beta}_0}\triangleq\begin{bmatrix}0 & \bar{\beta}_0 I\end{bmatrix}^T,\\
\hat{\Gamma}_{22}(k) &\triangleq \mathrm{diag}\left\{-\bar{Q}(k,l),-\frac{\gamma^2}{l+1}P_b,-\frac{\gamma^2}{l+1}I_l\otimes P_b,-\gamma^2 P_a\right\},\\
\mathcal{A}_{f_0}(k) &\triangleq I_2\otimes(\bar{A}_f(k)+HA(k)H^T),\ \mathcal{E}_{2K}\triangleq\bar{H}_K\tilde{\mathcal{E}}_{2l}(k),\\
\tilde{\mathcal{E}}_{2l}(k) &\triangleq \mathrm{diag}\left\{\hat{\mathcal{E}}_2(k-1),\hat{\mathcal{E}}_2(k-2),\ldots,\hat{\mathcal{E}}_2(k-l)\right\},\\
\hat{\Gamma}_{31}(k) &\triangleq \begin{bmatrix}\sqrt{\nu_0}H_0K(k)\hat{C}(k) & 0\\ \mathcal{A}_{f_0}(k)+\Lambda_{\bar{\beta}_0}K(k)\tilde{C}(k) & \mathcal{H}\\ \sqrt{\bar{\alpha}(1-\bar{\alpha})}\mathcal{D}_f(k) & 0\\ 0 & 0\end{bmatrix},\\
\hat{\Gamma}_{32}(k) &\triangleq \begin{bmatrix}0 & \sqrt{\nu_0}\mathcal{E}_K & 0 & 0\\ \Lambda_\beta H_K\check{C}_l(k) & \beta_0\mathcal{E}_K & \Lambda_\beta\mathcal{E}_{2K} & \hat{\mathcal{E}}_{1K}\\ 0 & 0 & 0 & 0\\ \bar{\Lambda}_\gamma H_K\tilde{C}_l(k) & 0 & \bar{\Lambda}_\gamma\mathcal{E}_{2K} & 0\end{bmatrix},\\
\hat{\Gamma}_{33}(k) &\triangleq \mathrm{diag}\{I_3\otimes -R(k+1),-\bar{R}(k+1)\},\ \hat{C}(k)\triangleq\begin{bmatrix}\bar{C}(k) & 0\end{bmatrix},\\
\tilde{C}(k) &\triangleq \begin{bmatrix}0 & \bar{C}(k)\end{bmatrix},\ \tilde{C}_l(k)\triangleq\mathrm{diag}\left\{\hat{C}(k-1),\hat{C}(k-2),\ldots,\hat{C}(k-l)\right\},\\
\check{C}_l(k) &\triangleq \mathrm{diag}\left\{\tilde{C}(k-1),\tilde{C}(k-2),\ldots,\tilde{C}(k-l)\right\},\\
\hat{\mathcal{E}}_1(k) &\triangleq \mathbf{1}_2\otimes\begin{bmatrix}\bar{E}_1(k) & 0\end{bmatrix},\ \hat{\mathcal{E}}_2(k)\triangleq\begin{bmatrix}0 & E_2^T(k)\end{bmatrix}^T,\ \hat{\mathcal{E}}_3(k)\triangleq\begin{bmatrix}0 & E_3(k)\end{bmatrix},\\
\bar{R}(k+1) &\triangleq \bar{P}^{-1}(k+1),\ H_K\triangleq I_l\otimes H_0K(k),\ \bar{H}_K\triangleq I_l\otimes\bar{K}(k),\\
\mathcal{E}_K &\triangleq \bar{K}(k)\hat{\mathcal{E}}_2(k),\ \hat{\mathcal{E}}_{1K}\triangleq\hat{\mathcal{E}}_1(k)+H_0K(k)\hat{\mathcal{E}}_3(k).
\end{aligned}
\tag{4.21}
$$

Theorem 4.2 *Consider the discrete time-varying nonlinear stochastic system (4.1) with the time-varying fault estimator (4.7). For the given disturbance attenuation level $\gamma > 0$, the positive definite matrices $P_a > 0$, $P_b > 0$ and $P_{ci} > 0$ $(i = -l, -l+1, \ldots, 0)$, the fault estimator $\hat{z}(k)$ $(k = 0, 1, \ldots, N-1)$ satisfies the performance criterion (4.10) if there exist families of positive scalars $\{\rho(k)\}_{k\in[0,N-1]}$, positive definite matrices $\{P(k)\}_{k\in[0,N]} > 0$, $\{Q(i,j)\}_{i\in[-l,N],j\in[1,l]} > 0$, $\{R(k)\}_{k\in[0,N]} > 0$ and real-valued matrices*

$K(k)_{k\in[0,N-1]}$ *satisfying*

$$\hat{\Gamma}(k) = \begin{bmatrix} \hat{\Gamma}_{11}(k) & * & * \\ 0 & \hat{\Gamma}_{22}(k) & * \\ \hat{\Gamma}_{31}(k) & \hat{\Gamma}_{32}(k) & \hat{\Gamma}_{33}(k) \end{bmatrix} < 0 \tag{4.22}$$

and the initial condition

$$\gamma^2 P_{c0} - P(0) > 0, \gamma^2 P_{-ci} - \sum_{j=i}^{l} Q(-i,j) > 0 \ (i = 1, 2, \ldots, l) \tag{4.23}$$

with the parameters updated by $P(k+1) = R^{-1}(k+1)$.

Proof *In order to avoid partitioning the positive define matrices* $\{P(k)\}_{k\in[0,N]}$ *and* $\{Q(i,j)\}_{i\in[-l,N],j\in[1,l]}$, *we rewrite the parameters in Theorem 4.1 in the following form:*

$$\begin{aligned} \mathcal{C}(k-s) &= \hat{C}(k-s), \mathcal{A}_f(k) = \mathcal{A}_{f_0}(k) + \Lambda_{\bar{\beta}_0} K(k)\tilde{C}(k), \\ \bar{\mathcal{C}}(k-s) &= H_0 K(k)\tilde{C}(k-s), \mathcal{E}_1(k) = \hat{\mathcal{E}}_1(k) + H_0 K(k)\hat{\mathcal{E}}_3(k), \\ \mathcal{E}_2(k-s) &= \bar{K}(k)\hat{\mathcal{E}}_2(k-s), \bar{K}(k) = I_2 \otimes K(k). \end{aligned} \tag{4.24}$$

Noticing (4.24) and using Lemma 2.1 (Schur Complement Lemma), (4.22) can be obtained by (4.11) after some straightforward algebraic manipulations. The proof of this theorem is now complete.

By means of Theorem 4.2, we can summarize the Finite-Horizon Fault Estimator Design (*FHFED*) algorithm as follows:

Algorithm *FHFED*:

Step 1.	Given the disturbance attenuation level γ, the positive definite matrices $P_a > 0$, $P_b > 0$ and $P_{ci} > 0$ $(i = -l, -l+1, \ldots, 0)$.
Step 2.	Set $k = 0$ and solve the matrix inequalities (4.23) and the recursive matrix inequalities (4.22) to obtain the values of matrices $P(0)$, $\sum_{j=i}^{l} Q(-i,j)$ $(i = 1, 2, \ldots, l)$, $R(1)$ and the estimator gain matrix $K(0)$.
Step 3.	Set $k = k+1$, update the matrices $P(k+1) = R^{-1}(k+1)$ and then obtain the estimator gain matrix $K(k)$ by solving the recursive matrix inequalities (4.22).
Step 4.	If $k < N$, then go to Step 3, else go to Step 5.
Step 5.	Stop.

Remark 4.6 *In Theorem 4.2, the finite-horizon fault estimator is designed by solving a series of recursive matrix inequalities where both the current system measurement and previous system states are employed to estimate the current system state. Such a recursive process is particularly useful for online real-time implementation. It can be observed from Algorithm FHFED that the main results established contain all the information of the addressed general systems including the time-varying systems parameters, the occurrence probabilities of the random faults as well as the statistics characteristics of the channel coefficients.*

4.2 Recursive Estimation of ROFs: the Finite-Horizon Case

In this section, we discuss the finite-horizon estimation problem of randomly occurring faults for a class of nonlinear systems whose parameters are all time-varying.

4.2.1 Problem Formulation

Consider the following class of discrete time-varying nonlinear stochastic systems defined on $k \in [0, N]$:

$$\begin{cases} x(k+1) = A(k)x(k) + g(k, x(k)) + \alpha_1(k)B_f(k)f(k) + B_d(k)d(k) \\ y(k) = C(k)x(k) + h(k, x(k)) + \alpha_2(k)D_f(k)f(k) + D_d(k)v(k) \\ x(0) = x_0 \end{cases} \tag{4.25}$$

where $x(k) \in \mathbb{R}^{n_x}$ represents the state vector; $y(k) \in \mathbb{R}^{n_y}$ is the measurement signal; $d(k) \in \mathbb{R}^{n_w}$, $v(k) \in \mathbb{R}^{n_v}$ and $f(k) \in \mathbb{R}^{n_l}$ are, respectively, the disturbance input, the measurement noises, and the fault signal. Moreover, it is assumed that $d(k)$, $v(k)$ and $f(k)$ belong to $l_2[0, \infty)$; and x_0 is an initial value that is unknown. $A(k)$, $B_f(k)$, $B_d(k)$, $C(k)$, $D_f(k)$ and $D_d(k)$ are known, real, time-varying matrices with appropriate dimensions.

We assume that the functions $g(k, x(k))$ and $h(k, x(k))$ with $g(k, 0) = 0$ and $h(k, 0) = 0$ are stochastic nonlinear functions described by their statistical characteristics as follows:

$$\mathbb{E}\left\{ \begin{bmatrix} g(k, x(k)) \\ h(k, x(k)) \end{bmatrix} \middle| x(k) \right\} = 0, \tag{4.26}$$

$$\mathbb{E}\left\{ \begin{bmatrix} g(k, x(k)) \\ h(k, x(k)) \end{bmatrix} \begin{bmatrix} g^{\mathrm{T}}(j, x(j)) & h^{\mathrm{T}}(j, x(j)) \end{bmatrix} \middle| x(k) \right\} = 0, \quad k \neq j \tag{4.27}$$

and

$$\mathbb{E}\left\{\left[\begin{array}{c} g(k,x(k)) \\ h(k,x(k)) \end{array}\right]\left[\begin{array}{cc} g^{\mathrm{T}}(k,x(k)) & h^{\mathrm{T}}(k,x(k)) \end{array}\right]\Bigg| x(k)\right\}$$
$$=\sum_{i=1}^{q}\pi_i\pi_i^{\mathrm{T}}\mathbb{E}\{x^{\mathrm{T}}(k)\Gamma_i x(k)\} := \sum_{i=1}^{q}\hat{\Theta}_i\mathbb{E}\{x^{\mathrm{T}}(k)\Gamma_i x(k)\} \tag{4.28}$$

where $\hat{\Theta}_i$ and Γ_i are known matrices with appropriate dimensions, and q is the number of independent state components.

Remark 4.7 *As pointed out in [127, 208], the nonlinearity description in (4.26)–(4.28) covers many well-studied nonlinearities in stochastic systems such as (1) linear system with state- and control-dependent multiplicative noise; (2) nonlinear systems with random vectors dependent on the norms of states and control input; and (3) nonlinear systems with a random sequence dependent on the sign of a nonlinear function of states and control inputs.*

It is assumed that the dynamic characteristics of the fault vector $f(k)$ are described as follows:

$$f(k+1) = A_f(k)f(k) \tag{4.29}$$

where $A_f(k)$ is a known matrix with appropriate dimensions. Note that the fault becomes a constant one when $A_f(k) \equiv I$.

The variables $\alpha_1(k)$ and $\alpha_2(k)$ in (4.25), which govern the random nature of the occurred faults, are Bernoulli distributed white sequences taking values on 0 or 1 with

$$\begin{aligned} &\text{Prob}\{\alpha_1(k)=1\}=\bar{\alpha}_1, \quad \text{Prob}\{\alpha_1(k)=0\}=1-\bar{\alpha}_1, \\ &\text{Prob}\{\alpha_2(k)=1\}=\bar{\alpha}_2, \quad \text{Prob}\{\alpha_2(k)=0\}=1-\bar{\alpha}_2 \end{aligned} \tag{4.30}$$

where $\bar{\alpha}_1 \in [0,1]$ and $\bar{\alpha}_2 \in [0,1]$ are known constants. It is assumed that $\alpha_1(k)$, $\alpha_2(k)$, $g(k,x(k))$ and $h(k,x(k))$ are unrelated each other.

Remark 4.8 *The proposed fault modelled by (4.29) may occur in a probabilistic way based on an individual probability distribution that can be specified a prior through statistical tests. The ROFs concept is to reflect the random fashion of the occurred faults in a network environment and therefore render more practical significance for the time-varying system (4.25). The statistical information of the ROF is used throughout the design of the fault estimators.*

We are now ready to state the fault estimation problem under consideration as follows: given a positive scalar γ, find the fault estimation $\hat{f}(k)$ $(k = 0, 1, \ldots, N-1)$ such that

$$\zeta(N) \triangleq \frac{\mathbb{E}\left\{\sum_{k=0}^{N-1} \|\hat{f}(k) - f(k)\|^2\right\}}{\mathbb{E}\left\{(x(0) - \hat{x}(0))^T P_A (x(0) - \hat{x}(0))\right\} + \zeta_1(N)} < \gamma^2, \tag{4.31}$$

$$\forall(\{d(k)\}, \{v(k)\}, x(0)) \neq 0,\ \zeta_1(N) = \sum_{k=0}^{N-1} \Big(d^T(k) P_B d(k) + v^T(k) P_C v(k)\Big)$$

holds, where P_A, P_B and P_C are known positive definite weighting matrices, and $\hat{x}(0)$ is the estimate of initial state $x(0)$. Without loss of generality, the initial state estimate $\hat{x}(0)$ is assumed to be zero.

Defining $\bar{x}(k) \triangleq \begin{bmatrix} x^T(k) & f^T(k) \end{bmatrix}^T$ and $z(k) \triangleq f(k)$, we rewrite (4.25) and (4.29) into the following augmented system:

$$\left\{\begin{array}{l} \bar{x}(k+1) = (\bar{A}(k) + \tilde{\alpha}_1(k)\bar{B}_f(k))\bar{x}(k) + Hg(k, x(k)) + \bar{B}_d(k)d(k) \\ y(k) = (\bar{C}(k) + \tilde{\alpha}_2(k)\bar{D}_f(k))\bar{x}(k) + h(k, x(k)) + D_d(k)v(k) \\ z(k) = L(k)\bar{x}(k) \\ \bar{x}(0) = \bar{x}_0 \end{array}\right. \tag{4.32}$$

where

$$\begin{aligned} \bar{A}(k) &\triangleq \begin{bmatrix} A(k) & \bar{\alpha}_1(k)B_f(k) \\ 0 & A_f(k) \end{bmatrix},\ \bar{B}_f(k) \triangleq \begin{bmatrix} 0 & B_f(k) \\ 0 & 0 \end{bmatrix}, \\ \bar{B}_d(k) &\triangleq \begin{bmatrix} B_d^T(k) & 0 \end{bmatrix}^T,\ \bar{D}_f(k) \triangleq \begin{bmatrix} 0 & D_f(k) \end{bmatrix}, \\ H &\triangleq \begin{bmatrix} I & 0 \end{bmatrix}^T,\ \bar{C}(k) \triangleq \begin{bmatrix} C(k) & \bar{\alpha}_2(k)D_f(k) \end{bmatrix}, \\ \tilde{\alpha}_1(k) &\triangleq \alpha_1(k) - \bar{\alpha}_1(k),\ \tilde{\alpha}_2(k) \triangleq \alpha_2(k) - \bar{\alpha}_2(k),\ L(k) \triangleq \begin{bmatrix} 0 & I \end{bmatrix}. \end{aligned}$$

In this section, the following fault estimator is adopted for the augmented system (4.32):

$$\left\{\begin{array}{l} \hat{\bar{x}}(k+1) = F(k)\hat{\bar{x}}(k) + G(k)y(k) \\ \hat{z}(k) = L(k)\hat{\bar{x}}(k) \end{array}\right. \tag{4.33}$$

where $\hat{\bar{x}}(k) \in \mathbb{R}^{n_x+n_l}$ is the estimate of the state $\bar{x}(k)$. We assume that $\hat{\bar{x}}(0) = 0$. Subsequently, the fault estimation problem can be reformulated as to find $F(k)$ and $G(k)$ such that (4.31) is satisfied.

Letting $e(k) \triangleq \bar{x}(k) - \hat{\bar{x}}(k)$, we have

$$\begin{aligned} e(k+1) = & \big(\bar{A}(k) - F(k) - G(k)\bar{C}(k)\big)\bar{x}(k) + \big(\tilde{\alpha}_1(k)\bar{B}_f(k) \\ & -\tilde{\alpha}_2(k)G(k)\bar{D}_f(k)\big)\bar{x}(k) + F(k)e(k) + \bar{B}_d(k)d(k) \\ & -G(k)D_d(k)v(k) + Hg(k, x(k)) - G(k)h(k, x(k)). \end{aligned} \tag{4.34}$$

Denoting $\eta(k) \triangleq \begin{bmatrix} \bar{x}^T(k) & e^T(k) \end{bmatrix}^T$, $\tilde{z}(k) \triangleq z(k) - \hat{z}(k)$ and $w(k) \triangleq$

$\left[d^T(k)\ \ v^T(k)\right]^T$, we have the following system to be investigated:

$$\begin{cases} \eta(k+1) = \big(\mathcal{A}(k) + \tilde{\alpha}_1(k)\mathcal{B}_f(k) + \tilde{\alpha}_2(k)\mathcal{D}_f(k)\big)\eta(k) \\ \qquad\qquad + \mathcal{H}(k)\mathcal{F}(k,x(k)) + \mathcal{G}(k)w(k) \\ \tilde{z}(k) = \mathcal{L}(k)\eta(k) \end{cases} \tag{4.35}$$

where

$$\begin{aligned} \mathcal{A}(k) &\triangleq \begin{bmatrix} \bar{A}(k) & 0 \\ \bar{A}(k) - F(k) - G(k)\bar{C}(k) & F(k) \end{bmatrix}, \ \mathcal{B}_f(k) \triangleq \begin{bmatrix} \bar{B}_f(k) & 0 \\ \bar{B}_f(k) & 0 \end{bmatrix}, \\ \mathcal{D}_f(k) &\triangleq \begin{bmatrix} 0 & 0 \\ -G(k)\bar{D}_f(k) & 0 \end{bmatrix}, \ \mathcal{H}(k) \triangleq \begin{bmatrix} H & 0 \\ H & -G(k) \end{bmatrix}, \ \mathcal{L}(k) \triangleq \begin{bmatrix} 0 & L(k) \end{bmatrix}, \\ \mathcal{G}(k) &\triangleq \begin{bmatrix} \bar{B}_d(k) & 0 \\ \bar{B}_d(k) & -G(k)D_d(k) \end{bmatrix}, \ \mathcal{F}(k,x(k)) \triangleq \begin{bmatrix} g(k,x(k)) \\ h(k,x(k)) \end{bmatrix}. \end{aligned}$$

Furthermore, the performance requirement (4.31) is rewritten as

$$\begin{aligned} J_N \ &\triangleq \ \sum_{k=0}^{N-1} \Big(\mathbb{E}\left\{\|\tilde{z}(k)\|^2\right\} - \gamma^2 w^T(k) P_\Psi w(k) \Big) - \gamma^2 \eta^T(0) R \eta(0) < 0, \\ &\qquad \forall(\{w(k)\}, \eta(0)) \neq 0 \end{aligned} \tag{4.36}$$

where $R \triangleq \text{diag}\{0, \bar{P}_A\}$, $\bar{P}_A \triangleq \text{diag}\{P_A, 0\}$ and $P_\Psi \triangleq \text{diag}\{P_B, P_C\}$.

Our objective of this section is to find the sequence of parameter matrices, $F(k)$ and $G(k)$, such that the dynamic system (4.35) satisfies the performance requirement (4.36).

4.2.2 Main Results

Lemma 4.1 *[2] Let matrices G, M and Γ be given with appropriate sizes. Then the following matrix equation*

$$GXM = \Gamma \tag{4.37}$$

has a solution X if and only if $GG^\dagger \Gamma M^\dagger M = \Gamma$. Moreover, any solution to (4.37) is represented by

$$X = G^\dagger \Gamma M^\dagger + Y - G^\dagger G Y M M^\dagger$$

where Y is a matrix with an appropriate size.

Theorem 4.3 *Consider the time-varying nonlinear stochastic system described by (4.25)–(4.29). For a given disturbance attenuation level $\gamma > 0$ and the positive definite matrices $P_B > 0$, $P_C > 0$ and $R > 0$, the fault estimator $\hat{f}(k)$*

$(k = 0, 1, \ldots, N-1)$ satisfies the performance criterion (4.31) if and only if the following discrete Riccati difference equation

$$\begin{cases} P(k) = \mathcal{A}^T(k)P(k+1)\mathcal{A}(k) + g_1^2(k)\mathcal{B}_f^T(k)P(k+1)\mathcal{B}_f(k) \\ \qquad + g_2^2(k)\mathcal{D}_f^T(k)P(k+1)\mathcal{D}_f(k) + \sum_{i=1}^{q}\hat{\Gamma}_i \cdot \mathrm{tr}\Big[\mathcal{H}^T(k) \\ \qquad \times P(k+1)\mathcal{H}(k)\hat{\Theta}_i\Big] + \mathcal{L}^T(k)\mathcal{L}(k) + \mathcal{A}^T(k)P(k+1) \\ \qquad \times \mathcal{G}(k)\Phi^{-1}(k)\mathcal{G}^T(k)P(k+1)\mathcal{A}(k) \\ P(N) = 0 \end{cases} \tag{4.38}$$

has a solution $(P(k), F(k), G(k))$ satisfying

$$\begin{cases} \Phi(k) = \gamma^2 P_\Psi - \mathcal{G}^T(k)P(k+1)\mathcal{G}(k) > 0 \\ P(0) < \gamma^2 R \end{cases} \tag{4.39}$$

where $\hat{\Gamma}_i \triangleq \begin{bmatrix} \bar{\Gamma}_i & 0 \\ 0 & 0 \end{bmatrix}$, $\bar{\Gamma}_i \triangleq \begin{bmatrix} \Gamma_i & 0 \\ 0 & 0 \end{bmatrix}$, $g_1(k) \triangleq \sqrt{\bar{\alpha}_1(k)(1-\bar{\alpha}_1(k))}$, $g_2(k) \triangleq \sqrt{\bar{\alpha}_2(k)(1-\bar{\alpha}_2(k))}$ and P_Ψ is defined in (4.36).

Proof *Sufficiency.* *Define $J(k) \triangleq \eta^T(k+1)P(k+1)\eta(k+1) - \eta^T(k)P(k)\eta(k)$. Noticing (4.35) and taking mathematical expectation as follows, we have*

$$\begin{aligned} & \mathbb{E}\{J(k)\} \\ = \; & \mathbb{E}\Big\{\eta^T(k+1)P(k+1)\eta(k+1) - \eta^T(k)P(k)\eta(k)\Big\} \\ = \; & \mathbb{E}\Big\{\Big(\big(\mathcal{A}(k) + \tilde{\alpha}_1(k)\mathcal{B}_f(k) + \tilde{\alpha}_2(k)\mathcal{D}_f(k)\big)\eta(k) + \mathcal{H}(k)\mathcal{F}(k, x(k)) \\ & + \mathcal{G}(k)w(k)\Big)^T P(k+1)\Big(\big(\mathcal{A}(k) + \tilde{\alpha}_1(k)\mathcal{B}_f(k) + \tilde{\alpha}_2(k)\mathcal{D}_f(k)\big)\eta(k) \\ & + \mathcal{H}(k)\mathcal{F}(k, x(k)) + \mathcal{G}(k)w(k)\Big) - \eta^T(k)P(k)\eta(k)\Big\}. \end{aligned} \tag{4.40}$$

Taking (4.28) into consideration, we have

$$\begin{aligned} & \mathbb{E}\left\{\mathcal{F}^T(k, x(k))\mathcal{H}^T(k)P(k+1)\mathcal{H}(k)\mathcal{F}(k, x(k))\right\} \\ = \; & \mathbb{E}\left\{\mathrm{tr}\left[\mathcal{H}^T(k)P(k+1)\mathcal{H}(k)\mathcal{F}(k, x(k))\mathcal{F}^T(k, x(k))\right]\right\} \\ = \; & \mathbb{E}\left\{\mathrm{tr}\left[\mathcal{H}^T(k)P(k+1)\mathcal{H}(k) \cdot \sum_{i=1}^{q}\hat{\Theta}_i x^T(k)\Gamma_i x(k)\right]\right\} \end{aligned}$$

$$= \mathbb{E}\left\{\eta^T(k)\sum_{i=1}^{q}\hat{\Gamma}_i \cdot \mathrm{tr}\left[\mathcal{H}^T(k)P(k+1)\mathcal{H}(k)\hat{\Theta}_i\right]\eta(k)\right\}. \quad (4.41)$$

Adding the following zero term

$$\|\tilde{z}(k)\|^2 - \gamma^2 w^T(k)P_\Psi w(k) - \Big(\|\tilde{z}(k)\|^2 - \gamma^2 w^T(k)P_\Psi w(k)\Big)$$

to the right side of (4.40) results in

$$\begin{aligned}
&\mathbb{E}\{J(k)\}\\
= \;& \mathbb{E}\Big\{\eta^T(k)\Big(\mathcal{A}^T(k)P(k+1)\mathcal{A}(k) + g_1^2(k)\mathcal{B}_f^T(k)P(k+1)\mathcal{B}_f(k) + g_2^2(k)\\
&\times\mathcal{D}_f^T(k)P(k+1)\mathcal{D}_f(k) + \sum_{i=1}^{q}\hat{\Gamma}_i \cdot \mathrm{tr}\left[\mathcal{H}^T(k)P(k+1)\mathcal{H}(k)\hat{\Theta}_i\right] + \mathcal{L}^T(k)\\
&\times\mathcal{L}(k) - P(k)\Big)\eta(k) + 2\eta^T(k)\mathcal{A}^T(k)P(k+1)\mathcal{G}(k)w(k) - w^T(k)\Phi(k)\\
&\times w(k) - \Big(\|\tilde{z}(k)\|^2 - \gamma^2 w^T(k)P_\Psi w(k)\Big)\Big\}. \qquad (4.42)
\end{aligned}$$

By the completing squares method, it is not difficult to see that

$$\begin{aligned}
&\mathbb{E}\{J(k)\}\\
= \;& \mathbb{E}\Big\{\eta^T(k)\Big(\mathcal{A}^T(k)P(k+1)\mathcal{A}(k) + g_1^2(k)\mathcal{B}_f^T(k)P(k+1)\mathcal{B}_f(k) + g_2^2(k)\\
&\times\mathcal{D}_f^T(k)P(k+1)\mathcal{D}_f(k) + \sum_{i=1}^{q}\hat{\Gamma}_i \cdot \mathrm{tr}\left[\mathcal{H}^T(k)P(k+1)\mathcal{H}(k)\hat{\Theta}_i\right] + \mathcal{L}^T(k)\\
&\times\mathcal{L}(k) - P(k) + \mathcal{A}^T(k)P(k+1)\mathcal{G}(k)\Phi^{-1}(k)\mathcal{G}^T(k)P(k+1)\mathcal{A}(k)\Big)\\
&\times\eta(k)\Big\} - \mathbb{E}\{(w(k)-w^*(k))^T\Phi(k)(w(k)-w^*(k))\} - \mathbb{E}\Big\{\|\tilde{z}(k)\|^2\\
&-\gamma^2 w^T(k)P_\Psi w(k)\Big\} \qquad (4.43)
\end{aligned}$$

where

$$w^*(k) \triangleq \Phi^{-1}(k)\mathcal{G}^T(k)P(k+1)\mathcal{A}(k)\eta(k). \quad (4.44)$$

Taking the sum on both sides of (4.43) from 0 *to* $N-1$, *we obtain*

$$\mathbb{E}\Big\{\sum_{k=0}^{N-1} J(k)\Big\} = \mathbb{E}\{\eta^T(N)P(N)\eta(N)\} - \mathbb{E}\{\eta^T(0)P(0)\eta(0)\}$$

$$
\begin{aligned}
&= -\sum_{k=0}^{N-1} \mathbb{E}\Big\{(w(k)-w^*(k))^T\Phi(k)(w(k)-w^*(k))\Big\} \\
&\quad -\sum_{k=0}^{N-1} \mathbb{E}\Big\{\|\tilde{z}(k)\|^2-\gamma^2 w^T(k)P_\Psi w(k)\Big\} \qquad (4.45)
\end{aligned}
$$

By noticing $\Phi(k) > 0$, $P(0) - \gamma^2 R < 0$, the final condition $P(N) = 0$ and $\eta(0) \neq 0$, we obtain

$$
\begin{aligned}
J_N &= \mathbb{E}\left\{\sum_{k=0}^{N-1}\Big(\|\tilde{z}(k)\|^2-\gamma^2 w^T(k)P_\Psi w(k)\Big)\right\} - \gamma^2\mathbb{E}\left\{\eta^T(0)R\eta(0)\right\} \\
&= \mathbb{E}\left\{\eta^T(0)(P(0)-\gamma^2 R)\eta(0)\right\} \\
&\quad -\sum_{k=0}^{N-1}\mathbb{E}\left\{(w(k)-w^*(k))^T\Phi(k)(w(k)-w^*(k))\right\} < 0 \qquad (4.46)
\end{aligned}
$$

which is equivalent to (4.36). Hence we finish completing the proof of sufficiency.

Necessity: *The proof follows directly from that of Theorem 1 in [183] and is therefore omitted.*

Remark 4.9 *So far, a necessary and sufficient condition has been obtained in Theorem 4.3 for the existence of the fault estimation performance of the dynamic system (4.35). That is, if there exists a solution $P(k)$ to (4.38) such that $\Phi(k) > 0$ and $P(0) < \gamma^2 R$, then the fault estimation $\hat{f}(k)$ ($k = 0, 1, \ldots, N-1$) satisfies the performance criterion (4.31). Moreover, according to (4.46), the worst-case disturbance can be expressed as $w^*(k) = \Phi^{-1}(k)\mathcal{G}^T(k)P(k+1)\mathcal{A}(k)\eta(k)$, and the performance objective $J_N = \eta^T(0)(P(0)-\gamma^2 R)\eta(0)$. In the main results, the occurring probability of the random faults is reflected. For probability 1, our results reduce to those for the traditional deterministic fault estimation problem and, in the case of probability 0, the fault estimation problem is no longer valid and our developed approach applies to the state estimation problem only. In the case that all the time-varying parameters become time-invariant (constant), (4.38)–(4.39) would become a rather standard Riccati difference equation whose feasibility has been widely investigated in the literature, see e.g. [95].*

In what follows, we aim to determine the gain matrices $F(k)$ and $G(k)$ of the desired fault estimator under the situation of the worst-case disturbance $w^*(k)$.

Theorem 4.4 *Consider the time-varying nonlinear stochastic system described by (4.25)–(4.29). Let a disturbance attenuation level $\gamma > 0$ and the positive* *definite matrices $P_B > 0$, $P_C > 0$ and $R > 0$ be given. For each $k = 0, 1, \ldots, N-1$, assume that the discrete Riccati difference equation (4.38) has a solution $(P(k), F(k), G(k))$ satisfying (4.39) and the following discrete Riccati difference equation:*

$$\begin{cases} Q(k) = \mathcal{A}_G^T(k)Q(k+1)\mathcal{A}_G(k) + g_1^2(k)\mathcal{B}_f^T(k)Q(k+1)\mathcal{B}_f(k) \\ \qquad + g_2^2(k)\mathcal{D}_f^T(k)Q(k+1)\mathcal{D}_f(k) + \sum_{i=1}^{q}\hat{\Gamma}_i \\ \qquad \cdot \operatorname{tr}\left[\mathcal{H}^T(k)Q(k+1)\mathcal{H}(k)\hat{\Theta}_i\right] + \mathcal{L}^T(k)\mathcal{L}(k) \\ \qquad - \mathcal{A}_G^T(k)Q(k+1)\Omega^{-1}(k)Q(k+1)\mathcal{A}_G(k), \\ Q(N) = 0, \\ \mathcal{N}(k) = \mathcal{N}(k)\mathcal{C}\dagger(k)\mathcal{C}(k) \end{cases} \tag{4.47}$$

has a solution $(Q(k), \mathcal{K}(k))$ satisfying

$$\Omega(k) = Q(k+1) + I > 0 \tag{4.48}$$

where

$$\begin{aligned} \mathcal{N}(k) &\triangleq -\bar{L}\Omega^{-1}(k)Q(k+1)\mathcal{A}_G(k), \quad \mathcal{K}(k) \triangleq \begin{bmatrix} F(k) & G(k) \end{bmatrix}, \\ \mathcal{C}(k) &\triangleq \begin{bmatrix} -I & I \\ -\bar{C}(k) & 0 \end{bmatrix}, \ \bar{A}(k) \triangleq \begin{bmatrix} \bar{A}(k) & 0 \\ \bar{A}(k) & 0 \end{bmatrix}, \\ \mathcal{A}_G(k) &\triangleq \bar{A}(k) + \mathcal{G}(k)\Phi^{-1}(k)\mathcal{G}^T(k)P(k+1)\mathcal{A}(k). \end{aligned} \tag{4.49}$$

Then, we can conclude that the fault estimation $\hat{f}(k)$ $(k = 0, 1, \ldots, N-1)$ satisfies the performance criterion (4.31) and the gain matrices of the fault estimator are given by

$$\begin{aligned} \mathcal{K}(k) &= \begin{bmatrix} F(k) & G(k) \end{bmatrix} = \mathcal{N}(k)\mathcal{C}\dagger(k) + Y(k) - Y(k)\mathcal{C}(k)\mathcal{C}\dagger(k), \\ &\qquad Y(k) \in \mathbb{R}^{n_x+n_l \times (n_x+n_l+n_y)}, \quad k = 1, 2, \ldots, N-1. \end{aligned} \tag{4.50}$$

Proof *We define a cost function as follows:*

$$J_N(w^*) \triangleq \mathbb{E}\left\{\sum_{k=0}^{N-1}(\|\tilde{z}(k)\|^2 + \|\Upsilon(k)\|^2)\right\} \tag{4.51}$$

where $\Upsilon(k) \triangleq \mathcal{K}(k)\mathcal{C}(k)\eta(k)$. Therefore, the original system (4.35) with the worst-case disturbance $w^*(k)$ can be rewritten as follows:

$$\begin{cases} \eta_{k+1} = (\mathcal{A}_G(k) + \tilde{\alpha}_1(k)\mathcal{B}_f(k) + \tilde{\alpha}_2(k)\mathcal{D}_f(k))\eta(k) + \mathcal{H}(k)\mathcal{F}(k, x(k)) + \tilde{\Upsilon}(k) \\ \tilde{z}(k) = \mathcal{L}(k)\eta(k) \end{cases} \tag{4.52}$$

where $\tilde{\Upsilon}(k) \triangleq \begin{bmatrix} 0 & \Upsilon(k)^T \end{bmatrix}^T$.

In order to obtain the parametric expression of $\mathcal{K}(k)$, *we define*

$$J_w(k) = \eta^T(k+1)Q(k+1)\eta(k+1) - \eta^T(k)Q(k)\eta(k). \tag{4.53}$$

It follows from (4.52) that

$$\begin{aligned}
&\mathbb{E}\{J_w(k)\} \\
= \;& \mathbb{E}\bigg\{\Big(\big(\mathcal{A}_G(k) + \tilde{\alpha}_1(k)\mathcal{B}_f(k) + \tilde{\alpha}_2(k)\mathcal{D}_f(k)\big)\eta(k) + \mathcal{H}(k)\mathcal{F}(k, x(k)) \\
&+ \tilde{\Upsilon}(k)\Big)^T Q(k+1)\Big(\big(\mathcal{A}_G(k) + \tilde{\alpha}_1(k)\mathcal{B}_f(k) + \tilde{\alpha}_2(k)\mathcal{D}_f(k)\big)\eta(k) \\
&+ \mathcal{H}(k)\mathcal{F}(k, x(k)) + \tilde{\Upsilon}(k)\Big) - \eta^T(k)Q(k)\eta(k)\bigg\} + \mathbb{E}\{\|\tilde{z}(k)\|^2 + \|\Upsilon(k)\|^2 \\
&- \|\tilde{z}(k)\|^2 - \|\Upsilon(k)\|^2\} \\
= \;& \mathbb{E}\bigg\{\eta^T(k)\Big(\mathcal{A}_G^T(k)Q(k+1)\mathcal{A}_G(k) + g_1^2(k)\mathcal{B}_f^T(k)Q(k+1)\mathcal{B}_f(k) \\
&+ g_2^2(k)\mathcal{D}_f^T(k)Q(k+1)\mathcal{D}_f(k) + \sum_{i=1}^{q}\hat{\Gamma}_i \cdot \mathrm{tr}\left[\mathcal{H}^T(k)Q(k+1)\mathcal{H}(k)\hat{\Theta}_i\right] \\
&+ \mathcal{L}^T(k)\mathcal{L}(k) - Q(k)\Big)\eta(k) + 2\eta^T(k)\mathcal{A}_G^T(k)Q(k+1)\tilde{\Upsilon}(k) \\
&+ \tilde{\Upsilon}^T(k)Q(k+1)\tilde{\Upsilon}(k) + \tilde{\Upsilon}^T(k)\tilde{\Upsilon}(k) - \Big(\|\tilde{z}(k)\|^2 + \|\Upsilon(k)\|^2\Big)\bigg\}.
\end{aligned} \tag{4.54}$$

By applying completing squares method again, we have

$$\begin{aligned}
&\mathbb{E}\{J_w(k)\} \\
=& \mathbb{E}\bigg\{\eta^T(k)\Big(\mathcal{A}_G^T(k)Q(k+1)\mathcal{A}_G(k) + g_1^2(k)\mathcal{B}_f^T(k)Q(k+1)\mathcal{B}_f(k) \\
&+ g_2^2(k)\mathcal{D}_f^T(k)Q(k+1)\mathcal{D}_f(k) + \sum_{i=1}^{q}\hat{\Gamma}_i \cdot \mathrm{tr}\left[\mathcal{H}^T(k)Q(k+1)\mathcal{H}(k)\hat{\Theta}_i\right] \\
&+ \mathcal{L}^T(k)\mathcal{L}(k) - Q(k) - \mathcal{A}_G^T(k)Q(k+1)\Omega^{-1}(k)Q(k+1)\mathcal{A}_G(k)\Big)\eta(k) \\
&+ (\tilde{\Upsilon}(k) - \tilde{\Upsilon}^*(k))^T\Omega(k)(\tilde{\Upsilon}(k) - \tilde{\Upsilon}^*(k))\bigg\} \\
&- \mathbb{E}\Big\{\|\tilde{z}(k)\|^2 + \|\Upsilon(k)\|^2\Big\}
\end{aligned} \tag{4.55}$$

where $\tilde{\Upsilon}^*(k) \triangleq -\Omega^{-1}(k)Q(k+1)\mathcal{A}_G(k)\eta(k)$. *Therefore, it is true that*

$$J_N(w^*)$$

$$
\begin{aligned}
&= \mathbb{E}\left\{\sum_{k=0}^{N-1}(\|\tilde{z}(k)\|^2 + \|\Upsilon(k)\|^2)\right\} \\
&= \mathbb{E}\Bigg\{\sum_{k=0}^{N-1}(\tilde{\Upsilon}(k) - \tilde{\Upsilon}^*(k))^T\Omega(k)(\tilde{\Upsilon}(k) - \tilde{\Upsilon}^*(k)) + \eta^T(0)Q(0)\eta(0) \\
&\quad -\eta^T(N)Q(N)\eta(N) + \sum_{k=0}^{N-1}\eta^T(k)\Big(\mathcal{A}_G^T(k)Q(k+1)\mathcal{A}_G(k) \\
&\quad +g_1^2(k)\mathcal{B}_f^T(k)Q(k+1)\mathcal{B}_f(k) + g_2^2(k)\mathcal{D}_f^T(k)Q(k+1)\mathcal{D}_f(k) \\
&\quad +\sum_{i=1}^{q}\hat{\Gamma}_i \cdot \mathrm{tr}\left[\mathcal{H}^T(k)Q(k+1)\mathcal{H}(k)\hat{\Theta}_i\right] + \mathcal{L}^T(k)\mathcal{L}(k) - Q(k) \\
&\quad -\mathcal{A}_G^T(k)Q(k+1)\Omega^{-1}(k)Q(k+1)\mathcal{A}_G(k)\Big)\eta(k)\Bigg\}.
\end{aligned}
$$

Under the zero final condition of $Q(N)$, in order to minimize the cost of $J_N(w^*)$, the best choice of $\mathcal{K}(k)$ is to satisfy the following condition:

$$
\begin{cases}
Q(k) = \mathcal{A}_G^T(k)Q(k+1)\mathcal{A}_G(k) + g_1^2(k)\mathcal{B}_f^T(k)Q(k+1)\mathcal{B}_f(k) \\
\qquad + g_2^2(k)\mathcal{D}_f^T(k)Q(k+1)\mathcal{D}_f(k) + \mathcal{L}^T(k)\mathcal{L}(k) \\
\qquad + \sum_{i=1}^{q}\hat{\Gamma}_i \cdot \mathrm{tr}\left[\mathcal{H}^T(k)Q(k+1)\mathcal{H}(k)\hat{\Theta}_i\right] \\
\qquad - \mathcal{A}_G^T(k)Q(k+1)\Omega^{-1}(k)Q(k+1)\mathcal{A}_G(k) \\
\mathcal{K}(k)\mathcal{C}(k) = -\bar{L}\Omega^{-1}(k)Q(k+1)\mathcal{A}_G(k)
\end{cases} \tag{4.56}
$$

where $\bar{L} \triangleq \begin{bmatrix} 0 & I \end{bmatrix}$. According to Lemma 4.1, it can be observed that the existence of a solution $\mathcal{K}(k)$ $(k = 0, 1, \ldots, N-1)$ to (4.56) is equivalent to the feasibility of

$$
-\bar{L}\Omega^{-1}(k)Q(k+1)\mathcal{A}_G(k)\mathcal{C}\dagger(k)\mathcal{C}(k) = -\bar{L}\Omega^{-1}(k)Q(k+1)\mathcal{A}_G(k)
$$

whose general solution is given by

$$
K_k = -\bar{L}\Omega^{-1}(k)Q(k+1)\mathcal{A}_G(k)\mathcal{C}\dagger(k) + Y(k) - Y(k)\mathcal{C}(k)\mathcal{C}\dagger(k)
$$

where $Y(k)$ is any matrix with dimension $(n_x + n_l) \times (n_x + n_l + n_y)$. The proof of this theorem is now complete.

By means of Theorem 4.4, we can summarize the Finite-Horizon Fault Estimator Design (*FHFED*) algorithm as follows:

Remark 4.10 *In this section, a coupled Riccati difference equation approach is proposed to solve the estimation problem of the randomly occurring faults*

Algorithm *FHFED*:	
Step 1.	Given the disturbance attenuation level γ, the positive definite matrices $P_B > 0$, $P_C > 0$ and $R > 0$, set $k = N - 1$.
Step 2.	Calculate $\Phi(k)$, $\Omega(k)$ and $\mathcal{N}(k)$ with known $P(k+1)$ and $Q(k+1)$ via the first equation of (4.39) and equations (4.48) and (4.49), respectively. Furthermore, the fault estimation gain matrix $\mathcal{K}(k)$ can be obtained by equation (4.50).
Step 3.	If $\mathcal{N}(k) = \mathcal{N}(k)\mathcal{C}\dagger(k)\mathcal{C}(k)$, then solve the first equation of (4.38) and (4.47) to get $P(k)$ and $Q(k)$, respectively, and go to the next step, else this algorithm is infeasible, stop.
Step 4.	If $k \neq 0$, $\Phi(k) > 0$ and $\Omega(k) > 0$, set $k = k - 1$ and go to *Step 2*, else go to the next step.
Step 5.	If $P(0) \geq r^2 R$ or $\Phi(k) \leq 0$ or $\Omega(k) \leq 0$, then this algorithm is infeasible, stop.

over a finite-horizon. The faults are allowed to occur dynamically governed by random variables with given probability laws. It can be observed from Algorithm FHFED that, in the estimator design procedure, all the important factors contributing to the system complexity have been reflected which include (1) the time-varying systems parameters; (2) the occurrence probabilities of the random faults; (3) statistics information about the stochastic nonlinearities; and (4) the prescribed disturbance attenuation level. The coupled RDE algorithm is backward recursive and therefore suitable for online application. The main theoretical contribution would be the establishment of the necessary and sufficient conditions for the existence of the desired finite-horizon H_∞ fault estimator.

4.3 Illustrative Examples

In this section, two simulation examples are given to demonstrate the approaches presented in this chapter.

4.3.1 Example 1

In this example, we use a nonlinear pendulum in a network environment to demonstrate the effectiveness and applicability of the proposed method in Section 4.1. Consider a pendulum system borrowed from [51]. It is assumed that two components of the system (that is, angle and angular velocity) are randomly perturbed by uncontrolled external forces. The equations of motion of the pendulum are described as follows:

$$\begin{aligned}
\dot{\theta}(t) &= \lambda\bar{\theta}(t) + \alpha(t)((1-\lambda)\bar{\theta}(t) + \lambda\theta(t)) \\
\dot{\bar{\theta}}(t) &= -\frac{g\sin(\theta(t)) + (b/lm)\bar{\theta}(t) + (aml/4)\bar{\theta}^2(t)\sin(2\theta(t))}{\frac{2}{3}l - \frac{a}{2}ml\cos^2(\theta(t))} \\
&\quad -(aml\lambda/4)w(t) \\
y(t) &= \sin(\theta(t)) + \lambda\bar{\theta}(t) + \lambda v(t)
\end{aligned} \tag{4.57}$$

where θ denotes the angle of the pendulum from the vertical, $\bar{\theta}$ is the angular velocity, $g = 9.8$ m/s^2 is the gravity constant, m is the mass of the pendulum, $a = 1/(m+M)$, M is the mass of the cart, l is the length of the pendulum, b is the damping coefficient of the pendulum around the pivot, and w and v are the disturbance applied to the cart and measurement noise, respectively. In this simulation, the pendulum parameters are chosen as $m = 2$ kg, $M = 8$ kg, $l = 0.5$ m and $b = 0.7$ Nm/s, and the retarded coefficient $\lambda = 0.6$.

Since the nonlinear pendulum system is in a network environment, wireless channels are known to be sensitive to fading effects which serve as one of the most dominant features in wireless communication links. Letting $x_1(t) = \theta(t)$, $x_2(t) = \bar{\theta}(t)$, considering the fading channel phenomenon and discretizing the plant with a sampling period 0.04 s, we obtain the following discrete-time system model to be investigated:

$$\begin{cases}
x(k+1) = g(k, x(k)) + \alpha(k)D_f(k)f(k) + E_1(k)w(k) \\
\tilde{y}(k) = C(k)x(k) + E_2(k)v(k) \\
y(k) = \sum_{s=0}^{l_k} \beta_s(k)\tilde{y}(k-s) + E_3(k)\xi(k).
\end{cases}$$

The system data are given as follows:

$$g(k, x(k)) = \begin{bmatrix} 0.48x_1(k) + 0.2x_2(k) + 0.12\sin(x_2(k)) \\ 0.03x_1(k) + 0.50x_2(k) \end{bmatrix},$$

$$D_f(k) = \begin{bmatrix} 0.4 + \sin(k) \\ 0.2 \end{bmatrix}, \quad E_1(k) = \begin{bmatrix} 0.2 \\ 0.5 \end{bmatrix},$$

$$E_3(k) = 0.1, \ C(k) = \begin{bmatrix} -0.2 + 0.1\sin(5k) & 0.5 \end{bmatrix}, \ E_2(k) = 0.3$$

where $x_i(k)$ $(i = 1, 2)$ is the ith element of $x(k)$. The probability of randomly occurring fault is taken as $\bar{\alpha} = 0.9$. In view of (4.58), the other system parameters can be obtained as follows:

$$A(k) = \begin{bmatrix} 0.48 & 0.2 \\ 0.03 & 0.50 \end{bmatrix}, \quad b(k) = 0.2.$$

The order of the fading model is $l = 1$ and the probability density functions of channel coefficients are as follows

$$\begin{cases} \varrho(\beta_0(k)) = 0.0005(e^{9.89\beta_0(k)} - 1), & 0 \leq \beta_0(k) \leq 1, \\ \varrho(\beta_1(k)) = 8.5017e^{-8.5\beta_1(k)}, & 0 \leq \beta_1(k) \leq 1. \end{cases}$$

It can be obtained that the mathematical expectation $\bar{\beta}_s$ and variance ν_s ($s = 0, 1$) are 0.8991, 0.1174, 0.0133 and 0.01364, respectively. The H_∞ performance level γ, the positive definite matrices P_a, P_b and P_{ci} ($i = -1, 0$) are chosen as $\gamma = 1$, $P_a = I$, $P_b = I$, $P_{c0} = P_{-c1} = 5I$, respectively. By applying Algorithm *FHFED*, the desired fault estimate parameters are obtained and listed in Table 4.1.

TABLE 4.1
Fault Estimate Parameters

k	**0**	**1**	**2**	**3**
$K(k)$	$\begin{bmatrix} 2.7544 \\ 2.7533 \\ 2.7552 \end{bmatrix}$	$\begin{bmatrix} 0.0021 \\ 0.0002 \\ -0.0008 \end{bmatrix}$	$\begin{bmatrix} -0.0067 \\ -0.0055 \\ -0.0301 \end{bmatrix}$	$\begin{bmatrix} -0.0020 \\ 0.0004 \\ -0.0011 \end{bmatrix}$
k	$\cdots$	**29**	**30**	
$K(k)$	$\cdots$	$\begin{bmatrix} -0.0405 \\ -0.0268 \\ 0.0097 \end{bmatrix}$	$\begin{bmatrix} -0.0161 \\ 0.0052 \\ -0.0165 \end{bmatrix}$	

From (4.10), we can obtain that

$$J(N) := \frac{\mathbb{E}\left\{ \sum_{k=0}^{N-1} \left(\|\tilde{z}(k)\|^2 \right) \right\}}{\mathbb{E}\left\{ \sum_{k=0}^{N-1} \left(\|\varpi(k)\|_{P_a}^2 + \|v(k)\|_{P_b}^2 \right) + \bar{\eta}(0) \right\}} < \gamma^2 \tag{4.58}$$

where $\bar{\eta}(0) = \sum_{i=-l}^{0} \eta^T(i) P_{ci} \eta(i)$. To illustrate the effectiveness of the designed fault estimator, we introduce the index $J(N)$ to reflect the actual fault estimation performance.

In the simulation, the initial value of the state is $x(0) = \begin{bmatrix} -0.55 & -0.16 \end{bmatrix}^T$ and the exogenous disturbance inputs are selected as $w(k) = 0.5e^{-2k} \sin(4k)$, $v(k) = 0.2e^{-4k} \cos(k)$ and $\xi(k) = \frac{4}{k+1} \cos(k)$. First, let the matrix $A_f(k) = -0.4I$. The fault to be estimated is $f(k) = 1$. Fig. 4.1 plots the simulation result on the fault signal and its estimate. Fig. 4.2 shows the evolution of the actual fault estimation performance in terms of the index $J(N)$ in (4.58), from which it can be seen that the index $J(N)$ ($N = 1, 2, ..., 30$) is always less than the prescribed upper bound 1. The simulation results confirm that the approach addressed in Section 4.1 provides a good performance of fault estimation.

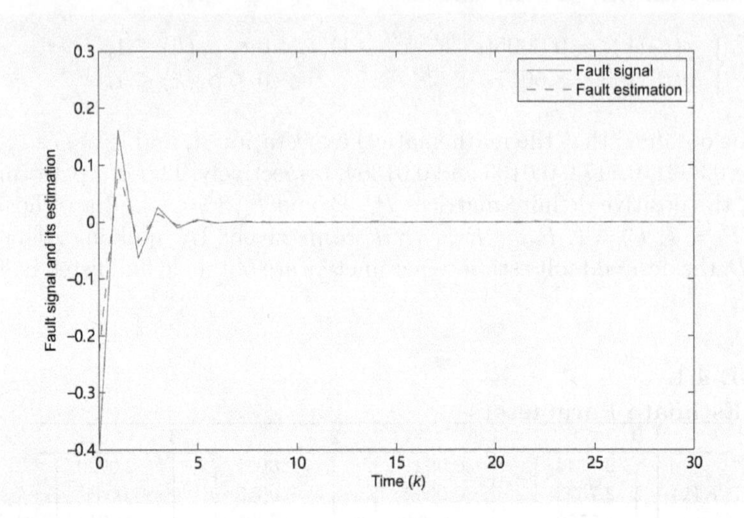

FIGURE 4.1
Fault signal and its estimate with $A_f(k) = -0.4I$.

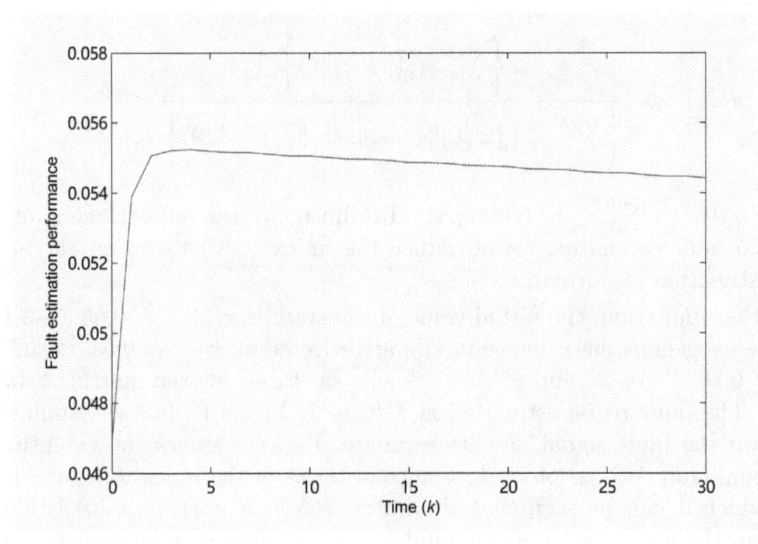

FIGURE 4.2
Fault estimation performance $J(N)$ with $A_f(k) = -0.4I$.

4.3.2 Example 2

In this example, we present a simulation example to illustrate the effectiveness of the finite-horizon estimation method of randomly occurring faults proposed in Section 4.2 for a class of nonlinear time-varying systems.

Consider a discrete time-varying system described by (4.25)–(4.29) with the following parameters over the finite time-horizon $[0, 50]$:

$$A(k) = \begin{bmatrix} 0 & -0.7 \\ 0.1 + 0.2\sin(3k) & -0.53 \end{bmatrix}, \quad C(k) = \begin{bmatrix} -0.2 + 0.1\sin(5k) & 0.5 \\ 0.1 & 1 \end{bmatrix},$$

$$B_f(k) = \begin{bmatrix} 0.4 \\ 0.2 \end{bmatrix}, \quad D_f(k) = \begin{bmatrix} 1 & 1 \end{bmatrix}^T, \quad B_d(k) = \begin{bmatrix} 0.2 & 0.5 \end{bmatrix}^T,$$

$$D_d(k) = \begin{bmatrix} 0.2 & 0.5 \end{bmatrix}^T, \quad L(k) = \begin{bmatrix} 0 & I \end{bmatrix},$$

$$g(k, x(k)) = \begin{bmatrix} 0.1 \\ 0.3 \end{bmatrix} \times \big(0.2x_1(k)\xi_1(k) + 0.3x_2(k)\xi_2(k)\big),$$

$$h(k, x(k)) = \begin{bmatrix} 0.1 \\ 0.1 \end{bmatrix} \times \big(0.2x_1(k)\xi_1(k) + 0.3x_2(k)\xi_2(k)\big)$$

where $x_i(k)$ $(i = 1, 2)$ is the ith element of $x(k)$, and $\xi_i(k)$ $(i = 1, 2)$ is zero mean, uncorrelated Gaussian white noise process with unity variance that is also uncorrelated with $d(k)$ and $v(k)$. It can be easily checked that the above class of stochastic nonlinearities satisfy

$$\mathbb{E}\left\{ \begin{bmatrix} g(k, x(k)) \\ h(k, x(k)) \end{bmatrix} \Big| x(k) \right\} = 0,$$

$$\mathbb{E}\left\{ \begin{bmatrix} g(k, x(k)) \\ h(k, x(k)) \end{bmatrix} \begin{bmatrix} g^T(k, x(k)) & h^T(k, x(k)) \end{bmatrix} \Big| x(k) \right\}$$

$$= \begin{bmatrix} 0.1 \\ 0.3 \\ 0.1 \\ 0.1 \end{bmatrix} \begin{bmatrix} 0.1 \\ 0.3 \\ 0.1 \\ 0.1 \end{bmatrix}^T \mathbb{E}\left\{ x^T(k) \begin{bmatrix} 0.04 & 0 \\ 0 & 0.09 \end{bmatrix} x(k) \right\}.$$

The H_∞ performance level γ, the positive definite matrices P_B, P_C and R are chosen as $\gamma = 1$, $P_B = I$, $P_C = I$ and $R = \text{diag}\{0, 0, 0, 2, 2, 0\}$, respectively. Let $\bar{\alpha}_1 = 0.9$ and $\bar{\alpha}_2 = 0.8$. Using the developed computational algorithm and Matlab (with the YALMIP 3.0), we can check the feasibility of the coupled recursive RDEs and obtain the desired fault estimate parameters which are listed in Table 4.2 from the time $k = 0$ to $k = 4$. In the simulation, the initial values of the states are $\bar{x}(0) = \begin{bmatrix} 0.2 & -0.6 & 0 \end{bmatrix}^T$ and $\hat{\bar{x}}(0) = \begin{bmatrix} 0 & 0 & 0 \end{bmatrix}^T$, the exogenous disturbance input is selected as $d(k) = 0.2\cos(k)$ and the measure noise is $v(k) = 0.3\sin(k)$.

First, let the matrix $A_f(k) = I$. We assume that the fault to be estimated is $f(k) = 2$. Fig. 4.3 plots the simulation result on the fault signal and its estimate. Fig. 4.4 shows the evolution of the actual fault estimation

TABLE 4.2
Fault Estimate Parameters

k	**1**	**2**
$F(k)$	$\begin{bmatrix} 0.3331 & 0 & 0 \\ 0 & 0.2220 & 0 \\ 0 & 0 & 0 \end{bmatrix}$	$\begin{bmatrix} 0.011 & 0 & 0 \\ 0 & 0.2220 & 0.02 \\ 0.509 & 0.001 & 0 \end{bmatrix}$
$G(k)$	$\begin{bmatrix} 0.0000 & 0.0000 \\ 0.0000 & 0.0000 \\ 2.0000 & -1.0000 \end{bmatrix}$	$\begin{bmatrix} 0.0155 & -0.0001 \\ -0.0001 & 0.0153 \\ -0.0046 & 0.0023 \end{bmatrix}$
k	**3**	**4**
$F(k)$	$\begin{bmatrix} 0.0061 & 0.0032 & 0.0011 \\ 0.0032 & 0.0107 & 0 \\ 0.0011 & 0 & 0.021 \end{bmatrix}$	$\begin{bmatrix} 0.0011 & 0 & -0.2 \\ 0.0026 & 0.0118 & 0.021 \\ 0 & 0.1001 & 0.001 \end{bmatrix}$
$G(k)$	$\begin{bmatrix} 0.0037 & -0.0002 \\ -0.0000 & 0.0031 \\ -1.0100 & 2.0252 \end{bmatrix}$	$\begin{bmatrix} -0.0046 & -0.0001 \\ -0.0001 & 0.0050 \\ 2.0212 & -1.0001 \end{bmatrix}$

performance in terms of the index $\zeta(N)$ in (4.31), from which it can be seen that the index $\zeta(N)$ $(N = 1, 2, ..., 50)$ is always less than the prescribed upper bound 1. Next, in order to examine the effects of the fault estimation over different time-intervals, we choose the matrix $A_f(k)$ as follows:

$$A_f(k) = \begin{cases} I, & 0 \le k \le 9 \\ 2I, & k = 10 \\ I, & 11 \le k \le 19 \\ -0.5I, & k = 20 \\ I, & \text{else.} \end{cases} \tag{4.59}$$

The fault to be estimated is $f(k) = 1$. Fig. 4.5 plots the fault signal and its estimate. The actual fault estimation performance is depicted in Fig. 4.6. It can be seen that the developed approach in Section 4.2 provides the desired performance for the addressed fault estimation problem.

4.4 Summary

In this chapter, the finite-horizon estimation problem of ROFs has been first studied for a class of nonlinear time-varying systems with fading channels. Some uncorrelated random variables have been introduced, respectively, to govern the fault occurrence probability and fading measurements. By employing the stochastic analysis techniques, some sufficient conditions have been provided to ensure that the dynamic system under consideration satisfies the fault estimation performance constraint. Moreover, the finite-horizon fault estimation problem has been investigated for a class of nonlinear stochastic

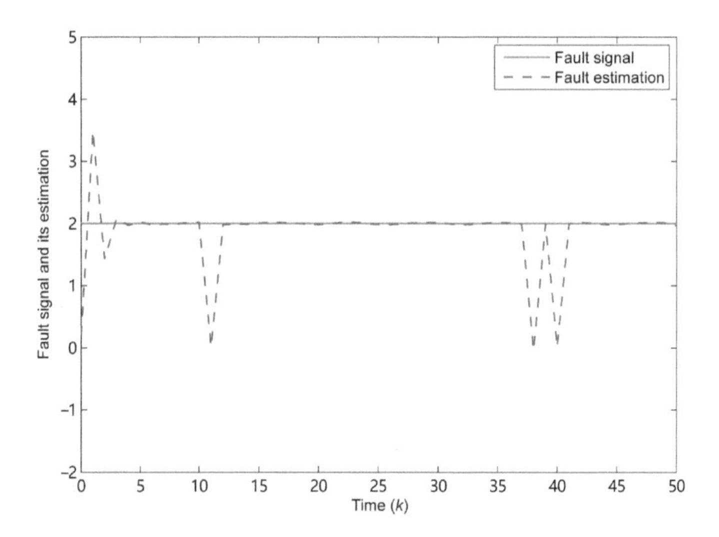

FIGURE 4.3
Fault signal and its estimate.

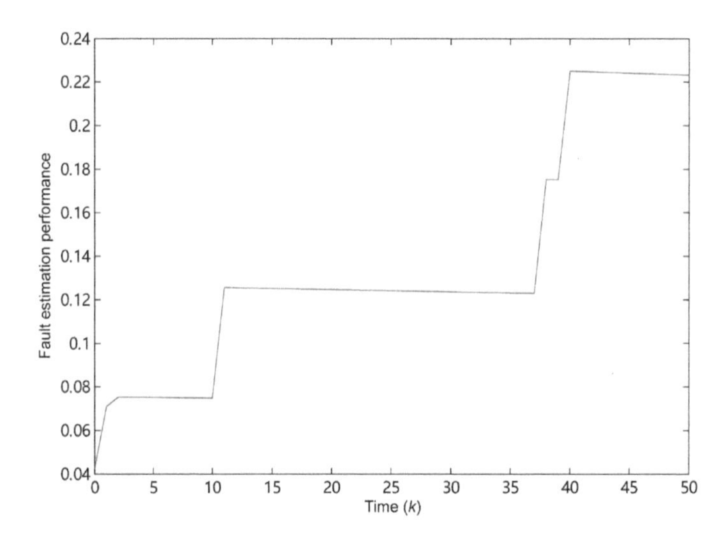

FIGURE 4.4
Fault estimation performance ($\zeta(k)$).

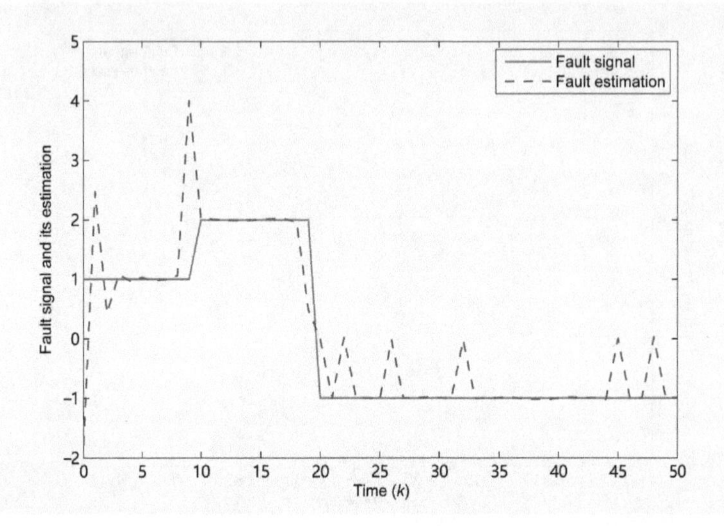

FIGURE 4.5
Fault signal and its estimate.

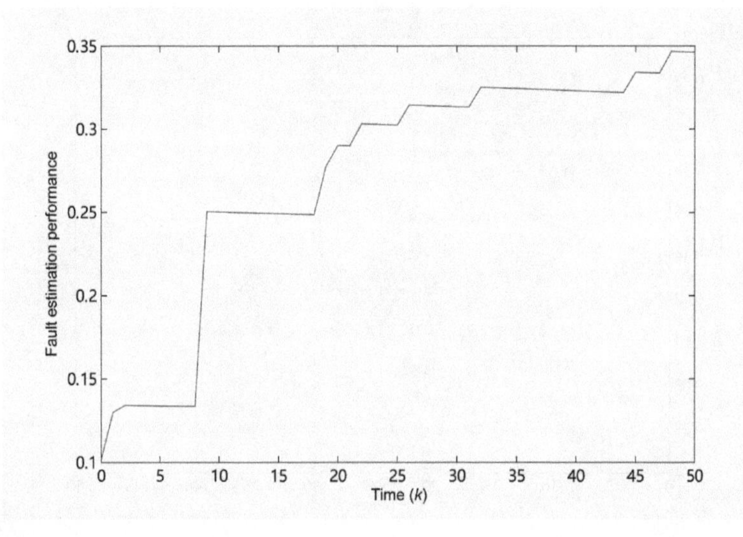

FIGURE 4.6
Fault estimation performance ($\zeta(k)$).

time-varying systems with ROFs. All the system parameters are time-varying and the stochastic nonlinearities under consideration could cover several classes of well-studied nonlinearities. The failures occur in a random way, and two sets of Bernoulli distributed white sequences have been introduced to govern the fault occurrence probability. The solvability of the addressed fault estimation problem has been dealt with by using the completing squares method and stochastic analysis techniques. The time-varying parameters of the fault estimator have been obtained by solving coupled backward recursive Riccati difference equations. Finally, two simulation examples have been provided to demonstrate the effectiveness of the proposed methods.

5

Set-Membership Filtering under Weighted Try-Once-Discard Protocol

Unlike the H_∞ filtering problems discussed in Chapter 2, this chapter is concerned with the set-membership filtering problem for a class of time-varying state-saturated systems with mixed time-delays under the communication protocol. Under the WTOD protocol, only the sensor node with the largest measurement difference is allowed to access the shared communication network at each transmission instant. The purpose of the problem addressed is to design a set of set-membership filters such that, in the simultaneous presence of mixed time-delays, state saturation, WTOD protocol and bounded noises, the filtering error dynamics is confined to certain ellipsoid regions. A sufficient condition is derived to guarantee the existence of the desired set-membership filters by means of the solutions to a set of recursive linear matrix inequalities. Subsequently, an optimization problem subject to certain inequality constraints is put forward to acquire the minimized ellipsoid in the sense of matrix trace. A simulation example is presented to illustrate the effectiveness of the proposed filter design scheme.

5.1 Problem Formulation

Consider a discrete time-varying state-saturated system with mixed time-delays of the following form:

$$\begin{cases} x(k+1) = \sigma(A(k)x(k)) + B(k)x(k-\tau_1) \\ \qquad\qquad + E(k)\sum_{i=1}^{\tau_2} \mu_i x(k-i) + F(k)w(k) \\ y(k) = C(k)x(k) + D(k)v(k) \\ x(i) = \phi_1(i), \ i = -\tau^*, -\tau^*+1, \ldots, 0 \end{cases} \tag{5.1}$$

where $x(k) \in \mathbb{R}^{n_x}$ and $y(k) \in \mathbb{R}^{n_y}$ are the system state to be estimated and the measurement output, respectively. The discrete time-delays τ_1 and τ_2 are known positive integers, and $\tau^* = \max\{\tau_1, \tau_2\}$. μ_i $(1 \le i \le \tau_2)$ stand for

DOI: 10.1201/9781003189497-5

the weight coefficients. $w(k) \in \mathbb{R}^{n_w}$ and $v(k) \in \mathbb{R}^{n_v}$ represent the process and measurement noises, respectively. $\phi_1(i)$ $(i = -\tau^*, -\tau^*+1, \ldots, 0)$ are the initial states of the system. $A(k)$, $B(k)$, $C(k)$, $D(k)$, $E(k)$ and $F(k)$ are real-valued matrices with compatible dimensions. The saturation function $\sigma(\cdot) : \mathbb{R}^{n_x} \longmapsto \mathbb{R}^{n_x}$ is defined as

$$\sigma(r) \triangleq \begin{bmatrix} \sigma_1(r_1) & \sigma_2(r_2) & \cdots & \sigma_{n_x}(r_{n_x}) \end{bmatrix}^T \tag{5.2}$$

with $\sigma_i(r_i) = \text{sign}(r_i)\min\{r_{i,\max}, |r_i|\}$, where $r_{i,\max}$ is the ith element of the vector $r_{\max}$ (i.e., the saturation level).

Assume that the noises $w(k)$ and $v(k)$ are confined to the following ellipsoidal sets:

$$\begin{cases} W(k) \triangleq \{w(k) : \ w^T(k)S^{-1}(k)w(k) \leq 1\} \\ V(k) \triangleq \{v(k) : \ v^T(k)R^{-1}(k)v(k) \leq 1\} \end{cases} \tag{5.3}$$

where $S(k) > 0$ and $R(k) > 0$ are known positive definite matrices with appropriate dimensions.

Define $y(k) \triangleq [y_1(k) \ \ y_2(k) \ \ \cdots \ \ y_{n_y}(k)]^T$, where $y_i(k)$ is the measurement output collected by the ith sensor. Denote by $\xi(k) \in \mathcal{R} \triangleq \{1, 2, \ldots, n_y\}$ the current transmission sensor node at time instant k, which can be determined by the following selection condition:

$$\xi(k) = \arg\max_{i \in \mathcal{R}} (y_i(k) - \bar{y}_i(k-1))^T Q_i (y_i(k) - \bar{y}_i(k-1)) \tag{5.4}$$

where $\bar{y}_i(k-1)$ denotes the last transmitted signal of sensor node i before time instant k and Q_i $(i \in \mathcal{R})$ are known positive definite weighted matrices.

It follows from (5.4) that $\bar{y}_i(k)$ $(k \in \mathbb{N}^+, i \in \mathcal{R})$ can be formulated as follows:

$$\bar{y}_i(k) = \begin{cases} y_i(k), & \text{if } \ i = \xi(k) \\ \bar{y}_i(k-1), & \text{otherwise.} \end{cases} \tag{5.5}$$

Then, by denoting $\bar{y}(k) \triangleq [\bar{y}_1(k) \ \ \bar{y}_2(k) \ \ \cdots \ \ \bar{y}_{n_y}(k)]^T$, the selection condition (5.4) can be rewritten as

$$\xi(k) = \arg\max_{i \in \mathcal{R}} (y(k) - \bar{y}(k-1))^T \bar{Q}_i (y(k) - \bar{y}(k-1)) \tag{5.6}$$

where $\bar{Q}_i \triangleq \bar{Q}\Omega_i$ with $\bar{Q} \triangleq \text{diag}\{Q_1, Q_2, \ldots, Q_{n_y}\}$ and $\Omega_i \triangleq \text{diag}\{\delta(i-1)I, \delta(i-2)I, \ldots, \delta(i-n_y)I\}$ $(i \in \mathcal{R})$ with $\delta(a)$ being a binary function that equals 1 when $a = 0$ and equals 0 otherwise. Accordingly, the concise form of $\bar{y}(k)$ can be obtained as

$$\begin{cases} \bar{y}(k) = \Omega_{\xi(k)} y(k) + (I - \Omega_{\xi(k)})\bar{y}(k-1) \\ \bar{y}(j) = \phi_2, \ j < 0 \end{cases} \tag{5.7}$$

where ϕ_2 is a known vector.

Before proceeding, we provide the following lemma.

Lemma 5.1 *[201]. If there exist diagonal matrices H_1 and H_2 satisfying $0 \leq H_1 < I \leq H_2$, then the saturation function $\sigma(A(k)x(k))$ in (5.1) can be written as follows:*

$$\sigma(A(k)x(k)) = H_1 A(k)x(k) + \psi(A(k)x(k)) \tag{5.8}$$

where $\psi(\cdot)$ is a nonlinear vector-valued function that satisfies the sector condition with $K_1 = 0$ and $K_2 = H$, in which $H = H_2 - H_1$, i.e., $\psi(A(k)x(k))$ satisfies the following inequality:

$$\psi^T(A(k)x(k))(\psi(A(k)x(k)) - HA(k)x(k)) \leq 0. \tag{5.9}$$

Now, letting $\bar{x}(k) \triangleq [x^T(k) \quad \bar{y}^T(k-1)]^T$ and $\bar{w}(k) \triangleq [w^T(k) \; v^T(k)]^T$, system (5.1) under the WTOD protocol can be represented as

$$\begin{cases} \bar{x}(k+1) = \bar{A}(k)\bar{x}(k) + \bar{I}\tilde{\psi}(k) + \bar{B}(k)\bar{x}(k-\tau_1) \\ \qquad\qquad + \bar{E}(k)\sum_{i=1}^{\tau_2} \mu_i \bar{x}(k-i) + \bar{F}(k)\bar{w}(k) \\ \bar{y}(k) = \bar{C}(k)\bar{x}(k) + \bar{D}(k)\bar{w}(k) \\ \bar{x}(i) = \bar{\phi}(i), \; i = -\tau^*, -\tau^*+1, \ldots, 0 \end{cases} \tag{5.10}$$

where

$$\bar{A}(k) \triangleq \begin{bmatrix} H_1 A(k) & 0 \\ \Omega_{\xi(k)} C(k) & I - \Omega_{\xi(k)} \end{bmatrix}, \; \bar{I} \triangleq \begin{bmatrix} I \\ 0 \end{bmatrix}, \; \tilde{\psi}(k) \triangleq \psi(\tilde{A}(k)\bar{x}(k)),$$
$$\bar{F}(k) \triangleq \mathrm{diag}\{F(k), \Omega_{\xi(k)} D(k)\}, \; \bar{B}(k) \triangleq \mathrm{diag}\{B(k), 0\}, \; \tilde{A}(k) \triangleq [A(k) \quad 0],$$
$$\bar{E}(k) \triangleq \mathrm{diag}\{E(k), 0\}, \; \bar{C}(k) \triangleq \begin{bmatrix} \Omega_{\xi(k)} C(k) & I - \Omega_{\xi(k)} \end{bmatrix},$$
$$\bar{D}(k) \triangleq \begin{bmatrix} 0 & \Omega_{\xi(k)} D(k) \end{bmatrix}, \; \bar{\phi}(i) \triangleq \begin{bmatrix} \phi_1^T(i) & \phi_2^T(i) \end{bmatrix}^T.$$

Based on the signal $\bar{y}(k)$, a protocol-based time-varying filter is constructed for the augmented system (5.10) with the following structure:

$$\begin{cases} \hat{x}(k+1) = \bar{A}(k)\hat{x}(k) + \bar{B}(k)\hat{x}(k-\tau_1) + \bar{E}(k)\sum_{i=1}^{\tau_2} \mu_i \hat{x}(k-i) \\ \qquad\qquad + K(k)(\bar{y}(k) - \bar{C}(k)\hat{x}(k)) \\ \hat{x}(i) = 0, \; i = -\tau^*, -\tau^*+1, \ldots, 0 \end{cases} \tag{5.11}$$

where $\hat{x}(k) \in \mathbb{R}^{n_x+n_y}$ is the state estimate of $\bar{x}(k)$ and $K(k) \in \mathbb{R}^{(n_x+n_y)\times n_y}$ is the filter parameter to be determined.

Denoting $e(k) \triangleq \bar{x}(k) - \hat{x}(k)$ as the filtering error, one has

$$\begin{aligned} e(k+1) =& (\bar{A}(k) - K(k)\bar{C}(k))e(k) + \bar{I}\tilde{\psi}(k) + \bar{E}(k)\sum_{i=1}^{\tau_2} \mu_i e(k-i) \\ &+ \bar{B}(k)e(k-\tau_1) + (\bar{F}(k) - K(k)\bar{D}(k))\bar{w}(k). \end{aligned} \tag{5.12}$$

The main purpose of the addressed problem is to design a sequence of filtering gains $\{K(k)\}_{k\in\mathbb{N}^+}$ such that, for a given sequence of positive definite matrices $\{P(k)\}_{k\in\mathbb{N}^+}$, the filtering error dynamics (5.12) satisfies the following ellipsoid constraint (also called $P(k)$-dependent constraint):

$$e^T(k)P^{-1}(k)e(k) \leq 1, \ k \in \mathbb{N}^+. \tag{5.13}$$

Assumption 5.1 *The initial conditions* $\bar{\phi}(i)$ $(i = -\tau^*, -\tau^*+1, \ldots, 0)$ *satisfy*

$$\bar{\phi}^T(i)P^{-1}(i)\bar{\phi}(i) \leq 1 \tag{5.14}$$

where $P(i)$ $(i = -\tau^*, -\tau^* + 1, \ldots, 0)$ *are known positive definite matrices.*

5.2 Main Results

5.2.1 Filter Design Subject to the $P(k)$-Dependent Constraint

In the following theorem, by employing the recursive matrix inequality technique, a sufficient condition is established under which the filtering error dynamics satisfies the $P(k)$-dependent constraint.

Theorem 5.1 *Consider the dynamic system (5.1), the WTOD protocol (5.6) and the time-varying filter (5.11). Let the sequence of positive definite constraint matrices* $\{P(k)\}_{k\in\mathbb{N}^+}$ *be given. Then, the dynamics of the time-varying system (5.12) meets the* $P(k)$*-dependent constraint condition (5.13) if there exist sequences of real-valued matrices* $\{K(k)\}_{k\in\mathbb{N}^+}$*, positive scalars* $\{\lambda_i(k)\}_{k\in\mathbb{N}^+}$ $(i \in \mathcal{R})$, $\{\alpha_l(k)\}_{k\in\mathbb{N}^+}$ $(l = 1, \ldots, 5)$ *and* $\{\beta_j(k)\}_{k\in\mathbb{N}^+}$ $(j = 1, \ldots, \tau_2)$ *satisfying the following recursive matrix inequality*

$$\begin{bmatrix} -\Theta(k) & \Pi^T(k) \\ * & -P(k+1) \end{bmatrix} \leq 0 \tag{5.15}$$

where

$$\begin{aligned} \Theta(k) \triangleq\ & \Gamma^T(k)\sum_{i=1}^{n_y}\lambda_i(k)(\bar{Q}(\Omega_i - \Omega_{\xi(k)}))\Gamma(k) + \alpha_5(k)\Upsilon(k) \\ & +\text{diag}\{1 - \alpha_1(k) - \alpha_2(k) - \sum_{j=1}^{\tau_2}\beta_j(k) - \alpha_3(k) \\ & -\alpha_4(k), \alpha_1(k)I, \alpha_2(k)I, \sum_{j=1}^{\tau_2}\beta_j(k)\mathcal{I}_j^T\mathcal{I}_j, \\ & \alpha_3(k)\bar{S}(k) + \alpha_4(k)\bar{R}(k), 0\}, \end{aligned}$$

$$\begin{aligned}
\Pi(k) &\triangleq \left[\begin{array}{llll} 0 & (\bar{A}(k)-K(k)\bar{C}(k))L(k) & \bar{B}(k)L(k-\tau_1) \\ \bar{E}_\tau(k) & \bar{F}(k)-K(k)\bar{D}(k) & \bar{I} \end{array}\right], \\
\bar{E}_\tau(k) &\triangleq \left[\begin{array}{lll} \mu_1\bar{E}(k)L(k-1) & \mu_2\bar{E}(k)L(k-2) & \cdots \\ \mu_{\tau_2}\bar{E}(k)L(k-\tau_2) \end{array}\right], \qquad (5.16) \\
\bar{S}(k) &\triangleq \mathrm{diag}\{S^{-1}(k),0\},\ \bar{R}(k)\triangleq \mathrm{diag}\{0,R^{-1}(k)\}, \\
\Gamma(k) &\triangleq \left[\begin{array}{cccccc} \tilde{C}(k)\hat{x}(k) & \tilde{C}(k)L(k) & 0 & 0 & \tilde{D}(k) & 0 \end{array}\right], \\
\tilde{C}(k) &\triangleq \left[\begin{array}{cc} C(k) & -I \end{array}\right], \tilde{D}(k)\triangleq\left[\begin{array}{cc} 0 & D(k) \end{array}\right], \\
\mathcal{I}_j &\triangleq [\underbrace{0,\ldots,0}_{j-1},I,\underbrace{0,\ldots,0}_{\tau_2-j}]\ (j=1,\ldots,\tau_2), \\
\Upsilon(k) &\triangleq \frac{1}{2}\left[\begin{array}{cccccc} 0 & 0 & 0 & 0 & 0 & \mathcal{A}_1(k)^T \\ 0 & 0 & 0 & 0 & 0 & \mathcal{A}_2(k)^T \\ 0 & 0 & 0 & 0 & 0 & 0 \\ 0 & 0 & 0 & 0 & 0 & 0 \\ 0 & 0 & 0 & 0 & 0 & 0 \\ \mathcal{A}_1(k) & \mathcal{A}_2(k) & 0 & 0 & 0 & 2I \end{array}\right], \\
\mathcal{A}_1(k) &\triangleq -H\tilde{A}(k)\hat{x}(k),\ \mathcal{A}_2(k)\triangleq -H\tilde{A}(k)L(k)
\end{aligned}$$

and $L(k)$ is the factorization of $P(k)$, i.e., $P(k)\triangleq L(k)L^T(k)$.

Proof *In order to prove this theorem, the mathematical induction method is employed, which can be divided into the initial step and the inductive step.*

Initial step: For $t=0$, it can be immediately obtained from the inequality (5.14) that

$$\begin{aligned} e^T(0)P^{-1}(0)e(0) &= \bar{x}^T(0)P^{-1}(0)\bar{x}(0) \\ &= \bar{\phi}^T(0)P^{-1}(0)\bar{\phi}(0) \le 1. \end{aligned} \tag{5.17}$$

Inductive step: Assume that $e^T(t)P^{-1}(t)e(t)\le 1$ is true for $0<t\le k$. Then, our purpose is to prove that the same requirement can be ensured for $t=k+1$. Based on the condition $e^T(t)P^{-1}(t)e(t)\le 1$ $(t\le k)$, it follows from [61] that there exists a vector $z(i)\in\mathbb{R}^{n_x+n_y}$ with $\|z(i)\|\le 1$ satisfying

$$e(i)=L(i)z(i) \tag{5.18}$$

where $L(i)$ is a factorization of $P(i)=L(i)L^T(i)$ for $i\in\{0,1,\ldots,k\}$.

Next, let us verify that $P(k+1)$ guarantees $e^T(k+1)P^{-1}(k+1)e(k+1)\le 1$ on the condition that the recursive inequality (5.15) is feasible.

For notational simplicity, define $\bar{z}_\tau(k)\triangleq[z^T(k-1)\ z^T(k-2)\ \cdots\ z^T(k-\tau_2)]^T$ and $\eta(k)\triangleq\left[1\ z^T(k)\ z^T(k-\tau_1)\ \bar{z}_\tau^T(k)\ \varpi^T(k)\ \tilde{\psi}^T(k)\right]^T$. Then, we reformulate $e(k+1)$ as

$$e(k+1)=\Pi(k)\eta(k) \tag{5.19}$$

where $\Pi(k)$ is defined in (5.16).

It follows from (5.9) that

$$\eta^T(k)\Upsilon(k)\eta(k) \leq 0 \tag{5.20}$$

with $\Upsilon(k)$ defined in (5.16). Moreover, it is inferred from (5.3), (5.18) and the selection principle of the WTOD that for any $j = 1, \ldots, \tau_2$, $i \in \mathcal{R}$,

$$\begin{cases} \|z(k)\|^2 \leq 1 \\ \|z(k-\tau_1)\|^2 \leq 1 \\ \|z(k-j)\|^2 \leq 1 \\ w^T(k)S^{-1}(k)w(k) \leq 1 \\ v^T(k)R^{-1}(k)v(k) \leq 1 \\ (y(k)-\bar{y}(k-1))^T\bar{Q}(\Omega_i - \Omega_{\xi(k)}) \\ \times (y(k)-\bar{y}(k-1)) \leq 0. \end{cases} \tag{5.21}$$

Combining (5.19), (5.20), (5.21) and S-Procedure, we conclude that, if there exist positive scalars $\alpha_l(k)$ $(l = 1, \ldots, 5)$, $\beta_j(k)$ $(j = 1, \ldots, \tau_2)$ and $\lambda_i(k)$ $(i \in \mathcal{R})$ such that

$$\Pi^T(k)P^{-1}(k+1)\Pi(k) - \Theta(k) \leq 0, \tag{5.22}$$

then $e^T(k+1)P^{-1}(k+1)e(k+1) \leq 1$ holds, where $\Theta(k)$ is defined in (5.16).

By means of the Schur complement, it is easily seen that (5.22) holds if the condition (5.15) is ensured, which implies that $e^T(k)P^{-1}(k)e(k) \leq 1$ is satisfied for $\forall k \in \mathbb{N}^+$, that is, the filtering error dynamics (5.12) satisfies the $P(k)$-dependent constraint. The proof is now complete.

5.2.2 Minimizing the Ellipsoids with Inequality Constraints

The following result is easily accessible from Theorem 5.1.

Corollary 5.1 *If there exist sequences of real-valued matrices $\{K(k)\}_{k\in\mathbb{N}^+}$, positive scalars $\{\lambda_i(k)\}_{k\in\mathbb{N}^+}$ $(i \in \mathcal{R})$, $\{\alpha_l(k)\}_{k\in\mathbb{N}^+}$ $(l = 1, \ldots, 5)$ and $\{\beta_j(k)\}_{k\in\mathbb{N}^+}$ $(j = 1, \ldots, \tau_2)$ solving the following optimization problem:*

$$\min_{\substack{P(k+1),K(k),\lambda_i(k)(i\in\mathcal{R}) \\ \alpha_l(k)(l=1,\ldots,5),\beta_j(k)(j=1,\ldots,\tau_2)}} \quad tr\{P(k+1)\} \tag{5.23}$$

subject to (5.15), then the ellipsoid determined by $P(k)$ with respect to the filtering error is minimized in the sense of the matrix trace.

According to Corollary 5.1, we can summarize the WTOD Protocol-Based Set-Membership Filter Design Algorithm (Algorithm 5.2.2) as follows.

WTOD Protocol-Based Set-Membership Filter Design Algorithm:

Step 1. Set $k = 0$ and give the initial constraint matrices $P(i)$ satisfying Assumption 5.1 for $(i = -\tau^*, -\tau^* + 1, \ldots, 0)$.

Step 2. Calculate $L(i)$ which is the factorization of $P(i) = L(i)L^T(i)$ for $k - \tau_2 \leq i \leq k$ and $i = k - \tau_1$.

Step 3. Solve the optimization problem (5.23) subject to (5.15). The filter parameter $K(k)$ and positive definite matrix $P(k+1)$ are derived according to the solution of the optimization problem.

Step 4. Set $k = k + 1$ and go to *Step 2.*

5.3 An Illustrative Example

In this section, we give an example to prove the effectiveness and applicability of the proposed method. The system model involves the experimental operation of an aircraft powered by two F-404 engines. Both engines are mounted close together in the aft fuselage. We are interested in tracking such an aircraft through wireless communications subject to state saturation and mixed time-delays under communication protocol. In this simulation, the nominal system matrix A and the measurement output matrix C are taken from the model of an F-404 aircraft engine system in [49]. We discretize them to obtain the following nominal system matrices:

$$A_1(k) = 1.5 + 0.01\sin(3k),\ A_2(k) = 0.413 + 0.01\cos(2k),\ \tau_1 = 3,\ \tau_2 = 2,$$

$$A_3(k) = -0.6 + 0.02\cos(2k),\ A(k) = \begin{bmatrix} A_1(k) & A_2(k) & A_3(k) \\ 1 & 0 & 0 \\ 0 & 1 & 0 \end{bmatrix},\ \tau_3 = 2,$$

$$B(k) = \begin{bmatrix} 0.01 & 0.08 & 0.11 \\ 0.3 & 0.04 & 0.04 \\ 0.8 & -0.02 & -0.02 \end{bmatrix},\ D(k) = \begin{bmatrix} 0 \\ 0.8 \\ 0 \end{bmatrix},\ F(k) = \begin{bmatrix} 0.5 \\ 1 \\ 2 \end{bmatrix},$$

$$E(k) = 0.1I,\ H = 0.3I,\ H_1 = 0.7I,\ C(k) = I,\ \mu_1 = \mu_2 = \mu_3 = 0.2.$$

The sensors of this system are grouped into 3 sensor nodes. The weight matrices of the WTOD protocol are taken to be $Q_1 = 0.8$, $Q_2 = 1.2$ and $Q_3 = 1.2$. The bounded noises are assumed to be $w(k) = 1.2\cos(0.2k)$ and $v(k) = 1.5\sin(0.2k)$, respectively. The matrices $S(k)$ and $R(k)$ are selected as $S(k) = R(k) = 0.25I$. The initial state and constraint matrices are chosen as follows: $\phi(l) = [2\ \ 2\ \ 2\ \ 1\ \ 1\ \ 1]^T$ and $P(l) = \text{diag}\{4, 4, 4, 1, 1, 1\}(l = -\max\{\tau_1, \tau_2, \tau_3\}, -\max\{\tau_1, \tau_2, \tau_3\} + 1, \ldots, 0)$.

The simulation results are shown in Figs. 5.1–5.3, which respectively

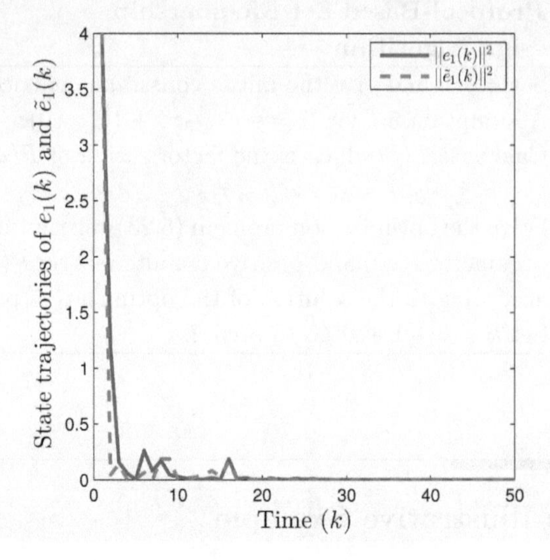

FIGURE 5.1
The norms of the filtering errors $e_1(k)$ and $\tilde{e}_1(k)$.

plot the norm of the filtering errors $e_i(k)$ and $\tilde{e}_i(k)$ with or without state saturations under the WTOD protocol, where $e_i(k)$ and $\tilde{e}_i(k)$ are the filtering errors of the ith element of the state. All the simulation results illustrate that the proposed filter design algorithm is effective.

5.4 Summary

In this chapter, the set-membership filtering problem has been investigated for a class of time-varying state-saturated systems with mixed time-delays under the WTOD protocol. A set of time-varying filters has been designed to obtain the estimation of the plant subject to the unknown but bounded noises and the WTOD protocol. A sufficient condition has been acquired for the designed filter to satisfy the prescribed $P(k)$-dependent constraint. Then, an optimization problem has been solved by optimizing the constraint ellipsoid of the estimation error subject to the WTOD protocol. Finally, a simulation example has been given to demonstrate the effectiveness of the proposed filter design algorithm.

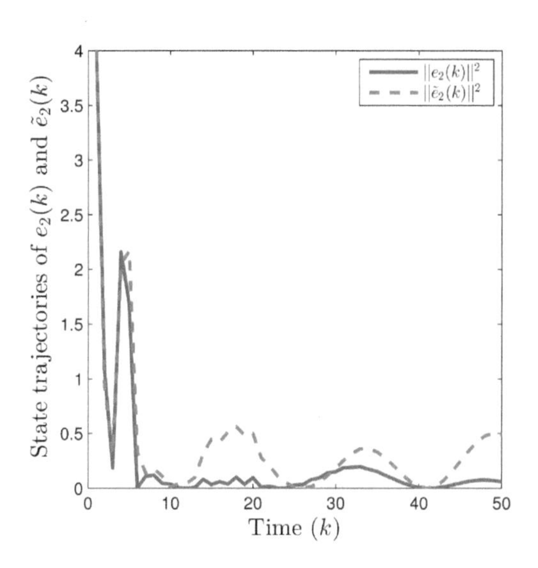

FIGURE 5.2
The norms of the filtering errors $e_2(k)$ and $\tilde{e}_2(k)$.

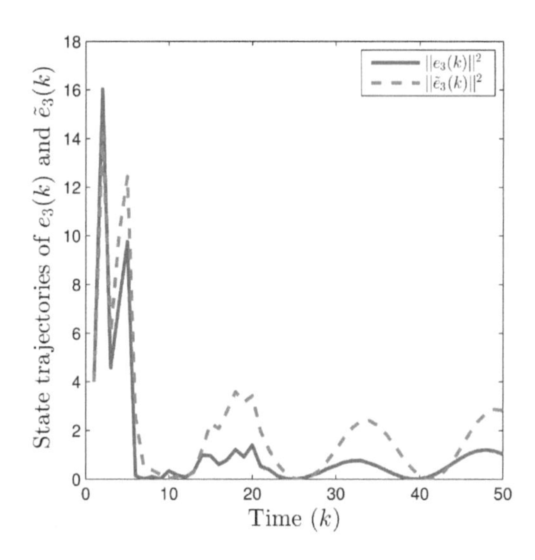

FIGURE 5.3
The norms of the filtering errors $e_3(k)$ and $\tilde{e}_3(k)$.

6

Distributed Estimation over Sensor Network

For extending the estimation method presented in Chapter 4, in this chapter, the distributed estimation/filtering problem is investigated for a class of time-varying systems over sensor networks. Firstly, we consider the systems over sensor networks with redundant channels and randomly switching topologies (RSTs). The Markovian jumping parameters are introduced to describe the switching topologies for sensor networks. Besides, in view of the realities of the data transmission, the redundant channels are used to improve the brittle network communication quality. The aim of the addressed distributed estimation problem is to design a class of distributed state estimators so that the state estimation error dynamics satisfies the given average H_∞ performance constraint. Secondly, we also concern the non-fragile distributed fault estimation problem for a class of time-varying systems subject to RONs and ROGVs. The occurrences of RONs and ROGVs are governed by random variables obeying certain probabilistic distributions on the interval $[0,1]$. Through stochastic analysis techniques, sufficient conditions are derived that guarantee the existence of the desired fault estimators. The explicit parameterization of the estimator parameters is obtained by solving a set of recursive matrix inequalities via a standard software package. The aim to design distributed fault estimators such that the dynamics of the estimation error achieves the prescribed level for H_∞ disturbance attenuation and rejection over a given finite horizon is finished. Thirdly, the finite-horizon filtering problem is discussed for nonlinear systems over sensor networks whose topologies are changeable subject to the RR protocol. The phenomenon of switching topologies for sensor networks is considered due to the weak connections between the node and its neighbours. The RR communication strategy is employed to save the limited bandwidth and reduce the network resource consumption. It is our target to design distributed filters so that the dynamics of augmented error system satisfies the given level of average H_∞ performance. We get the sufficient conditions in order to guarantee the existence of the desired distributed filters whose parameter matrices would be attained by a recursive method in terms of a set of matrix inequalities. Finally, several simulation examples are given to illustrate the effectiveness of the proposed estimator/filter design schemes.

DOI: 10.1201/9781003189497-6

6.1 Finite-Horizon Distributed State Estimation with RSTs and RCs

6.1.1 Problem Formulation

In this section, it is assumed that a sensor network has n sensor nodes which are distributed in space according to s randomly switching network topologies represented by directed graphs $\mathcal{G}^{(l)} = (\mathcal{V}^{(l)}, \mathcal{E}^{(l)}, \mathcal{A}^{(l)})$ $(l = 1, 2, \ldots, s)$ of order n with the set of nodes $\mathcal{V}^{(l)} = \{1, 2, \ldots, n\}$, the set of edges $\mathcal{E}^{(l)} \subseteq \mathcal{V}^{(l)} \times \mathcal{V}^{(l)}$, and the weighted adjacency matrices $\mathcal{A}^{(l)} = [a_{ij}^{(l)}]$ $(l = 1, 2, \ldots, s)$ with nonnegative adjacency elements $a_{ij}^{(l)}$, and l stands for the numbers of weighted adjacency matrices. An edge of $\mathcal{G}^{(l)}$ is denoted by ordered pair (i, j). The adjacency elements associated with the edges of the graph are positive, i.e., $a_{ij}^{(l)} > 0 \Longleftrightarrow (i, j) \in \mathcal{E}^{(l)}$ $(l = 1, 2, \ldots, s)$ which means that for network topology l $(l = 1, 2, \ldots, s)$, sensor i can obtain information from sensor j. Also, assume that $a_{ii}^{(l)} = 1$ for all $i \in \mathcal{V}^{(l)}$ and $l = 1, 2, \ldots, s$, and therefore (i, i) can be regarded as an additional edge. The set of neighbours of node $i \in \mathcal{V}^{(l)}$ plus the node itself are denoted by $\mathcal{N}_i^{(l)} = \{j \in \mathcal{V}^{(l)} : (i, j) \in \mathcal{E}^{(l)}\}$.

Let a finite time-horizon be denoted by $[0, N] := \{0, 1, 2, \ldots, N\}$. Consider a target plant described by the following discrete-time nonlinear time-varying stochastic system on $k \in [0, N]$:

$$\begin{cases} x(k+1) = A(k)x(k) + \beta(k)f(x(k)) + G(k)w(k) \\ \quad z(k) = M(k)x(k) \end{cases} \tag{6.1}$$

where $x(k) \in \mathbb{R}^{n_x}$ represents the state vector which cannot be observed directly; $z(k) \in \mathbb{R}^{n_z}$ is the output to be estimated; $w(k) \in \mathbb{R}^{n_w}$ denotes the noise signal belonging to $l_2[0, N-1]$. $f(\cdot)$ is a continuously vector-valued function.

In this section, the model of sensor node i is given as follows:

$$\begin{aligned} y_i(k) =& \alpha_{i1}(k)C_{i1}(k)x(k) + \sum_{p=2}^{r} \left\{ \prod_{q=1}^{p-1} (1 - \alpha_{iq}(k))\alpha_{ip}(k)C_{ip}(k)x(k) \right\} \\ &+ D_i(k)v(k), \ i = 1, 2, \ldots, n, \ p = 1, 2, \ldots, r \end{aligned} \tag{6.2}$$

where $y_i(k) \in \mathbb{R}^{n_y}$ is the output measured by sensor i and $v(k) \in l_2[0, N-1]$ is the external disturbance. Moreover, all the mentioned matrices (i.e., $A(k)$, $G(k)$, $M(k)$ and $C_{ip}(k)$) are known, real and time-varying with appropriate dimensions. The stochastic variables $\alpha_{ip}(k)$ $(i = 1, 2, \ldots, n$ and $p = 1, 2, \ldots, r)$, which describe the packet dropout phenomena of the ith channel, take values of 0 or 1 with

$$\text{Prob}\{\alpha_{ip}(k) = 1\} = \bar{\alpha}_{ip}, \quad \text{Prob}\{\alpha_{ip}(k) = 0\} = 1 - \bar{\alpha}_{ip} \tag{6.3}$$

where $\bar{\alpha}_{ip}$ are known non-negative constants. It is assumed that all the random variables $\alpha_{ip}(k)$ are mutually independent.

Remark 6.1 *As proposed in (6.2), the novel scheme called redundant channels which is used to improve the reliability of data transmission, depicts a new method to judge whether the packet dropout happen by the random variables a_{ip} ($i = 1, 2, \ldots, n$ and $p = 1, 2, \ldots, r$). For sensor node i, if $a_{i1} = 1$, it means that the first channel is capable of transmitting the information and the other channels would not work, or the current channel drops the information and another channel would be activated to deliver data. The phenomenon of packet dropouts happens when all the channels fail to transmit the data. Based on above discussions, packet dropout rate would be greatly reduced by using the redundant channels scheme.*

Assumption 6.1 *The nonlinear function $f(\cdot)$ satisfies $f(0) = 0$ and the following sector-bounded conditions:*

$$[f(x) - f(y) - L_1(x-y)]^T[f(x) - f(y) - L_2(x-y)] \leq 0, \ \forall x, \ y \in \mathbb{R}^{n_x} \quad (6.4)$$

where $L_1, L_2 \in \mathbb{R}^{n_x \times n_x}$ are real matrices of appropriate dimensions.

The stochastic variable $\beta(k)$ is introduced to account for the probabilistic nature of the occurrence of the nonlinearity, where $\beta(k)$ is a Bernoulli distributed white sequence taking values on 0 or 1 with

$$\begin{cases} \text{Prob}\{\beta(k) = 1\} = \bar{\beta} \\ \text{Prob}\{\beta(k) = 0\} = 1 - \bar{\beta} \end{cases}$$

where $\bar{\beta} \in [0, 1]$ is a known constant.

In this section, the following state estimator structure is adopted on sensor node i:

$$\begin{cases} \hat{x}_i(k+1) = \sum_{l=1}^{s} \sum_{j \in \mathcal{N}_i^{(l)}} \delta(\tau(k), l) a_{ij}^{(l)} K_{ij}^{(l)}(k) \hat{x}_j(k) \\ \qquad\qquad + \sum_{l=1}^{s} \sum_{j \in \mathcal{N}_i^{(l)}} \delta(\tau(k), l) a_{ij}^{(l)} H_{ij}^{(l)}(k) y_j(k) \\ \hat{z}_i(k) = M(k) \hat{x}_i(k) \end{cases} \quad (6.5)$$

where $\hat{x}_i(k) \in \mathbb{R}^{n_x}$ is the state estimate on sensor node i and $\hat{z}_i(k) \in \mathbb{R}^{n_z}$ is the estimate of $z(k)$ on sensor node i. Here, $K_{ij}^{(l)}(k)$ and $H_{ij}^{(l)}(k)$ ($l = 1, 2, \ldots, s$) are the estimator gain parameters on node i to be determined. $\delta(\cdot, \cdot)$ is the Kronecker delta function, i.e.,

$$\delta(a, b) = \begin{cases} 0, & \text{if } a \neq b \\ 1, & \text{if } a = b \end{cases}$$

The stochastic variable $\tau(k)$ is introduced to describe the phenomenon of randomly switching topologies at time instant k. In the present section, it is assumed that the sequence $\{\tau(k)\}$ is a discrete-time homogeneous Markovian chain taking values in the finite state space

$$S = \{1, 2, \ldots, s\} \tag{6.6}$$

and $\Xi = [\lambda_{m_k n_{k+1}}]$ is the transition probability matrix with its entities defined as

$$\lambda_{m_k n_{k+1}} = \text{Prob}\{\tau(k+1) = n_{k+1} | \tau(k) = m_k\} \tag{6.7}$$

where m_k is the mode at time instant k and n_{k+1} is the mode at time instant $k+1$.

Remark 6.2 *The state estimator (6.5) represents a quite general model structure and establishes the communications between sensor node i and its neighbouring nodes, in which the sensor nodes are distributed over a spatial region. It can be observed from (6.5) that the topology of sensor network may switch in a probabilistic way by means of their intensity. It is worth mentioning that, by recurring to the Kronecker delta function, 'Markov jumping' assumption on $\{\tau(k)\}$ is able to describe randomly switching topologies adequately.*

For convenience of later analysis, denote

$$\begin{aligned}
&\hat{x}(k) \triangleq \begin{bmatrix} \hat{x}_1^T(k) & \hat{x}_2^T(k) & \cdots & \hat{x}_n^T(k) \end{bmatrix}^T, \quad \mathcal{F}(I_2\eta(k)) \triangleq \mathbf{1}_n \otimes f(x(k)), \\
&\bar{w}(k) \triangleq \mathbf{1}_n \otimes w(k), \quad \bar{v}(k) \triangleq \mathbf{1}_n \otimes v(k), \quad w_v(k) \triangleq \begin{bmatrix} \bar{w}^T(k) & \bar{v}^T(k) \end{bmatrix}^T, \\
&\bar{x}(k) \triangleq \mathbf{1}_n \otimes x(k), \quad y(k) \triangleq \begin{bmatrix} y_1^T(k) & y_2^T(k) & \cdots & y_n^T(k) \end{bmatrix}^T, \\
&\bar{z}(k) \triangleq \mathbf{1}_n \otimes z(k), \quad \hat{z}(k) \triangleq \begin{bmatrix} \hat{z}_1^T(k) & \hat{z}_2^T(k) & \cdots & \hat{z}_n^T(k) \end{bmatrix}^T, \\
&\bar{A}(k) \triangleq I_n \otimes A(k), \quad \bar{G}(k) \triangleq I_n \otimes G(k), \\
&\bar{C}_1(k) \triangleq \text{diag}\{C_{11}(k), C_{21}(k), \ldots, C_{n1}(k)\}, \\
&\bar{C}_p(k) \triangleq \text{diag}\{C_{1p}(k), C_{2p}(k), \ldots, C_{np}(k)\}, \\
&\tilde{A}_1(k) \triangleq \text{diag}\{\alpha_{11}(k), \alpha_{21}(k), \ldots, \alpha_{n1}(k)\}, \\
&\bar{A}_1 \triangleq \text{diag}\{\bar{\alpha}_{11}, \bar{\alpha}_{21}, \ldots, \bar{\alpha}_{n1}\}, \quad \bar{M}(k) \triangleq I_n \otimes M(k), \\
&\tilde{A}_p(k) \triangleq \text{diag}\Bigg\{ \prod_{q=1}^{p-1}(1 - \alpha_{1q}(k)\alpha_{1p}(k), \prod_{q=1}^{p-1}(1 - \alpha_{2q}(k)\alpha_{2p}(k), \ldots, \\
&\qquad \prod_{q=1}^{p-1}(1 - \alpha_{nq}(k)\alpha_{np}(k) \Bigg\}, \\
&\bar{A}_p \triangleq \text{diag}\Bigg\{ \prod_{q=1}^{p-1}(1 - \bar{\alpha}_{1q})\bar{\alpha}_{1p}, \prod_{q=1}^{p-1}(1 - \bar{\alpha}_{2q})\bar{\alpha}_{2p}, \ldots, \prod_{q=1}^{p-1}(1 - \bar{\alpha}_{nq})\bar{\alpha}_{np} \Bigg\},
\end{aligned}$$

$$\bar{K}^{(l)}(k) \triangleq [\bar{K}_{ij}^{(l)}(k)]_{n\times n} \quad \text{with}$$

$$\bar{K}_{ij}^{(l)}(k) \triangleq \begin{cases} a_{ij}^{(l)} K_{ij}^{(l)}(k), & i=1,2,\ldots,n;\ j\in\mathcal{N}_i^{(l)};\ l=1,2,\ldots,s \\ 0, & i=1,2,\ldots,n;\ j\notin\mathcal{N}_i^{(l)};\ l=1,2,\ldots,s \end{cases}$$

$$\bar{H}^{(l)}(k) \triangleq [\bar{H}_{ij}^{(l)}(k)]_{n\times n} \quad \text{with}$$

$$\bar{H}_{ij}^{(l)}(k) \triangleq \begin{cases} a_{ij}^{(l)} H_{ij}^{(l)}(k), & i=1,2,\ldots,n;\ j\in\mathcal{N}_i^{(l)};\ l=1,2,\ldots,s \\ 0, & i=1,2,\ldots,n;\ j\notin\mathcal{N}_i^{(l)};\ l=1,2,\ldots,s. \end{cases}$$

Obviously, since $a_{ij}^{(l)}=0$ when $j\notin\mathcal{N}_i$ and $l\in S$, $\bar{K}^{(l)}(k)$ and $\bar{H}^{(l)}(k)$ can be expressed as

$$\bar{K}^{(l)}(k)\in\mathscr{T}_{n_x\times n_x}, \quad \bar{H}^{(l)}(k)\in\mathscr{T}_{n_x\times n_y} \tag{6.8}$$

where $\mathscr{T}_{p\times q}\triangleq\left\{\bar{T}=[T_{ij}]\in\mathbb{R}^{np\times nq}\mid T_{ij}\in\mathbb{R}^{p\times q},\quad T_{ij}=0\ \text{ if } j\notin\mathcal{N}_i^{(l)}\right\}$.

Letting $\eta(k)\triangleq\begin{bmatrix}\bar{x}^T(k) & \hat{x}^T(k)\end{bmatrix}^T$ and $\tilde{z}(k)\triangleq\bar{z}(k)-\hat{z}(k)$, the following augmented system is obtained that governs the estimation dynamics for the sensor network:

$$\begin{cases} \eta(k+1) = \mathcal{A}(k,\tau(k))\eta(k) + \mathcal{B}_1(k,\tau(k))\eta(k) + \mathcal{C}_1(k,\tau(k))\eta(k) \\ \qquad\qquad + \sum_{p=2}^{r}\mathcal{B}_p(k,\tau(k))\eta(k) + \sum_{p=2}^{r}\mathcal{C}_p(k,\tau(k))\eta(k) \\ \qquad\qquad + \mathcal{D}(k,\tau(k))w_v(k) + (\bar{\beta}I_1+\tilde{\beta}(k)I_1)\mathcal{F}(I_2\eta(k)) \\ \tilde{z}(k) = \mathcal{M}(k)\eta(k) \end{cases} \tag{6.9}$$

where

$$\mathcal{A}(k,\tau(k))\triangleq\operatorname{diag}\{\bar{A}(k),\sum_{l=1}^{s}\delta(\tau(k),l)\bar{K}^{(l)}(k)\}, \quad \bar{L}_1\triangleq I_n\otimes L_1,$$

$$\mathcal{B}_1(k,\tau(k))\triangleq\begin{bmatrix}0 & 0\\ \sum_{l=1}^{s}\delta(\tau(k),l)\bar{H}^{(l)}(k)\bar{A}_1\bar{C}_1(k) & 0\end{bmatrix},$$

$$\mathcal{C}_1(k,\tau(k))\triangleq\begin{bmatrix}0 & 0\\ \check{\mathcal{C}}_1(k,\tau(k)) & 0\end{bmatrix}, \quad \bar{L}_2\triangleq I_n\otimes L_2,$$

$$\check{\mathcal{C}}_1(k,\tau(k))\triangleq\sum_{l=1}^{s}\delta(\tau(k),l)\bar{H}^{(l)}(k)(\tilde{A}_1(k)-\bar{A}_1)\bar{C}_1(k),$$

$$\mathcal{B}_p(k,\tau(k))\triangleq\begin{bmatrix}0 & 0\\ \check{\mathcal{B}}_p(k,\tau(k)) & 0\end{bmatrix}, \quad \mathcal{C}_p(k,\tau(k))\triangleq\begin{bmatrix}0 & 0\\ \check{\mathcal{C}}_p(k,\tau(k)) & 0\end{bmatrix}, \tag{6.10}$$

$$\check{\mathcal{B}}_p(k,\tau(k))\triangleq\sum_{l=1}^{s}\delta(\tau(k),l)\bar{H}^{(l)}(k)\bar{A}_p\bar{C}_p(k),$$

$$\mathcal{D}(k,\tau(k))\triangleq\operatorname{diag}\left\{\bar{G}(k),\sum_{l=1}^{s}\delta(\tau(k),l)\bar{H}^{(l)}(k)\bar{D}(k)\right\},$$

$$\check{C}_p(k,\tau(k)) \triangleq \sum_{l=1}^{s}\delta(\tau(k),l)\bar{H}^{(l)}(k)(\tilde{A}_p(k)-\bar{A}_p)\bar{C}_p(k),$$

$$I_1 \triangleq \begin{bmatrix} I & 0 \end{bmatrix}^T, \quad \mathcal{M}(k) \triangleq \begin{bmatrix} \bar{M}(k) & -\bar{M}(k) \end{bmatrix}, \quad I_2 \triangleq \begin{bmatrix} I & 0 \end{bmatrix}.$$

Before proceeding further, the following definition is introduced.

Definition 6.1 *For a given disturbance attenuation level $\gamma > 0$ and some given positive definite matrices $S_i > 0$ $(0 \le i \le n)$, the state estimation error $\tilde{z}_i(k) = z(k) - \hat{z}_i(k)$ is said to satisfy the average H_∞ performance constraints if the following inequality holds:*

$$\begin{aligned}\frac{1}{n}\sum_{i=1}^{n}\mathbb{E}\{\|\tilde{z}_i\|^2_{[0,N-1]}\} <\gamma^2\Big\{ & \|w\|^2_{[0,N-1]} + \|v\|^2_{[0,N-1]} + \frac{1}{n}\sum_{i=1}^{n}(x(0) \\ & - \hat{x}_i(0))^T S_i(\tau(k))(x(0)-\hat{x}_i(0))\Big\}.\end{aligned} \tag{6.11}$$

The average H_∞ performance constraints (6.11) can be rewritten as follows:

$$J \triangleq \mathbb{E}\{\|\tilde{z}\|^2_{[0,N-1]}\} - \gamma^2\{\|w_v\|^2_{[0,N-1]} + e^T(0)R(\tau(k))e(0)\} < 0 \tag{6.12}$$

where $e(0) \triangleq \bar{x}(0) - \hat{x}(0)$ and $R(\tau(k)) \triangleq \mathrm{diag}\{S_1(\tau(k)), S_2(\tau(k)), \ldots, S_n(\tau(k))\}$.

The target in this section is to acquire the state estimators parameters $K_{ij}^{(l)}(k)$ and $H_{ij}^{(l)}(k)$ such that the estimation error system (6.9) satisfies the performance constraint (6.12).

6.1.2 Main Results

A. Analysis of H_∞ Performances

In this subsection, we need to investigate the estimation error analysis for system (6.1) with n sensors whose topology is determined by the graphs $\mathcal{G}^{(l)} = (\mathcal{V}^{(l)}, \mathcal{E}^{(l)}, \mathcal{A}^{(l)})$.

Theorem 6.1 *Assume the parameters of designed state estimators $K_{ij}^{(l)}(k)$ and $H_{ij}^{(l)}(k)$ in (6.5) are given. Define u_N and l_0 are respectively the mode at time instant N and 0. For a positive scalar $\gamma > 0$ and a sequence of positive definite matrices $S_i(m_k) > 0$ $(i = 1, 2, \ldots, n)$, the average H_∞ performance requirement defined in (6.12) is achieved for all nonzero $w_v(k)$, if there exists a sequence of positive definite matrices $\{P(k, m_k)\}_{0 \le k \le N+1}$ $(1 < i < n)$ with the initial condition $\eta^T(0)P(0, l_0)\eta(0) \le \gamma^2 \bar{e}^T(0)R(l_0)\bar{e}(0)$ satisfying the following recursive matrix inequalities:*

$$\Upsilon(k,m_k) = \begin{bmatrix} \Upsilon_{11}(k,m_k) & * & * \\ \Upsilon_{21}(k,m_k) & \Upsilon_{22}(k,m_k) & * \\ \Upsilon_{31}(k,m_k) & \Upsilon_{32}(k,m_k) & \Upsilon_{33}(k,m_k) \end{bmatrix} < 0 \tag{6.13}$$

where

$$
\begin{aligned}
\Upsilon_{11}(k,m_k) \triangleq\ & \mathcal{A}_1{}^T(k,m_k)\tilde{P}(k+1,m_k)\mathcal{A}_1(k,m_k) + \bar{\mathcal{C}}_1^T(k,m_k) \\
& \times \tilde{P}(k+1,m_k)\bar{\mathcal{C}}_1(k,m_k) + \sum_{p=2}^{r}(\bar{\mathcal{C}}_p(k,m_k))^T\tilde{P}(k+1,m_k) \\
& \times (\bar{\mathcal{C}}_p(k,m_k)) - P(k,m_k) + \mathcal{M}^T(k)\mathcal{M}(k) - \varepsilon_1\mathcal{L}_1 \\
& + (r-1)\mathcal{B}_1^T(k)\tilde{P}(k+1,m_k)\mathcal{B}_1(k) \\
& + (r-1)\sum_{p=2}^{r}\bar{\mathcal{B}}_p^T(k,m_k)\tilde{P}(k+1,m_k)\bar{\mathcal{B}}_p(k,m_k), \\
\Upsilon_{22}(k,m_k) \triangleq\ & (\bar{\beta}^2+g)I_1^T\tilde{P}(k+1,m_k)I_1 - \varepsilon_1 I, \\
\Upsilon_{31}(k,m_k) \triangleq\ & \mathcal{D}^T(k,m_k)\tilde{P}(k+1,m_k)\mathcal{A}_1(k,m), \\
\Upsilon_{33}(k,m_k) \triangleq\ & \mathcal{D}^T(k,m_k)\tilde{P}(k+1,m_k)\mathcal{D}(k,m_k) - \gamma^2 I, \\
\Upsilon_{21}(k,m_k) \triangleq\ & \bar{\beta}I_1^T\tilde{P}(k+1,m_k)\mathcal{A}_1(k,m_k) - \varepsilon_1\mathcal{L}_2, \qquad (6.14)\\
\Upsilon_{32}(k,m_k) \triangleq\ & \bar{\beta}\mathcal{D}^T(k,m_k)\tilde{P}(k+1,m_k)I_1, \\
\mathcal{A}_1(k,m_k) \triangleq\ & \mathcal{A}(k,m_k) + \mathcal{B}_1(k,m_k) + \sum_{p=2}^{r}\mathcal{B}_p(k,m_k), \\
\tilde{P}(k+1,m_k) \triangleq\ & \sum_{n_{k+1}=1}^{s}\lambda_{m_k n_{k+1}}P(k+1,n_{k+1}|m_k), \quad g \triangleq \bar{\beta}(1-\bar{\beta}), \\
\mathcal{L}_1 \triangleq\ & \frac{I_2^T(\bar{L}_1^T\bar{L}_2 + \bar{L}_2^T\bar{L}_1)I_2}{2}, \quad \mathcal{L}_2 \triangleq -\frac{(\bar{L}_1+\bar{L}_2)I_2}{2}, \quad \tilde{\beta} \triangleq \beta(k) - \bar{\beta}, \\
\bar{\mathcal{C}}_p(k,m_k) \triangleq\ & \begin{bmatrix} 0 & 0 \\ \bar{\mathcal{C}}_p^1(k,m_k) & 0 \end{bmatrix}, \quad \bar{\mathcal{C}}_1(k,m_k) \triangleq \begin{bmatrix} 0 & 0 \\ \bar{\mathcal{C}}_1^1(k,m_k) & 0 \end{bmatrix}, \\
\bar{\mathcal{C}}_1^1(k,m_k) \triangleq\ & \sum_{l=1}^{s}\delta(m_k,l)\bar{H}^{(l)}(k)\sqrt{\bar{A}_1(I-\bar{A}_1)}\bar{C}_1(k), \\
\bar{\mathcal{C}}_p^1(k,m_k) \triangleq\ & \sum_{l=1}^{s}\delta(m_k,l)\bar{H}^{(l)}(k)\sqrt{\bar{A}_p(I-\bar{A}_p)}\bar{C}_p(k).
\end{aligned}
$$

$P(k+1, n_{k+1}|m_k)$ is the positive definite matrix presented in Lyapunov function with the mode n_{k+1} at time instant $k+1$ given the mode m_k at time instant k.

Proof *Define*

$$J(k,m_k) \triangleq \eta^T(k+1)\tilde{P}(k+1,m_k)\eta(k+1) - \eta^T(k)P(k,m_k)\eta(k). \qquad (6.15)$$

From (6.9) we have

$$\mathbb{E}\{J(k,m_k)\}$$

$$
\begin{aligned}
&= \mathbb{E}\Big\{\eta^T(k+1)\sum_{n_{k+1}=1}^{s}\lambda_{m_k n_{k+1}}P(k+1,n_{k+1}|m_k)\eta(k+1) \\
&\quad -\eta^T(k)P(k,m_k)\eta(k)\Big\} \\
&= \mathbb{E}\{\eta^T(k+1)\tilde{P}(k+1,m_k)\eta(k+1)-\eta^T(k)P(k,m_k)\eta(k)\} \\
&= \mathbb{E}\Big\{\Big(\mathcal{A}_1(k,m_k)\eta(k)+\mathcal{C}_1(k,m_k)\eta(k)+\sum_{p=2}^{r}\mathcal{C}_p(k,m_k)\eta(k) \\
&\quad +\mathcal{D}(k,m_k)w_v(k)+(\bar{\beta}I_1+\tilde{\beta}(k)I_1)\mathcal{F}(I_2\eta(k))\Big)^T\tilde{P}(k+1,m_k) \\
&\quad \times\Big((\mathcal{A}_1(k,m_k)\eta(k)+\mathcal{C}_1(k,m_k)\eta(k)+\sum_{p=2}^{r}\mathcal{C}_p(k,m_k)\eta(k) \\
&\quad +\mathcal{D}(k,m_k)w_v(k)+(\bar{\beta}I_1+\tilde{\beta}(k)I_1)\mathcal{F}(I_2\eta(k))\Big)\Big\} \qquad (6.16)\\
&\quad -\eta^T(k)P(k,m_k)\eta(k) \\
&= \mathbb{E}\Big\{(\mathcal{A}_1(k,m_k)\eta(k)+\mathcal{D}(k,m_k)w_v(k) \\
&\quad +\bar{\beta}I_1\mathcal{F}(I_2\eta(k)))^T\tilde{P}(k+1,m_k)(\mathcal{A}_1(k,m_k)\eta(k) \\
&\quad +\mathcal{D}(k,m_k)w_v(k)+\bar{\beta}I_1\mathcal{F}(I_2\eta(k)) \\
&\quad +\Big(\mathcal{C}_1(k,m_k)\eta(k)+\sum_{p=2}^{r}\mathcal{C}_p(k,m_k)\eta(k)\Big)^T\tilde{P}(k+1,m_k) \\
&\quad +\Big(\mathcal{C}_1(k,m_k)\eta(k)+\sum_{p=2}^{r}\mathcal{C}_p(k,m_k)\eta(k)\Big) \\
&\quad +g(I_1\mathcal{F}(I_2\eta(k)))^T\tilde{P}(k+1,m_k)(I_1\mathcal{F}(I_2\eta(k))) \\
&\quad -\eta^T(k)P(k,m_k)\eta(k)\Big\}.
\end{aligned}
$$

Also, we have from (6.4) that

$$[\mathcal{F}(I_2\eta(k))-\bar{L}_1I_2\eta(k)]^T[\mathcal{F}(I_2\eta(k))-\bar{L}_2I_2\eta(k)]\le 0. \qquad (6.17)$$

Adding the zero term $\tilde{z}^T(k)\tilde{z}(k)-\gamma^2w_v^T(k)w_v(k)-\tilde{z}^T(k)\tilde{z}(k)+\gamma^2w_v^T(k)w_v(k)$ *to* $\mathbb{E}(J(k))$ *results in*

$$\mathbb{E}\{J(k,m_k)\}=\mathbb{E}\{\zeta^T(k)\bar{\Upsilon}(k,m_k)\zeta(k)-\tilde{z}^T(k)\tilde{z}(k)+\gamma^2w_v^T(k)w_v(k)\} \qquad (6.18)$$

where

$$\bar{\Upsilon}(k,m_k)\triangleq\begin{bmatrix}\bar{\Upsilon}_{11}(k,m_k) & * & * \\ \bar{\Upsilon}_{21}(k,m_k) & \bar{\Upsilon}_{22}(k,m_k) & * \\ \bar{\Upsilon}_{31}(k,m_k) & \bar{\Upsilon}_{32}(k,m_k) & \bar{\Upsilon}_{33}(k,m_k)\end{bmatrix},$$

$$
\begin{aligned}
&\bar{\Upsilon}_{11}(k,m_k) \triangleq \Upsilon_{11}(k,m_k)+\varepsilon_1\mathcal{L}_1, \quad \bar{\Upsilon}_{21}(k,m_k) \triangleq \Upsilon_{21}(k,m_k)+\varepsilon_1\mathcal{L}_2, \\
&\bar{\Upsilon}_{22}(k,m_k) \triangleq \Upsilon_{22}(k,m_k)+\varepsilon_1 I, \quad \bar{\Upsilon}_{31}(k,m_k) \triangleq \Upsilon_{31}(k,m_k), \\
&\bar{\Upsilon}_{32}(k,m_k) \triangleq \Upsilon_{32}(k,m_k), \quad \bar{\Upsilon}_{33}(k,m_k) \triangleq \Upsilon_{33}(k,m_k), \\
&\zeta(k) \triangleq \begin{bmatrix}\eta^T(k) & \mathcal{F}^T(I_2\eta(k)) & w_v^T(k)\end{bmatrix}^T.
\end{aligned} \tag{6.19}
$$

Moreover, it follows from the constraint (6.12) that

$$
\begin{aligned}
&\mathbb{E}\{J(k,m_k)\} \\
\leq&\mathbb{E}\{\zeta^T(k)\bar{\Upsilon}(k,m_k)\zeta(k)-[\mathcal{F}(I_2\eta(k))-\bar{L}_1I_2\eta(k)]^T \\
&\times[\mathcal{F}(I_2\eta(k))-\bar{L}_2I_2\eta(k)]-\tilde{z}^T(k)\tilde{z}(k)+\gamma^2w_v^T(k)w_v(k)\} \\
=&\mathbb{E}\{\zeta^T(k)\Upsilon(k,m_k)\zeta(k)-\tilde{z}^T(k)\tilde{z}(k)+\gamma^2w_v^T(k)w_v(k)\}.
\end{aligned} \tag{6.20}
$$

Summing up (6.20) on both sides from 0 to $N-1$ with respect to k, we can obtain:

$$
\begin{aligned}
&\sum_{k=0}^{N-1}\mathbb{E}\{J(k,m_k)\} \\
=&\mathbb{E}\{\eta^T(N)\tilde{P}(N,u_N)\eta(N)\}-\eta^T(0)P(0,l_0)\eta(0) \\
\leq&\mathbb{E}\Bigg\{\sum_{k=0}^{N-1}\zeta^T(k)\Upsilon(k,m_k)\zeta(k)\Bigg\}-\mathbb{E}\Bigg\{\sum_{k=0}^{N-1}(\tilde{z}^T(k)\tilde{z}(k) \\
&-\gamma^2w_v^T(k)w_v(k))\Bigg\}.
\end{aligned} \tag{6.21}
$$

Therefore, the average H_∞ performance index defined in (6.11) is given by

$$
\begin{aligned}
J\leq&\mathbb{E}\Bigg\{\sum_{k=0}^{N-1}\zeta^T(k)\Upsilon(k,m_k)\zeta(k)\Bigg\}-\mathbb{E}\{\eta^T(N)\tilde{P}(N,u_N)\eta(N)\} \\
&+\eta^T(0)P(0,l_0)\eta(0)-\gamma^2e^T(0)R(l_0)e(0).
\end{aligned} \tag{6.22}
$$

Noting that $\tilde{P}(N,u_N)>0$ and the initial condition $\eta^T(0)P(0,l_0)\eta(0)\leq\gamma^2e^T(0)R(l_0)e(0)$, we have $J<0$ when (6.13) holds.

For convenience of later analysis, we denote

$$
\begin{aligned}
\Omega_{11}(k,m_k)\triangleq&\begin{bmatrix}-P(k,m_k)-\varepsilon_1\mathcal{L}_1 & * & * \\ -\varepsilon_1\mathcal{L}_2 & -\varepsilon_1 I & * \\ 0 & 0 & -\gamma^2 I\end{bmatrix}, \\
\Omega_{22}(k,m_k)\triangleq&\mathrm{diag}\{-I,-\tilde{P}^{-1}(k+1,m_k),-\tilde{P}^{-1}(k+1,m_k), \\
&-I_{2r}\otimes\tilde{P}^{-1}(k+1,m_k)\},
\end{aligned}
$$

$$\Omega_{21}(k,m_k) \triangleq \begin{bmatrix} \mathcal{M}(k) & 0 & 0 \\ \mathcal{A}_1(k,m_k) & \bar{\beta}I_1 & \mathcal{D}(k,m_k) \\ 0 & \tilde{\beta}I_1 & 0 \\ \Omega_{2241}(k,m_k) & 0 & 0 \\ \Omega_{2251}(k,m_k) & 0 & 0 \end{bmatrix}, \tag{6.23}$$

$$\Omega_{2241}(k,m_k) \triangleq \begin{bmatrix} \bar{\mathcal{C}}_1^T(k,m_k) & \cdots & \bar{\mathcal{C}}_p^T(k,m_k) \end{bmatrix}^T,$$

$$\Omega_{2251}(k,m_k) \triangleq \begin{bmatrix} \varrho\bar{\mathcal{B}}_1^T(k,m_k) & \cdots & \varrho\bar{\mathcal{B}}_p^T(k,m_k) \end{bmatrix}^T, \ \varrho \triangleq \sqrt{r-1}.$$

Theorem 6.2 *Given a positive scalar $\gamma > 0$ and positive definite matrices $S_i(m_k) = S_i^T(m_k) > 0$ $(1 \le i \le n)$. For the target plant (6.1) with randomly varying nonlinearities and the time-varying state estimators with switching topologies, the addressed finite-horizon state estimator design problem is solved if there exist positive definite matrices $\{P(k,m_k)\}_{0\le k\le N+1}$ satisfying the initial condition*

$$\eta^T(0)P(0,l_0)\eta(0) \leqslant \gamma^2\bar{e}^T(0)R(l_0)\bar{e}(0) \tag{6.24}$$

and a positive scalar ε_1 such that for all $0 \le k \le N$ the following recursive matrix inequalities hold:

$$\Omega(k,m_k) = \begin{bmatrix} \Omega_{11}(k,m_k) & * \\ \Omega_{21}(k,m_k) & \Omega_{22}(k,m_k) \end{bmatrix} < 0. \tag{6.25}$$

Proof *On the basis of the previous analysis on the H_∞ performance, (6.13) is equal to*

$$\begin{aligned}
&\Omega(k,m_k) \\
&= \begin{bmatrix} 0 & \tilde{\beta}I_1^T & 0 \end{bmatrix}^T \tilde{P}(k+1,m_k) \begin{bmatrix} 0 & \tilde{\beta}I_1 & 0 \end{bmatrix} + \begin{bmatrix} \bar{\mathcal{C}}_1^T(k,m_k) & 0 & 0 \end{bmatrix}^T \\
&\quad \times \tilde{P}(k+1,m_k) \begin{bmatrix} \bar{\mathcal{C}}_1(k,m_k) & 0 & 0 \end{bmatrix} + \sum_{p=2}^{r} \begin{bmatrix} \bar{\mathcal{C}}_p^T(k,m_k) & 0 & 0 \end{bmatrix}^T \\
&\quad \times \tilde{P}(k+1,m_k) \begin{bmatrix} \bar{\mathcal{C}}_p(k,m_k) & 0 & 0 \end{bmatrix} + \begin{bmatrix} \varrho\bar{\mathcal{B}}_1^T(k,m_k) & 0 & 0 \end{bmatrix}^T \\
&\quad \times \tilde{P}(k+1,m_k) \begin{bmatrix} \varrho\bar{\mathcal{B}}_1(k,m_k) & 0 & 0 \end{bmatrix} + \sum_{p=2}^{r} \begin{bmatrix} \varrho\bar{\mathcal{B}}_p^T(k,m_k) & 0 & 0 \end{bmatrix}^T \\
&\quad \times \tilde{P}(k+1,m_k) \begin{bmatrix} \varrho\bar{\mathcal{B}}_p(k,m_k) & 0 & 0 \end{bmatrix} \\
&\quad + \begin{bmatrix} \mathcal{A}_1^T(k,m_k) & \bar{\beta}I_1^T & \mathcal{D}^T(k,m_k) \end{bmatrix}^T \tilde{P}(k+1,m_k) \\
&\quad \times \begin{bmatrix} \mathcal{A}_1(k,m_k) & \bar{\beta}I_1 & \mathcal{D}(k,m_k) \end{bmatrix} \\
&\quad + \begin{bmatrix} \mathcal{M}^T(k) & 0 & 0 \end{bmatrix}^T I \begin{bmatrix} \mathcal{M}(k) & 0 & 0 \end{bmatrix} + \Omega_{11}(k,m_k) < 0.
\end{aligned} \tag{6.26}$$

In terms of the aforementioned equalities and Lemma 2.1, we can easily obtain (6.25) by (6.26). The proof of this theorem is now complete.

The performance analysis of the dynamic system (6.9) has been derived in Theorem 6.1 and Theorem 6.2, then we are devoted to designing the distributed state estimator for a class of time-varying systems over sensor networks with redundant channels and switching topologies.

B. Design of State Estimator

In the following theorem, we will develop a new method to design the state estimator for the sensor network with redundant channels and switching topologies over a finite horizon.

Theorem 6.3 *Given a positive scalar $\gamma > 0$ and positive definite matrices $S_i(m_k) = S_i^T(m_k) > 0$ $(1 \leq i \leq n)$. For the target plant (6.1) with randomly varying nonlinearities and sensor network with switching topologies, the addressed finite-horizon state estimation problem is solved if there exist positive definite matrices $\{\check{\Psi}(k, m_k)\}_{0 \leq k \leq N+1}$ and $\{\check{\Phi}(k, m_k)\}_{0 \leq k \leq N+1}$, a positive scalar ε_1 and two sets of matrices $K_{ij}^{(l)}(k) \in \mathscr{T}_{n_x \times n_x}$ $(l \in S)$ and $H_{ij}^{(l)}(k) \in \mathscr{T}_{n_x \times n_y}$ $(l \in S)$ satisfying the initial condition*

$$\eta^T(0)\text{diag}\{\Psi(0, l_0), \Phi(0, l_0)\}\eta(0) \leqslant \gamma^2 \bar{e}^T(0) R(l_0) \bar{e}(0) \tag{6.27}$$

and the following RLMIs:

$$\Gamma(k, m_k) = \begin{bmatrix} \Gamma_{11}(k, m_k) & * \\ \Gamma_{21}(k, m_k) & \Gamma_{22}(k, m_k) \end{bmatrix} < 0 \tag{6.28}$$

with the parameters updated by

$$\check{\Psi}^{-1}(k+1, m_k) = \sum_{n_{k+1}=1}^{s} \lambda_{m_k n_{k+1}} \Psi(k+1, n_{k+1}|m_k),$$

$$\check{\Phi}^{-1}(k+1, m_k) = \sum_{n_{k+1}=1}^{s} \lambda_{m_k n_{k+1}} \Phi(k+1, n_{k+1}|m_k)$$

where

$$\begin{aligned}
\Gamma_{11}(k, m_k) &\triangleq \begin{bmatrix} \Gamma_{11}^1(k, m_k) & * \\ 0 & \Gamma_{11}^2(k, m_k) \end{bmatrix}, \\
\Gamma_{11}^{11}(k, m_k) &\triangleq -\frac{\varepsilon_1(\bar{L}_1^T \bar{L}_2 + \bar{L}_2^T \bar{L}_1)}{2} - \Psi(k, m_k), \\
\Gamma_{11}^1(k, m_k) &\triangleq \begin{bmatrix} \Gamma_{11}^{11}(k, m_k) & 0 & * \\ 0 & -\Phi(k, m_k) & * \\ \frac{\varepsilon_1(\bar{L}_1 + \bar{L}_2)}{2} & 0 & -\varepsilon_1 I \end{bmatrix}, \\
\Gamma_{11}^2(k, m_k) &\triangleq \begin{bmatrix} -\gamma^2 I & * \\ 0 & -\gamma^2 I \end{bmatrix}, \\
\Gamma_{22}(k, m_k) &\triangleq \text{diag}\{-I, I_{2r+2} \otimes \Psi_\Phi(k+1, m_k)\},
\end{aligned} \tag{6.29}$$

$$\Gamma_{21}(k,m_k) \triangleq \begin{bmatrix} \mathcal{M}(k) & 0 & 0 \\ \mathcal{A}_1(k,m_k) & \bar{\beta}I & \mathcal{D}(k,m_k) \\ 0 & \tilde{\beta}I_1 & 0 \\ \Gamma_{2241}(k,m_k) & 0 & 0 \\ \Gamma_{2251}(k,m_k) & 0 & 0 \end{bmatrix},$$

$$\Gamma_{2241}(k,m_k) \triangleq \Omega_{2241}(k,m_k), \quad \Gamma_{2251}(k,m_k) \triangleq \Omega_{2251}(k,m_k),$$
$$\Psi_{\Phi}(k+1,m_k) \triangleq \mathrm{diag}\{-\check{\Psi}(k+1,m_k), -\check{\Phi}(k+1,m_k)\}.$$

Proof *Suppose that the following variables can be decomposed:*

$$\begin{aligned} P(k,m_k) &= \mathrm{diag}\{\Psi(k,m_k), \Phi(k,m_k)\}, \\ \tilde{P}(k+1,m_k) &= \mathrm{diag}\{\bar{\Psi}(k+1,m_k), \bar{\Phi}(k+1,m_k)\}, \\ \bar{\Phi}(k+1,m_k) &= \sum_{n_{k+1}=1}^{s} \lambda_{m_k n_{k+1}} \Phi(k+1, n_{k+1}|m_k), \\ \bar{\Psi}(k+1,m_k) &= \sum_{n_{k+1}=1}^{s} \lambda_{m_k n_{k+1}} \Psi(k+1, n_{k+1}|m_k), \\ \tilde{P}^{-1}(k+1,m_k) &= \mathrm{diag}\{\check{\Psi}(k+1,m_k), \check{\Phi}(k+1,m_k)\}. \end{aligned} \tag{6.30}$$

In terms of the aforementioned equalities and Lemma 2.1, we can easily obtain (6.28) by (6.25). Therefore, according to Theorem 6.2, the proof of this theorem is now complete.

By means of Theorem 6.3, we can summarize the Distributed State Estimation Design (*DSED*) algorithm as follows.

Algorithm *DSED*	
Step 1.	Set the average H_∞ performance index γ, the matrix $R(l_0) > 0$, the initial states $x(0)$ and $\hat{x}_i(0)$. Determine the initial value for the matrices $\Psi(0,l_0)$ and $\Phi(0,l_0)$ according to the initial condition (6.27) and set $k = 0$.
Step 2.	Compute the positive definite matrices $\check{\Psi}(k,m_k)$ and $\check{\Phi}(k,m_k)$, the matrices $K_{ij}^{(l)}(k)$ and $H_{ij}^{(l)}(k)$ at the sampling instant k by solving the RLMIs (6.28).
Step 3.	Set $k = k+1$ and acquire $\Psi(k,m_k)$ and $\Phi(k,m_k)$ by the formula (6.29). If $k < N$, then go to Step 2, or go to Step 4.
Step 4.	Stop.

Remark 6.3 *In Theorem 6.3, the state estimator gains have been presented by calculating the proposed recursive linear matrix inequalities (6.28). It is*

worthy to note that, the main novelties of this section lie in (1) the redundant channels and switching topologies schemes are simultaneously introduced in the sensor network for the first time and (2) the distributed estimation framework is established to deal with the measurements from the coupling sensor network.

6.2 Non-Fragile Distributed Fault Estimation: the Finite-Horizon Case

In this section, we discuss the non-fragile distributed fault estimation problem for a class of time-varying systems subject to RONs and ROGVs.

6.2.1 Problem Formulation

For the sensor network considered in this section, the sensor nodes are deployed according to a fixed network topology represented by a directed graph $\mathcal{G} = (\mathcal{V}, \mathcal{E}, \mathcal{A})$ of order n with the set of nodes $\mathcal{V} = 1, 2, \ldots, n$, the set of edges $\mathcal{E} \in \mathcal{V} \times \mathcal{V}$, and the weighted adjacency matrix $\mathcal{A} = [a_{ij}]$ having nonnegative adjacency element a_{ij}. Every edge of $\mathcal{G}$ is denoted by an ordered pair (i, j). The adjacency elements associated with the edges of the graph are positive, i.e., $a_{ij} > 0 \Longleftrightarrow (i, j) \in \mathcal{E}$, which means that sensor i can obtain information from sensor j. Also, we assume that $a_{ii} = 1$ for all $i \in \mathcal{V}$, and therefore (i, i) can be regarded as an additional edge. The set of neighbours of node $i \in \mathcal{V}$ plus the node itself are denoted by $\mathcal{N}_i = \{j \in \mathcal{V} : (i, j) \in \mathcal{E}\}$.

Consider a target plant described by the following discrete-time nonlinear time-varying stochastic system on a finite-horizon $k \in [0, N]$:

$$x(k+1) = A(k)x(k) + \alpha(k)h(x(k)) + D(k)w(k) + G(k)f(k) \tag{6.31}$$

where $x(k) \in \mathbb{R}^{n_x}$ is the system state, $w(k) \in \mathbb{R}^{n_v}$ is the disturbance input belonging to $l_2[0, N)$, and $f(k) \in \mathbb{R}^l$ is the fault to be detected. For the ith sensor, the model is described by

$$y_i(k) = C_i(k)x(k) + E_i(k)v(k) + H_i(k)f(k), \quad i = 1, 2, \ldots, n \tag{6.32}$$

where $y_i(k) \in \mathbb{R}^{n_y}$ is the output measured by sensor i and $v(k) \in l_2[0, N)$ is an external disturbance. Moreover, all the matrices mentioned above (i.e., $A(k), D(k), G(k)$, $C_i(k)$, $E_i(k)$ and $H_i(k)$) are known, real, time-varying matrices with appropriate dimensions.

The stochastic variable $\alpha(k) \in \mathbb{R}$, which accounts for the phenomenon of RONs, is a Bernoulli-distributed white sequence with its probability distribution given as follows:

$$\text{Prob}\{\alpha(k) = 1\} = \bar{\alpha}, \qquad \text{Prob}\{\alpha(k) = 0\} = 1 - \bar{\alpha} \tag{6.33}$$

where $0 \leq \bar{\alpha} \leq 1$ is a known constant which can be obtained by statistical experiments.

The nonlinear vector-valued function $h : \mathbb{R}^{n_x} \rightarrow \mathbb{R}^{n_x}$ with $h(0) = 0$ is supposed to be continuous and satisfies the following sector-bounded condition:

$$[h(x) - h(y) - \Psi(x-y)]^T[h(x) - h(y) - \Omega(x-y)] \leq 0 \tag{6.34}$$

for all $x, y \in \mathbb{R}^{n_x}$, where Ψ and Ω are real matrices of appropriate dimensions.

In this section, the following estimator structure is adopted on sensor node i:

$$\begin{cases} \hat{x}_i(k+1) = \sum\limits_{j\in\mathcal{N}_i} a_{ij}(K_{ij}(k) + \sigma_{1k}\Delta K_{ij}(k))\hat{x}_i(k) \\ \qquad\qquad + \sum\limits_{j\in\mathcal{N}_i} a_{ij}(H_{ij}(k) + \sigma_{2k}\Delta H_{ij}(k)) \times y_i(k) \\ r_i(k) = \sum\limits_{j\in\mathcal{N}_i} a_{ij}L_{ij}(k)(y_i(k) - C_i(k)\hat{x}_i(k)) \end{cases}$$

where $\hat{x}_i(k) \in \mathbb{R}^{n_x}$ is the state estimate on sensor node i and $r_i(k) \in \mathbb{R}^l$ is the so-called residual that is compatible with the fault vector $f(k)$. The matrices $K_{ij}(k)$, $H_{ij}(k)$ and $L_{ij}(k)$ are estimator parameters on node i to be determined. The stochastic variables σ_{1k} and σ_{2k}, which govern the occurrence of the estimator gain variations, have the expectations $\bar{\sigma}_{1k}$, $\bar{\sigma}_{2k}$ and variances $\tilde{\sigma}_{1k}^2$, $\tilde{\sigma}_{2k}^2$, respectively. $\Delta K_{ij}(k)$ and $\Delta H_{ij}(k)$ quantify the estimator gain variations and are assumed to the following multiplicative forms:

$$\Delta K_{ij}(k) = K_{ij}(k)H_aF_a(k)E_a, \; \Delta H_{ij}(k) = H_{ij}(k)H_bF_b(k)E_b \tag{6.35}$$

where H_a, H_b, E_a and E_b are known matrices with appropriate dimensions, and $F_a(k)$, $F_b(k)$ are unknown matrices satisfying $F_a^T(k)F_a(k) \leq I$, $F_b^T(k)F_b(k) \leq I$.

Remark 6.4 *Due to a variety of reasons such as rounding errors and finite precision in engineering practice, designed estimators may not be implemented accurately and the parameter fluctuations are often unavoidable. As such, the norm-bounded implementation errors described by $\Delta K_{ij}(k)$ and $\Delta H_{ij}(k)$ are utilized to cater for the gain variations in this section.*

For convenience of later analysis, we denote

$$\begin{aligned} &\bar{K}(k) \triangleq [\bar{K}_{ij}(k)]_{n\times n} \quad \text{with} \quad \bar{K}_{ij}(k) \triangleq \begin{cases} a_{ij}K_{ij}(k), & i = 1, 2, \ldots, n; \; j \in \mathcal{N}_i \\ 0, & i = 1, 2, \ldots, n; \; j \notin \mathcal{N}_i \end{cases} \\ &\bar{H}(k) \triangleq [\bar{H}_{ij}(k)]_{n\times n} \quad \text{with} \quad \bar{H}_{ij}(k) \triangleq \begin{cases} a_{ij}H_{ij}(k), & i = 1, 2, \ldots, n; \; j \in \mathcal{N}_i \\ 0, & i = 1, 2, \ldots, n; \; j \notin \mathcal{N}_i \end{cases} \\ &\bar{L}(k) \triangleq [\bar{L}_{ij}(k)]_{n\times n} \quad \text{with} \quad \bar{L}_{ij}(k) \triangleq \begin{cases} a_{ij}L_{ij}(k), & i = 1, 2, \ldots, n; \; j \in \mathcal{N}_i \\ 0, & i = 1, 2, \ldots, n; \; j \notin \mathcal{N}_i. \end{cases} \end{aligned} \tag{6.36}$$

Obviously, since $a_{ij} = 0$ when $j \notin \mathcal{N}_i$, $\bar{K}(k)$, $\bar{H}(k)$ and $\bar{L}(k)$ are sparse matrices which can be expressed as

$$\bar{K}(k) \in \mathscr{T}_{n_x \times n_x}, \quad \bar{H}(k) \in \mathscr{T}_{n_x \times n_y}, \quad \bar{L}(k) \in \mathscr{T}_{l \times n_y} \tag{6.37}$$

where $\mathscr{T}_{p\times q} \triangleq \{\mathcal{T} = [T_{ij}] \in \mathbb{R}^{np\times nq} | T_{ij} \in \mathbb{R}^{p\times q}, T_{ij} = 0 \text{ if } j \notin \mathcal{N}_i\}$.

Letting $\eta(k) \triangleq \begin{bmatrix}\bar{x}^T(k) & \hat{x}^T(k)\end{bmatrix}^T$, $\xi(k) \triangleq [\bar{\omega}^T(k) \quad \bar{v}^T(k) \quad \bar{f}^T(k)]]^T$, $\tilde{r}_i(k) \triangleq r_i(k) - f(k)$, $\tilde{r}(k) \triangleq \begin{bmatrix}\tilde{r}_1^T(k) & \cdots & \tilde{r}_n^T(k)\end{bmatrix}^T$, $\bar{\omega}(k) \triangleq \mathbf{1}_n \otimes \omega(k)$, $\bar{v}(k) \triangleq \mathbf{1}_n \otimes v(k)$, $\bar{f}(k) \triangleq \mathbf{1}_n \otimes f(k)$, $\bar{x}(k) \triangleq \mathbf{1}_n \otimes x(k)$ and $\hat{x}(k) \triangleq \begin{bmatrix}\hat{x}_1^T(k) & \cdots & \hat{x}_n^T(k)\end{bmatrix}^T$, we obtain the following augmented system:

$$\begin{cases} \eta(k+1) = (A_{11}(k) + \bar{\sigma}_{2k}A_{12}(k) + \bar{\sigma}_{1k}A_{13}(k))\eta(k) \\ \qquad\qquad + (\tilde{\sigma}_{2k}A_{12}(k) + \tilde{\sigma}_{1k}A_{13}(k))\eta(k) + (B_{11}(k) + \bar{\sigma}_{2k}B_{12}(k))\xi(k) \\ \qquad\qquad + \tilde{\sigma}_{2k}B_{12}\xi(k) + (\bar{\alpha}I_1 + \tilde{\alpha}I_1)\bar{h}(I_2\eta(k)) \\ \tilde{r}(k) = D_1\eta(k) + D_2\xi(k) \end{cases}$$

where

$$\begin{aligned}
&A_{11}(k) \triangleq \begin{bmatrix}\bar{A}(k) & 0 \\ \bar{H}(k)\bar{C}(k) & \bar{K}(k)\end{bmatrix}, \ A_{12}(k) \triangleq \begin{bmatrix}0 & 0 \\ \Delta\bar{H}(k)\bar{C}(k) & 0\end{bmatrix}, \\
&A_{13}(k) \triangleq \begin{bmatrix}0 & 0 \\ 0 & \Delta\bar{K}(k)\end{bmatrix}, \ B_{11}(k) \triangleq \begin{bmatrix}\bar{D}(k) & 0 & \bar{G}(k) \\ 0 & \bar{H}(k)\bar{E}(k) & \bar{H}(k)\tilde{H}(k)\end{bmatrix}, \\
&D_1(k) \triangleq \begin{bmatrix}\bar{L}(k)\bar{C}(k) & -\bar{L}(k)\bar{C}(k)\end{bmatrix}, \ \bar{C}(k) \triangleq \mathrm{diag}\{C_1(k), C_2(k), \ldots, C_n(k)\}, \\
&B_{12}(k) \triangleq \begin{bmatrix}0 & 0 & 0 \\ 0 & \Delta\bar{H}(k)\bar{E}(k) & \Delta\bar{H}(k)\tilde{H}(k)\end{bmatrix}, \ \bar{D}(k) \triangleq I_n \otimes D(k), \\
&D_2(k) \triangleq \begin{bmatrix}0 & \bar{L}(k)\bar{E}(k) & \bar{L}(k)\tilde{H}(k) - I_n\end{bmatrix}, \ \bar{\Psi} \triangleq I_n \otimes \Psi, \ \bar{\Omega} \triangleq I_n \otimes \Omega, \\
&\bar{E}(k) \triangleq \mathrm{diag}\{E_1(k), E_2(k), \ldots, E_n(k)\}, \ \bar{G}(k) \triangleq I_n \otimes G(k), \ \tilde{\alpha} \triangleq \alpha(k) - \bar{\alpha}, \\
&\tilde{H}(k) \triangleq \mathrm{diag}\{H_1(k), H_2(k), \ldots, H_n(k)\}, \ \bar{h}(I_2\eta(k)) \triangleq \mathbf{1}_n \otimes h(x(k)), \\
&\bar{A}(k) \triangleq I_n \otimes A(k).
\end{aligned} \tag{6.38}$$

Furthermore, it follows from (6.34) that

$$[\bar{h}(I_2\eta(k)) - \bar{\Psi}I_2\eta(k)]^T[\bar{h}(I_2\eta(k)) - \bar{\Omega}I_2\eta(k)] < 0. \tag{6.39}$$

Let the disturbance attenuation level $\gamma > 0$ and all nonzero $w(k)$ and $v(k)$ be given. The objective of this section is to design the parameters ($K_{ij}(k)$, $H_{ij}(k)$ and $L_{ij}(k)$) for each non-fragile fault estimator (6.35) such that, under zero initial conditions, the fault estimation error $\tilde{r}(k)$ in (6.38) satisfies the following condition:

$$J \triangleq \frac{1}{n}\mathbb{E}\left\{\sum_{k=0}^{N-1} \|\tilde{r}(k)^2\|\right\} - \gamma^2\left\{\sum_{k=0}^{N-1} \|\xi(k)\|^2 + \bar{e}^T(0)R\bar{e}(0)\right\} \leq 0 \tag{6.40}$$

where

$$\bar{e}(0) \triangleq \bar{x}(0) - \hat{x}(0), \quad R \triangleq \mathrm{diag}\{S_1, S_2, \ldots, S_n\}. \tag{6.41}$$

6.2.2 Main Results

In this subsection, we will investigate the distributed fault estimation problem for the sensor network whose topology is given by the graph $\mathcal{G} = (\mathcal{V}, \mathcal{E}, \mathcal{A})$.

The following lemma will be needed in establishing our main results.

Lemma 6.1 *[147] Let $P \triangleq diag\{P_1, P_2, \ldots, P_n\}$ with $P_i \in \mathbb{R}^{p\times p}$ $(1 \le i \le n)$ being invertible matrices. If $X = PW$ for $W \in \mathbb{R}^{np\times nq}$, then we have $W \in \mathscr{T}_{p\times q} \Longleftrightarrow X \in \mathscr{T}_{p\times q}$.*

We are now in a position to provide our main results.

Theorem 6.4 *Assume that the fault estimation parameters $K_{ij}(k)$, $H_{ij}(k)$ and $L_{ij}(k)$ in (6.35) are given. For a positive scalar $\gamma > 0$ and a sequence of positive definite matrices $S_i > 0$ $(i = 1, 2, \ldots, n)$, the average $\mathcal{H}_\infty$ performance requirement defined in (6.40) is achieved for all nonzero $\xi(k)$ if there exists a sequence of positive definite matrices $\{P(k)\}_{0\le k\le N+1}$ satisfying the following recursive matrix inequalities:*

$$\Pi(k) = \begin{bmatrix} \Pi_{11}(k) & * & * \\ \Pi_{21}(k) & \Pi_{22}(k) & * \\ \Pi_{31}(k) & \Pi_{32}(k) & \Pi_{33}(k) \end{bmatrix} < 0, \quad \text{for } 0 \le k \le N \tag{6.42}$$

with the initial condition $\eta^T(0)P(0)\eta(0) \le \gamma^2\bar{e}^T(0)R\bar{e}(0)$, where

$$\begin{aligned}
\Pi_{11}(k) \triangleq\ & (A_{11}(k) + \bar{\sigma}_{2k}A_{12}(k) + \bar{\sigma}_{1k}A_{13}(k))^T P(k+1)(A_{11}(k) \\
& + \bar{\sigma}_{2k}A_{12}(k) + \bar{\sigma}_{1k}A_{13}(k)) + \frac{1}{n}D_1^T(k)D_1(k) + \tilde{\sigma}_{2k}^2 A_{12}^T(k) \\
& \times P(k+1)A_{12}(k) + \tilde{\sigma}_{1k}^2 A_{13}^T(k)P(k+1)A_{13}(k) \\
& - \frac{1}{2}\lambda I_2^T(\bar{\Psi}^T\bar{\Omega} + \bar{\Omega}^T\bar{\Psi})I_2 - P(k), \\
\Pi_{21}(k) \triangleq\ & (B_{11}(k) + \bar{\sigma}_{2k}B_{12}(k))^T P(k+1)(A_{11}(k) + \bar{\sigma}_{2k}A_{12}(k) \\
& + \bar{\sigma}_{1k}A_{13}(k)) + \frac{1}{n}D_2^T(k)D_1(k) + \tilde{\sigma}_{2k}^2 B_{12}^T(k)P(k+1)A_{12}(k), \\
\Pi_{22}(k) \triangleq\ & (B_{11}(k) + \bar{\sigma}_{2k}B_{12}(k))^T P(k+1)(B_{11}(k) + \bar{\sigma}_{2k}B_{12}(k)) \\
& + \tilde{\sigma}_{2k}^2 B_{12}^T(k)P(k+1)B_{12}(k) + \frac{1}{n}D_2^T(k)D_2(k) - \gamma^2 I, \\
\Pi_{31}(k) \triangleq\ & \bar{\alpha}I_1^T P(k+1)(A_{11}(k) + \bar{\sigma}_{2k}A_{12}(k) + \bar{\sigma}_{1k}A_{13}(k)) + \frac{1}{2}\lambda(\bar{\Omega} + \bar{\Psi})I_2, \\
I_2 \triangleq\ & \begin{bmatrix} I & 0 \end{bmatrix}, \quad \Pi_{32}(k) \triangleq \bar{\alpha}I_1^T P(k+1)(B_{11}(k) + \bar{\sigma}_{2k}B_{12}(k)), \\
I_1 \triangleq\ & \begin{bmatrix} I & 0 \end{bmatrix}^T, \quad \Pi_{33}(k) \triangleq (\bar{\alpha}^2 + g)I_1^T P(k+1)I_1 - \lambda I, \quad g \triangleq \bar{\alpha}(1-\bar{\alpha}).
\end{aligned} \tag{6.43}$$

Proof *Define*

$$J(k) \triangleq \eta^T(k+1)P(k+1)\eta(k+1) - \eta^T(k)P(k)\eta(k). \tag{6.44}$$

It follows from (6.38) that

$$
\begin{aligned}
&\mathbb{E}\{J(k)\} \\
=&\mathbb{E}\left\{[(A_{11}(k)+\bar{\sigma}_{2k}A_{12}(k)+\bar{\sigma}_{1k}A_{13}(k))\eta(k)+(B_{11}(k)+\bar{\sigma}_{2k}B_{12}(k))\xi(k)\right. \\
&+\bar{\alpha}I_1\bar{h}(I_2\eta(k))]^T\ P(k+1)\left[(A_{11}(k)+\bar{\sigma}_{2k}A_{12}(k)+\bar{\sigma}_{1k}A_{13}(k))\eta(k)\right. \\
&\left.+(B_{11}(k)+\bar{\sigma}_{2k}B_{12}(k))\xi(k)+\bar{\alpha}I_1\bar{h}(I_2\eta(k))\right]+\tilde{\sigma}_{2k}^2\eta^T(k)A_{12}^T(k)P(k+1) \\
&\times A_{12}(k)\eta(k)+\tilde{\sigma}_{1k}^2\eta^T(k)A_{13}^T(k)P(k+1)A_{13}(k)\eta(k) \qquad (6.45)\\
&+\tilde{\sigma}_{2k}^2\xi^T(k)B_{12}^T(k)P(k+1)B_{12}(k)\xi(k)+\tilde{\alpha}^2\bar{h}^T(I_2\eta(k))I_1^T \\
&\times P(k+1)I_1\bar{h}(I_2\eta(k))+\tilde{\sigma}_{2k}^2\eta^T(k)A_{12}^T(k)P(k+1) \\
&\left.\times B_{12}(k)\xi(k)+\tilde{\sigma}_{2k}^2\xi^T(k)B_{12}^T(k)P(k+1)A_{12}(k)\eta(k)\ -\eta^T(k)P(k)\eta(k)\right\}.
\end{aligned}
$$

Adding the zero term $\frac{1}{n}\mathbb{E}\{\|\tilde{r}(k)\|^2\}-\gamma^2\|\xi(k)\|^2-\frac{1}{n}\mathbb{E}\left\{\|\tilde{r}(k)\|^2\right\}+\gamma^2\|\xi(k)\|^2$ *to* $\mathbb{E}\{J(k)\}$ *results in*

$$\mathbb{E}\left\{J(k)\right\}=\mathbb{E}\left\{\tilde{\eta}^T(k)\bar{\Pi}(k)\tilde{\eta}(k)-\frac{1}{n}\mathbb{E}\{\|\tilde{r}(k)\|^2\}+\gamma^2\|\xi(k)\|^2\right\} \qquad (6.46)$$

where

$$
\begin{aligned}
\bar{\Pi}(k) &\triangleq \begin{bmatrix} \Pi_{11}(k)+\frac{1}{2}\lambda I_2^T(\bar{\Psi}^T\bar{\Omega}+\bar{\Omega}^T\bar{\Psi})I_2 & * & * \\ \Pi_{21}(k) & \Pi_{22}(k) & * \\ \Pi_{31}(k)-\frac{1}{2}\lambda(\bar{\Omega}+\bar{\Psi})I_2 & \Pi_{32}(k) & \Pi_{33}(k)+\lambda I \end{bmatrix}, \qquad (6.47)\\
\tilde{\eta}(k) &\triangleq \begin{bmatrix}\eta^T(k) & \xi^T(k) & \bar{h}^T(I_2\eta(k))\end{bmatrix}^T.
\end{aligned}
$$

Moreover, it follows from the condition (6.39) that

$$\mathbb{E}\{J(k)\}\leq\mathbb{E}\left\{\tilde{\eta}^T(k)\Pi(k)\tilde{\eta}(k)-\frac{1}{n}\mathbb{E}\left\{\|\tilde{r}(k)\|^2\right\}+\gamma^2\|\xi(k)\|^2\right\}. \qquad (6.48)$$

Summing up (6.48) on both sides from 0 *to* $N-1$ *with respect to* k*, we obtain*

$$
\begin{aligned}
\sum_{k=0}^{N-1}\mathbb{E}\Big\{J(k)\Big\}\leq&\mathbb{E}\Big\{\sum_{k=0}^{N-1}\tilde{\eta}^T(k)\Pi(k)\tilde{\eta}(k)\Big\} \\
&-\left(\frac{1}{n}\mathbb{E}\{\sum_{k=0}^{N-1}\|\tilde{r}(k)\|^2\}-\gamma^2\sum_{k=0}^{N-1}\|\xi(k)\|^2\right). \qquad (6.49)
\end{aligned}
$$

Hence, the average $\mathcal{H}_\infty$ *performance index defined in (6.40) is given by*

$$
\begin{aligned}
J<&\mathbb{E}\Big\{\sum_{k=0}^{N-1}\tilde{\eta}^T(k)\Pi(k)\tilde{\eta}(k)\Big\}-\mathbb{E}\left\{\eta^T(N)P(N)\eta(N)\right\}+\eta^T(0)P(0)\eta(0) \\
&-\gamma^2\bar{e}^T(0)R\bar{e}(0). \qquad (6.50)
\end{aligned}
$$

Noting that $P(N) > 0$ and $\eta^T(0)P(0)\eta(0) \leq \gamma^2\bar{e}^T(0)R\bar{e}(0)$, we have $J < 0$ when (6.42) holds, which ends the proof.

After the performance analysis of the fault estimator in Theorem 6.4, we are now ready to design the parameters for the distributed fault estimators. The solutions of the distributed fault estimator design problem over sensor networks are provided in the following theorem.

Theorem 6.5 *Given a positive scalar $\gamma > 0$ and positive definite matrices $S_i = S_i^T > 0$ $(1 \leq i \leq n)$. For the target plant (6.31) with randomly varying nonlinearities and sensor network (6.32), the fault estimator design problem is solved if there exist positive definite matrices $P_i(k)_{0\leq k\leq N+1}$ $(1 \leq i \leq n)$, a family of matrices $Q(k)_{0\leq k\leq N} \in \mathscr{T}_{2nn_x\times n(n_x+n_y)}$, $\bar{L}(k)_{0\leq k\leq N} \in \mathscr{T}_{nn_y\times nn_y}$ and a positive constant scalar μ satisfying the initial condition*

$$\eta^T(0)P(0)\eta(0) \leqslant \gamma^2\bar{e}^T(0)R\bar{e}(0) \tag{6.51}$$

and the recursive linear matrix inequalities

$$\Sigma(k) = \begin{bmatrix} \Sigma_{11}(k) & * & * & * \\ \Sigma_{21}(k) & \Sigma_{22}(k) & * & * \\ 0 & \Sigma_{32}^T(k) & -\mu I & * \\ \mu\Sigma_{41}(k) & 0 & 0 & -\mu I \end{bmatrix} < 0 \tag{6.52}$$

for all $0 \leq k \leq N$, where

$$\Sigma_{11}(k) \triangleq \begin{bmatrix} -P(k) - \frac{1}{2}\lambda I_2^T(\bar{\Psi}^T\bar{\Omega} + \bar{\Omega}^T\bar{\Psi})I_2 & * & * \\ 0 & -\gamma^2 I & * \\ \frac{1}{2}\lambda(\bar{\Omega} + \bar{\Psi})I_2 & 0 & -\lambda I \end{bmatrix},$$

$$\Sigma_{41}(k) \triangleq \begin{bmatrix} \bar{E}_b\hat{C}(k) & 0 & 0 \\ 0 & \bar{E}_b\hat{\xi}(k) & 0 \\ \bar{E}_a I_4 & 0 & 0 \end{bmatrix},$$

$$\Sigma_{21}(k) \triangleq \begin{bmatrix} \sqrt{\frac{1}{n}}\bar{L}(k)\bar{C}_1(k) & \sqrt{\frac{1}{n}}(\bar{L}(k)\hat{\xi}(k) - \hat{\xi}_1) & 0 \\ 0 & 0 & P(k+1)\sqrt{g}I_1 \\ P(k+1)A_{110}(k) + Q(k)R_0(k) & P(k+1)B_{110}(k) + Q(k)R_1(k) & P(k+1)\bar{\alpha}I_1 \\ 0 & 0 & 0 \\ 0 & 0 & 0 \end{bmatrix},$$

$$\Sigma_{22}(k) \triangleq \mathrm{diag}\{-I, -P(k+1), -P(k+1), -P(k+1), -P(k+1)\},$$

$$\Sigma_{32}(k) \triangleq \begin{bmatrix} 0 & 0 & 0 \\ 0 & 0 & 0 \\ \bar{\sigma}_{2k}Q(k)I_5\bar{H}_b & \bar{\sigma}_{2k}Q(k)I_5\bar{H}_b & \bar{\sigma}_{1k}Q(k)I_3\bar{H}_a \\ \tilde{\sigma}_{2k}Q(k)I_5\bar{H}_b & \tilde{\sigma}_{2k}Q(k)I_5\bar{H}_b & 0 \\ 0 & 0 & \tilde{\sigma}_{1k}Q(k)I_3\bar{H}_a, \end{bmatrix}, \tag{6.53}$$

$$A_{110}(k) \triangleq \begin{bmatrix} \bar{A}(k) & 0 \\ 0 & 0 \end{bmatrix}, \quad R_1(k) \triangleq \begin{bmatrix} 0 & \bar{E}(k) & \tilde{H}(k) \\ 0 & 0 & 0 \end{bmatrix},$$

$$I_3 \triangleq \begin{bmatrix} 0 & I \end{bmatrix}^T, \quad \hat{C}(k) \triangleq \begin{bmatrix} \bar{C}(k) & 0 \end{bmatrix}, \quad I_4 \triangleq \begin{bmatrix} 0 & I \end{bmatrix},$$

$$\hat{\xi}(k) \triangleq \begin{bmatrix} 0 & \bar{E}(k) & \tilde{H}(k) \end{bmatrix}, \quad \bar{C}_1(k) \triangleq \begin{bmatrix} \bar{C}(k) & -\bar{C}(k) \end{bmatrix},$$

$$R_0(k) \triangleq \begin{bmatrix} \bar{C}(k) & 0 \\ 0 & I \end{bmatrix}, \quad B_{110}(k) \triangleq \begin{bmatrix} \bar{D}(k) & 0 & \bar{G}(k) \\ 0 & 0 & 0 \end{bmatrix},$$

$$\bar{H}_b \triangleq I_n \otimes H_b, \quad \bar{E}_a \triangleq I_n \otimes E_a, \quad \bar{E}_b \triangleq I_n \otimes E_b,$$

$$\bar{H}_a \triangleq I_n \otimes H_a, \quad I_5 \triangleq \begin{bmatrix} I & 0 \end{bmatrix}^T, \quad \hat{\xi}_1 \triangleq \begin{bmatrix} 0 & 0 & I \end{bmatrix}.$$

Moreover, if the above inequalities are feasible, the matrices $\bar{H}(k)$ and $\bar{K}(k)$ are given as follows:

$$\begin{bmatrix} \bar{H}(k) & \bar{K}(k) \end{bmatrix} = (I_3^T P(k+1) I_3)^{-1} I_3^T Q(k). \tag{6.54}$$

Accordingly, the desired filter parameters K_{ij} and H_{ij} $(i = 1, 2, \ldots, n;\ j \in \mathcal{N}_i)$ can be obtained from (6.36).

Proof *In order to reduce unnecessary conservatism, we rewrite the parameters in Theorem 6.4 in the following form:*

$$A_{11}(k) = A_{110}(k) + I_3X(k)R_0(k), \quad B_{11}(k) = B_{110}(k) + I_3X(k)R_1(k) \tag{6.55}$$

where

$$X(k) \triangleq \begin{bmatrix} \bar{H}(k) & \bar{K}(k) \end{bmatrix}. \tag{6.56}$$

According to the Lemma 2.1, the inequality (6.42) holds if and only if

$$\Gamma_1(k) = \begin{bmatrix} \Sigma_{11}(k) & * \\ \Gamma(k) & \Sigma_{22}^{-1}(k) \end{bmatrix} < 0 \tag{6.57}$$

where

$$\Gamma(k) \triangleq \begin{bmatrix} \sqrt{\frac{1}{n}}D_1(k) & \sqrt{\frac{1}{n}}(\bar{L}(k)\hat{\xi}(k) - \hat{\xi}_1(k)) & 0 \\ 0 & 0 & \sqrt{g}I_1 \\ A_{11}(k) + \bar{\sigma}_{2k}A_{12}(k) + \bar{\sigma}_{1k}A_{13}(k) & B_{11}(k) + \bar{\sigma}_{2k}B_{12}(k) & \bar{\alpha}I_1 \\ \tilde{\sigma}_{2k}A_{12}(k) & \tilde{\sigma}_{2k}B_{12}(k) & 0 \\ \tilde{\sigma}_{1k}A_{13}(k) & 0 & 0 \end{bmatrix}. \tag{6.58}$$

Then, by using certain congruence transformation, the above inequality holds if and only if the following is true:

$$\Sigma_0(k) + \Sigma_{320}(k)F(k)\Sigma_{410}(k) + \Sigma_{410}^T(k)F^T(k)\Sigma_{320}^T(k) < 0 \tag{6.59}$$

where

$$\begin{aligned}
&\Sigma_0(k) \triangleq \begin{bmatrix} \Sigma_{11}(k) & * \\ \Sigma_{21}(k) & \Sigma_{22}(k) \end{bmatrix}, \quad \Sigma_{320}(k) \triangleq \begin{bmatrix} 0 & \Sigma_{32}^T(k) \end{bmatrix}^T, \\
&\Sigma_{410}(k) \triangleq \begin{bmatrix} \Sigma_{41}(k) & 0 \end{bmatrix}, \quad \bar{F}_b(k) \triangleq I_n \otimes F_b(k), \\
&\bar{F}_a(k) \triangleq I_n \otimes F_a(k), \quad F(k) \triangleq \text{diag}\{\bar{F}_b(k), \bar{F}_b(k), \bar{F}_a(k)\}.
\end{aligned} \tag{6.60}$$

Let $P(k) \triangleq \text{diag}\{P_1(k), P_2(k), \ldots, P_n(k)\}$. *Noticing* $P(k+1)I_3X(k) = Q(k)$, *from Lemma 6.1, it is easy to verify that the sparsity is maintained for* $X(k)$. *Applying Lemma 3.1 and after some straightforward algebraic manipulations, it can be shown that (6.52) holds if and only if (6.59) is true, and the proof of this theorem follows readily from that of Theorem 6.4.*

By means of Theorem 6.5, we can summarise the Non-fragile Fault Estimator Design ($NFED$) algorithm as follows.

Algorithm $NFED$

Step 1. Given the average H_∞ performance index γ, the positive definite matrix R, the initial conditions $x(0)$, $\hat{x}_i(0)$ and $y_i(0)$, select the initial value for matrix $P(0)$ which satisfies the initial condition (6.51) and set $k = 0$.

Step 2. Obtain the positive definite matrix $P(k+1)$ and matrix $Q(k)$ for the sampling instant k by solving the recursive linear matrix inequalities (6.52).

Step 3. Derive the distributed estimator parameter matrices $\bar{K}(k)$ and $\bar{H}(k)$ by solving (6.54). Accordingly, the desired fault estimator parameters $K_{ij}(k)$, $H_{ij}(k)$ and $L_{ij}(k)$ $(i = 1, 2, \ldots, n, j \in \mathcal{N}_i)$ can be obtained from (6.36).

Step 4. If $k < N$, then go to Step 2, else go to Step 5.

Step 5. Stop.

Remark 6.5 *In Theorem 6.5, a recursive algorithm is proposed for the recursive design of the fault estimator parameters, and the main results are quite comprehensive that covers the following phenomena frequently encountered in engineering practice: (1) randomly occurring nonlinearities in the target plant; (2) coupling relationship between the sensor nodes* *subject to a given topology; and (3) random gain variations involved in the estimator structures of the individual sensors. Furthermore, the average H_∞ performance is achieved that ensures the disturbance rejection capability over a finite time-horizon. The non-fragile distributed fault estimators are designed by resorting to recursive linear matrix inequalities that can be solved by standard software package.*

6.3 Distributed Filtering with RSTs under the RR Protocol

6.3.1 Problem Formulation

In this section, we suppose that the networked system contains n sensor nodes. Besides, all the sensors are distributed in space in the light of 2 switching topologies denoted by directed graphs $\mathbb{G}^{(r)} = (\mathbb{V}^{(r)}, \mathbb{E}^{(r)}, \mathbb{A}^{(r)})$ $(r = 1, 2)$, in which the number of nodes $\mathbb{V}^{(r)}$ is defined by $\{1, 2, \ldots, n\}$, the number of edges is described as $\mathbb{E}^{(r)} \subseteq \mathbb{V}^{(r)} \times \mathbb{V}^{(r)}$, and $\mathbb{A}^{(r)} = [a_{ij}^{(r)}]$ $(r = 1, 2)$ denotes the weighted adjacency matrices with nonnegative elements $a_{ij}^{(r)}$, and r is the number of weighted adjacency matrices. (i, j) stands for an edge of $\mathbb{G}^{(r)}$. Defining that all the elements in the adjacency matrices are positive, i.e., $a_{ij}^{(r)} > 0$ $(r = 1, 2)$, which shows that for topology r $(r = 1, 2)$, the information could be transmitted from sensor j to sensor i. Also, supposing that $a_{ii}^{(r)} = 1$ for all $i \in \mathbb{V}^{(r)}$ and $r = 1, 2$, and therefore (i, i) would be defined as a new edge for $\mathbb{E}^{(r)}$. The number of neighbours is described by $\mathbb{N}_i^{(r)} = \{j \in \mathbb{V}^{(r)} : (i, j) \in \mathbb{E}^{(r)}\}$, which contains the sensor i itself.

In this section, we define the time horizon as $[0, N-1] := \{0, 1, 2, \ldots, N-1\}$. Then, the target system on instant $u \in [0, N-1]$ is considered as follows:

$$\begin{cases} x(u+1) = A(u)x(u) + h(x(u)) + B(u)w(u) \\ \quad z(u) = L(u)x(u) \end{cases} \tag{6.61}$$

where the state vector is indicated by $x(u) \in \mathbb{R}^{n_x}$; $w(u) \in \mathbb{R}^{n_w}$ which is owned by $l_2[0, N-1]$ stands for the noise signal and the nonlinear function is defined by $h(x(u))$.

The nonlinear function $h : \mathbb{R}^{n_x} \to \mathbb{R}^{n_x}$ is a continuous function which satisfies the sector boundary condition shown as follows:

$$[h(x) - h(y) - M_1(x-y)]^T [h(x) - h(y) - M_2(x-y)] \leq 0 \tag{6.62}$$

where M_1 and M_2 are matrices that have proper dimensions.

The measurement output from the ith sensor node is expressed by

$$y_i(u) = C_i(u)x(u) + D_i(u)v(u) \tag{6.63}$$

where $y_i(u) \in \mathbb{R}^{n_y}$, and $v(u) \in l_2[0, N-1]$ denotes the external disturbance. Besides, $C_i(u)$, $B(u)$, $D_i(u)$, $A(u)$ and $L(u)$ are given matrices with homologous dimensions.

Remark 6.6 *Due to the limited bandwidth of network, data collisions occur inevitably in the process of data transmission. In order to resolve this problem, it is a common method that only one sensor has the access to the communication channel at the data delivery instant. The RR method would be suitably used to determine which sensor delivers the information via SNs at a certain instant in such a situation. Considering the characteristic that the RR method is a static protocol employed by sensor networks, every sensor has equal opportunity to transmit the information.*

Suppose that the ith sensor delivers the data for the first time at the transmission instant i. When all the sensors have been given the access to the network one after one in a loop, which means that a round is completed, then, the transmission sequence in the $(\gamma+1)$th $(\gamma = 0, 1, 2, \ldots)$ round can be described as

$$y_1(\gamma n+1), y_2(\gamma n+2), y_3(\gamma n+3), \ldots, y_n(\gamma n+n). \tag{6.64}$$

(6.64) describes that the output $y_i(u)$, from the node i, is delivered at u only when

$$\mathrm{mod}(u-i, n) = 0 \tag{6.65}$$

holds.

$\tilde{y}(u)$ is defined as $\begin{bmatrix} \tilde{y}_1^T(u) & \tilde{y}_2^T(u) & \cdots & \tilde{y}_n^T(u) \end{bmatrix}^T \in \mathbb{R}^{n_y n}$. According to the RR protocol, the measurement $\tilde{y}_i(u)$ $(i = 1, 2, \ldots, n)$ is updated as follows:

$$\tilde{y}_i(u) = \begin{cases} y_i(u), & \text{if } \mathrm{mod}(u-i, n) = 0 \text{ and } u > 0 \\ \tilde{y}_i(u-1), & \text{otherwise.} \end{cases} \tag{6.66}$$

In order to simplify the expression, we assume that $\tilde{y}_i(q) = 0$ for $q = -(n-1), -(n-2), \ldots, -1, 0$. The update matrix is defined as

$$\Gamma_i \triangleq \mathrm{diag}\{\vartheta(i-1)I, \vartheta(i-2)I, \ldots, \vartheta(i-n)I\} \tag{6.67}$$

where $\vartheta(\cdot) \in \{0, 1\}$ denotes the Kronecker delta function. In view of the communication process presented by (6.66), $\tilde{y}(u)$ can be described as follows:

$$\tilde{y}(u) = \Gamma_{\theta(u)} y(u) + (I - \Gamma_{\theta(u)})\tilde{y}(u-1) \tag{6.68}$$

where $\theta(u) \triangleq \mathrm{mod}(u-1, n) + 1$ stands for the selected node which can deliver data to the filter.

Let

$$\begin{aligned}
&\bar{x}(u) \triangleq \left[\tilde{x}^T(u) \quad \tilde{y}^T(u-1)\right]^T, \quad w_v(u) \triangleq \left[\bar{w}^T(u) \quad \bar{v}^T(u)\right]^T, \\
&\bar{w}(u) \triangleq \mathbf{1}_n \otimes w(u), \quad \bar{v}(u) \triangleq \mathbf{1}_n \otimes v(u), \\
&\tilde{x}(u) \triangleq \mathbf{1}_n \otimes x(u), \quad \bar{z}(u) \triangleq \mathbf{1}_n \otimes z(u), \\
&H(\bar{x}(u)) \triangleq \left[h^T(\tilde{x}(u)), 0\right]^T, \ h(\tilde{x}(u)) \triangleq \mathbf{1}_n \otimes h(x(u)).
\end{aligned}$$

Then the time-varying system is formulated as follows:

$$\begin{cases}
\bar{x}(u+1) = \mathbb{A}(u)\bar{x}(u) + H(\bar{x}(u)) + \mathbb{B}(u)w_v(u) \\
\tilde{y}(u) = \mathbb{C}_1(u)\bar{x}(u) + \mathbb{D}_1(u)w_v(u) \\
\bar{z}(u) = \bar{L}(u)\bar{x}(u)
\end{cases} \tag{6.69}$$

where

$$\begin{aligned}
&\mathbb{A}(u) \triangleq \begin{bmatrix} \bar{A}(u) & 0 \\ \Gamma_{\theta(u)}\bar{C}(u) & I - \Gamma_{\theta(u)} \end{bmatrix}, \ \mathbb{B}(u) \triangleq \mathrm{diag}\{\bar{B}(u), \Gamma_{\theta(u)}\bar{D}(u)\}, \\
&\bar{D}_1(u) \triangleq \mathrm{diag}\{D_1(u), D_2(u), \ldots, D_n(u)\}, \ \bar{A}(u) \triangleq I_n \otimes A(u), \\
&\bar{L}(u) \triangleq \left[\tilde{L}(u), 0\right], \ \tilde{L}(u) \triangleq I_n \otimes L(u), \ \bar{B}(u) \triangleq I_n \otimes B(u), \\
&\mathbb{D}_1 \triangleq \left[0 \quad \Gamma_{\theta(u)}\bar{D}(u)\right], \ \mathbb{C}_1(u) \triangleq \left[\Gamma_{\theta(u)}\bar{C}(u) \quad I_n - \Gamma_{\theta(u)}\right], \\
&\bar{C}(u) \triangleq \mathrm{diag}\{C_1(u), C_2(u), \ldots, C_n(u)\}.
\end{aligned} \tag{6.70}$$

Then, the distributed filter is designed for sensor node i as follows:

$$\begin{cases}
\hat{x}_i(u+1) = A(u)\hat{x}_i(u) + \sum_{j\in\mathbb{N}_i} \pi(u)a_{ij}^{(1)}K_{ij}^{(1)}(u)(\tilde{y}_j(u) - \vartheta(\theta(u)-j) \\
\qquad \times (C_j(u)\hat{x}_j(u)) - (1-\vartheta(\theta(u)-j))\hat{y}_j(u-1)) \\
\qquad + \sum_{j\in\mathbb{N}_i}(1-\pi(u))a_{ij}^{(2)}K_{ij}^{(2)}(u)(\tilde{y}_j(u) - \vartheta(\theta(u)-j) \\
\qquad \times (C_j(u)\hat{x}_j(u)) - (1-\vartheta(\theta(u)-j))\hat{y}_j(u-1)) + h(\hat{x}_i(u)) \\
\hat{y}_i(u) = \vartheta(\theta(u)-i)(C_i(u)\hat{x}_i(u)) + (1-\vartheta(\theta(u)-i))\hat{y}_i(u-1)) \\
\qquad + \sum_{j=i}\pi(u)a_{ij}^{(1)}H_{ij}^{(1)}(u)(\tilde{y}_j(u) - \vartheta(\theta(u)-j)(C_j(u)\hat{x}_j(u)) \\
\qquad - (1-\vartheta(\theta(u)-j))\hat{y}_j(u-1)) \sum_{j=i}(1-\pi(u))a_{ij}^{(2)}H_{ij}^{(2)}(u) \\
\qquad \times (\tilde{y}_j(u) - \vartheta(\theta(u)-j)(C_j(u)\hat{x}_j(u)) \\
\qquad - (1-\vartheta(\theta(u)-j))\hat{y}_j(u-1)) \\
\hat{z}_i(u) = L(u)\hat{x}_i(u)
\end{cases} \tag{6.71}$$

where the state estimate from node i is denoted by $\hat{x}_i(u) \in \mathbb{R}^{n_x}$, the estimation of the measurement signal is shown by $\hat{y}_i(u) \in \mathbb{R}^{n_y}$, the estimation of $z(u)$ is described by $\hat{z}_i(u)$, and $h(\hat{x}_i(u))$ is the estimation of $h(x(u))$. Here, $K_{ij}^{(r)}(u)$ and $H_{ij}^{(r)}(u)$ $(r = 1, 2)$ are the filter parameters which need to be designed.

In order to present the phenomenon of switching topologies, the stochastic variable $\pi(u) \in \mathbb{R}$, whose values are 0 or 1, is introduced as follows:

$$\text{Prob}\{\pi(u) = 1\} = \bar{\pi}, \qquad \text{Prob}\{\pi(u) = 0\} = 1 - \bar{\pi} \tag{6.72}$$

where $\bar{\pi}(0 \le \bar{\pi} \le 1)$ is a known constant which can be obtained by statistical experiments.

Remark 6.7 *The filter (6.71) reflects the spatial nature of the distributed estimation problem, in which the information from both the sensor node i and its neighbours is used for the updating. It can be observed from (6.71) that the topology of SN may switch in a probabilistic way by means of their intensity. It is worth mentioning that, with the help from the stochastic variable $\pi(u)$, it is able to describe the randomly switching topologies in an adequate way.*

Denote

$$\begin{aligned}
\breve{x}(u) \triangleq & \begin{bmatrix} \hat{x}_1^T(u) & \hat{x}_2^T(u) & \cdots & \hat{x}_n^T(u) & \hat{y}_1^T(u-1) & \hat{y}_2^T(u-1) & \cdots \end{bmatrix}\\
& \hat{y}_n^T(u-1) \Big]^T, \; \hat{z}(u) \triangleq \begin{bmatrix} \hat{z}_1^T(u) & \hat{z}_2^T(u) & \cdots & \hat{z}_n^T(u) \end{bmatrix}^T,\\
\bar{K}^{(r)}(u) \triangleq & [\bar{K}_{ij}^{(r)}(u)]_{n \times n} \quad \text{with}\\
& \bar{K}_{ij}^{(r)}(u) \triangleq \begin{cases} a_{ij}^{(r)} K_{ij}^{(r)}(u), & i = 1, 2, \ldots, n; \; j \in \mathbb{N}_i^{(r)}; \; r = 1, 2 \\ 0, & i = 1, 2, \ldots, n; \; j \notin \mathbb{N}_i^{(r)}; \; r = 1, 2 \end{cases}\\
\bar{H}^{(r)}(u) \triangleq & [\bar{H}_{ij}^{(r)}(u)]_{n \times n} \quad \text{with}\\
& \bar{H}_{ij}^{(r)}(u) \triangleq \begin{cases} a_{ij}^{(r)} H_{ij}^{(r)}(u), & i = 1, 2, \ldots, n; \; j = i; \; r = 1, 2 \\ 0, & i = 1, 2, \ldots, n; \; j \neq i; \; r = 1, 2. \end{cases}
\end{aligned}$$

It is easy to see that, because $a_{ij}^{(r)} = 0$ when $j \notin \mathbb{N}_i^{(r)}$. $\bar{K}^{(r)}(u)$ and $\bar{H}^{(r)}(u)$ can be showed as

$$\bar{K}^{(r)}(u) \in \mathscr{T}_{n_x \times n_y}, \quad \bar{H}^{(r)}(u) \in \mathscr{T}_{n_y \times n_y} \tag{6.73}$$

where $\mathscr{T}_{p \times q} \triangleq \left\{ \bar{T} = [T_{ij}] \in \mathbb{R}^{np \times nq} \mid T_{ij} \in \mathbb{R}^{p \times q}, \; T_{ij} = 0 \;\text{ if } j \notin \mathbb{N}_i^{(r)} \right\}$.

Letting $e(u) \triangleq \bar{x}(u) - \breve{x}(u)$ and $\tilde{z}(u) \triangleq \bar{z}(u) - \hat{z}(u)$, the following augmented system is obtained for the sensor network:

$$\begin{cases} e(u+1) = \mathbb{A}(u, \pi(u))e(u) + \mathbb{C}(u, \pi(u))w_v(u) + H(e(u)) \\ \tilde{z}(u) = \bar{L}(u)e(u) \end{cases} \tag{6.74}$$

where

$$
\begin{aligned}
&\mathbb{A}(u,\pi(u)) \triangleq \begin{bmatrix} \mathbb{A}_{11}(u,\pi(u)) & \mathbb{A}_{12}(u,\pi(u)) \\ \mathbb{A}_{21}(u,\pi(u)) & \mathbb{A}_{22}(u,\pi(u)) \end{bmatrix}, \\
&\mathbb{C}(u,\pi(u)) \triangleq \begin{bmatrix} \bar{B}(u) & -\bar{K}(u,\pi(u))\Gamma_{\theta(u)}\bar{D}(u) \\ 0 & \Gamma_{\theta(u)}\bar{D}(u) - \bar{H}(u,\pi(u))\Gamma_{\theta(u)}\bar{D}(u) \end{bmatrix}, \\
&\mathbb{A}_{12}(u,\pi(u)) \triangleq -\bar{K}(u,\pi(u))(I - \Gamma_{\theta(u)}), \\
&\mathbb{A}_{11}(u,\pi(u)) \triangleq \bar{A}(u) - \bar{K}(u,\pi(u))\Gamma_{\theta(u)}\bar{C}(u), \\
&\mathbb{A}_{21}(u,\pi(u)) \triangleq \Gamma_{\theta(u)}\bar{C}(u) - \bar{H}(u,\pi(u))\Gamma_{\theta(u)}\bar{C}(u), \\
&\mathbb{A}_{22}(u,\pi(u)) \triangleq (I - \Gamma_{\theta(u)}) - \bar{H}(u,\pi(u))(I - \Gamma_{\theta(u)}), \\
&H(e(u)) \triangleq H(\bar{x}(u)) - H(\hat{x}(u)),\ H(\hat{x}(u)) \triangleq \begin{bmatrix} h^T(\hat{x}(u)) & 0 \end{bmatrix}^T, \\
&\bar{K}(u,\pi(u)) \triangleq \pi(u)\bar{K}^{(1)}(u) + (1-\pi(u))\bar{K}^{(2)}(u), \\
&h(\hat{x}(u)) \triangleq \begin{bmatrix} h^T(\hat{x}_1(u)) & h^T(\hat{x}_2(u)) & \cdots & h^T(\hat{x}_n(u)) \end{bmatrix}^T, \\
&\bar{H}(u,\pi(u)) \triangleq \pi(u)\bar{H}^{(1)}(u) + (1-\pi(u))\bar{H}^{(2)}(u).
\end{aligned} \tag{6.75}
$$

The following definition is introduced in order to proceed further.

Definition 6.2 *We suppose that a disturbance attenuation level $\gamma > 0$ and a matrix $R(u) > 0$ are given, the H_∞ performance index satisfies the following inequality, namely:*

$$
J \triangleq \mathbb{E}\{\|\tilde{z}(u)\|^2_{[0,N-1]}\} - \gamma^2\{\|w_v(u)\|^2_{[0,N-1]} + e^T(0)R(u)e(0)\} < 0 \tag{6.76}
$$

where $e(0)$ is defined as $\bar{x}(0) - \check{x}(0)$.

In this section, we are committed to obtaining the filter gains $K_{ij}^{(r)}(u)$ and $H_{ij}^{(r)}(u)$ such that the error dynamics matches the preset performance constraint (6.76).

6.3.2 Main Results

Now we need to investigate the filtering problem for SNs whose topologies are described by $\mathbb{G}^{(r)} = (\mathbb{V}^{(r)}, \mathbb{E}^{(r)}, \mathbb{A}^{(r)})$ $(r = 1, 2)$.

The following theorem gives a sufficient condition under which the augmented system (6.74) satisfies the performance constraint (6.76).

Theorem 6.6 *In view of the time-varying system (6.61) and SNs whose topologies are changeable under RR protocol. Assuming the parameters of the designed filter $K_{ij}^{(r)}(u)$ and $H_{ij}^{(r)}(u)(r = 1, 2)$ are given, the filtering dynamic system (6.74) satisfies the given performance constraint (6.76) if there exists a scalar $\epsilon > 0$ and a class of positive definite matrices $\{\tilde{\mathbb{P}}(u)\}_{0 \le u \le N+1}$ with* *the initial condition* $\mathbb{P}(0) \leq \gamma^2 R(0)$ *satisfying*

$$\Theta(u,\pi(u)) = \begin{bmatrix} \Theta_{11}(u,\pi(u)) & * & * \\ \Theta_{21}(u,\pi(u)) & \Theta_{22}(u,\pi(u)) & * \\ \Theta_{31}(u,\pi(u)) & \Theta_{32}(u,\pi(u)) & \Theta_{33}(u,\pi(u)) \end{bmatrix} < 0 \quad \text{for } 0 \leq u \leq N-1 \tag{6.77}$$

where

$$\begin{aligned}
\Theta_{11}(u,\pi(u)) &\triangleq \mathbb{A}^T(u,\pi(u))\tilde{\mathbb{P}}(u+1)\mathbb{A}(u,\pi(u)) - \mathbb{P}(u) + \bar{L}^T(u)\bar{L}(u) - \epsilon\tilde{M}, \\
\Theta_{21}(u,\pi(u)) &\triangleq \tilde{\mathbb{P}}(u+1)\mathbb{A}(u,\pi(u)) + \epsilon\bar{M}, \ \mathbb{M}_1 \triangleq \mathrm{diag}\{\bar{M}_1, 0\}, \\
\Theta_{22}(u,\pi(u)) &\triangleq \tilde{\mathbb{P}}(u+1) - \gamma^2 I - \epsilon I, \ \mathbb{M}_2 \triangleq \mathrm{diag}\{\bar{M}_2, 0\}, \ \bar{M}_1 \triangleq I_n \otimes M_1, \\
\Theta_{31}(u,\pi(u)) &\triangleq \mathbb{C}^T(u,\pi(u))\tilde{\mathbb{P}}(u+1)\mathbb{A}(u,\pi(u)), \ \bar{M}_2 \triangleq I_n \otimes M_2, \\
\Theta_{33}(u,\pi(u)) &\triangleq \mathbb{C}^T(u,\pi(u))\tilde{\mathbb{P}}(u+1)\mathbb{C}(u,\pi(u)), \ \bar{M} \triangleq \frac{\mathbb{M}_2 + \mathbb{M}_1}{2}, \\
\Theta_{32}(u,\pi(u)) &\triangleq \mathbb{C}^T(u,\pi(u))\tilde{\mathbb{P}}(u+1), \ \tilde{M} \triangleq \frac{\mathbb{M}_1^T\mathbb{M}_2 + \mathbb{M}_2^T\mathbb{M}_1}{2}, \\
\tilde{\mathbb{P}}(u+1) &\triangleq \bar{\pi}\mathbb{P}(u+1|\pi(u)=1) + (1-\bar{\pi})\mathbb{P}(u+1|\pi(u)=0).
\end{aligned}$$

Here, $\mathbb{P}(u+1|\pi(u))$ denotes a certain positive definite matrix at the instant $u+1$ given the stochastic variable $\pi(u)$.

Proof *Define*

$$J(u) \triangleq e^T(u+1)\tilde{\mathbb{P}}(u+1)e(u+1) - e^T(u)\mathbb{P}(u)e(u). \tag{6.78}$$

Through the calculation of the expectation for function (6.78), we can achieve that

$$\begin{aligned}
\mathbb{E}\{J(u)\} &= \mathbb{E}\Big\{e^T(u+1)\tilde{\mathbb{P}}(u+1)e(u+1) - e^T(u)\mathbb{P}(u)e(u)\Big\} \\
&= \mathbb{E}\Bigg\{\Big(\mathbb{A}(u,\pi(u))e(u) + \mathbb{C}(u,\pi(u))w_v(u) + H(e(u))\Big)^T \tilde{\mathbb{P}}(u+1) \\
&\quad \times \Big(\mathbb{A}(u,\pi(u))e(u) + \mathbb{C}(u,\pi(u))w_v(u) + H(e(u))\Big) \\
&\quad - e^T(u)\mathbb{P}(u)e(u)\Bigg\}.
\end{aligned} \tag{6.79}$$

Then, adding $\tilde{z}^T(u)\tilde{z}(u) - \gamma^2\bar{w}_v^T(u)\bar{w}_v(u) - \tilde{z}^T(u)\tilde{z}(u) + \gamma^2\bar{w}_v^T(u)\bar{w}_v(u)$ *to* $\mathbb{E}\{J(u)\}$, *we have*

$$\mathbb{E}\{J(u)\} \leq \mathbb{E}\{\eta^T(u)\Theta(u,\pi(u))\eta(u) - \tilde{z}^T(u)\tilde{z}(u) + \gamma^2\bar{w}^T(u)\bar{w}(u)\} \tag{6.80}$$

where

$$\eta(u) \triangleq \left[e^T(u) \quad H^T(e(u)) \quad w_v^T(u)\right]^T. \tag{6.81}$$

Summing up the equation (6.80) from 0 to $N-1$, one has the following inequality:

$$\begin{aligned} J &< \mathbb{E}\left\{\sum_{u=0}^{N-1} \eta^T(u)\Theta(u,\pi(u))\eta(u)\right\} - \mathbb{E}\{e^T(N)\tilde{\mathbb{P}}(N)e(N)\} \\ &\quad + e^T(0)\mathbb{P}(0)e(0) - \gamma^2 e^T(0)R(0)e(0). \end{aligned} \tag{6.82}$$

It is remarkable that the positive definite matrix $\tilde{\mathbb{P}}(N) > 0$ and the initial condition $\mathbb{P}(0) \leq \gamma^2 R(0)$, we have $J < 0$ when (6.77) holds.

As discussed in Theorem 6.6, the performance analysis has been conducted for the presented system (6.74). Then, we are motivated to achieve the filter gains. In the next theorem, we would deal with this problem.

Theorem 6.7 *Assume that a scalar $\mu > 0$ and a matrix $R(0) = R^T(0) > 0$ are given. For the target plant (6.61) over sensor networks, the proposed time-varying filtering problem is solved if there exist matrices $\check{\Omega}(u)$ and $\check{\Upsilon}(u)$ $(0 \leq u \leq N+1)$, a scalar $\epsilon > 0$ and a set of gains $K_{ij}^{(r)}(u) \in \mathscr{T}_{n_x \times n_y}$ and $H_{ij}^{(r)}(u) \in \mathscr{T}_{n_y \times n_y}$ $(r = 1, 2)$ satisfying the condition*

$$\operatorname{diag}\{\Omega(0), \Upsilon(0)\} \leqslant \gamma^2 R(0) \tag{6.83}$$

and the following RLMIs:

$$\Pi(u,\pi(u)) = \begin{bmatrix} \Pi_{11}(u,\pi(u)) & * \\ \Pi_{21}(u,\pi(u)) & \Pi_{22}(u,\pi(u)) \end{bmatrix} < 0 \tag{6.84}$$

with the parameters updated by

$$\begin{aligned} \check{\Omega}^{-1}(u+1) &= \bar{\pi}\Omega(u+1|\pi(u)=1) + (1-\bar{\pi})\Omega(u+1|\pi(u)=0), \\ \check{\Upsilon}^{-1}(u+1) &= \bar{\pi}\Upsilon(u+1|\pi(u)=1) + (1-\bar{\pi})\Upsilon(u+1|\pi(u)=0), \end{aligned}$$

where

$$\begin{aligned} \Pi_{11}(u,\pi(u)) &\triangleq \begin{bmatrix} \Pi_{11}^1(u,\pi(u)) & * \\ 0 & \Pi_{11}^2(u,\pi(u)) \end{bmatrix} \\ \Pi_{11}^{11}(u,\pi(u)) &\triangleq -\frac{\epsilon(\bar{M}_1^T\bar{M}_2 + \bar{M}_2^T\bar{M}_1)}{2} - \Omega(u), \\ \Pi_{11}^1(u,\pi(u)) &\triangleq \begin{bmatrix} \Gamma_{11}^{11}(u,\pi(u)) & 0 & * \\ 0 & -\Upsilon(u) & * \\ \frac{\epsilon(\bar{M}_1+\bar{M}_2)}{2} & 0 & -\epsilon I \end{bmatrix}, \end{aligned} \tag{6.85}$$

$$\Pi_{11}^2(u,\pi(u)) \triangleq \begin{bmatrix} -\gamma^2 I & * \\ 0 & -\gamma^2 I \end{bmatrix}, \ \Pi_{22}(u,\pi(u)) \triangleq \text{diag}\{-I, \Omega_\Upsilon(u+1)\},$$

$$\Pi_{21}(u,\pi(u)) \triangleq \begin{bmatrix} \bar{L}(u) & 0 & 0 \\ \mathbb{A}(u,\pi(u)) & I & \mathbb{C}(u,\pi(u)) \end{bmatrix},$$

$$\Omega_\Upsilon(u+1) \triangleq \text{diag}\{-\check{\Omega}(u+1), -\check{\Upsilon}(u+1)\}.$$

Proof *Assume that the positive definite matrix $\mathbb{P}(u)$ could be decomposed as follows:*

$$\begin{aligned} &\mathbb{P}(u) = \text{diag}\{\Omega(u), \Upsilon(u)\}, \ \tilde{\mathbb{P}}^{-1}(u+1) = \text{diag}\{\check{\Omega}(u+1), \check{\Upsilon}(u+1)\}, \\ &\bar{\Upsilon}(u+1) = \bar{\pi}\Upsilon(u+1|\pi(u)=1) + (1-\bar{\pi})\Upsilon(u+1|\pi(u)=0), \\ &\bar{\Omega}(u+1) = \bar{\pi}\Omega(u+1|\pi(u)=1) + (1-\bar{\pi})\Omega(u+1|\pi(u)=0), \\ &\bar{\mathbb{P}}(u+1) = \text{diag}\{\bar{\Omega}(u+1), \bar{\Upsilon}(u+1)\}. \end{aligned} \tag{6.86}$$

Then, considering Lemma 2.1, it is easy to achieve (6.84) by some matrix calculation. Therefore, the proof of this theorem is now completed.

By using Theorem 6.7, we can easily carry out the Distributed Filter Design (DFD) algorithm in the following table.

Algorithm DFD	
Step 1.	Select suitable values for the performance index γ , the matrix $R(0) > 0$, the states $x(0)$, $\hat{x}_i(0)$ and $\tilde{y}_i(-1)$ when time instant $u = 0$. Choose values for the matrices $\Omega(0)$ and $\Upsilon(0)$ under the condition (6.83) as well as set the time instant $u = 0$.
Step 2.	Calculate the matrices $\check{\Omega}(u) > 0$, $\check{\Upsilon}(u) > 0$. By solving the RLMIs (6.84), the gain matrices of $K_{ij}^{(r)}(u)$ and $H_{ij}^{(r)}(u)$ would be obtained at time instant u .
Step 3.	Set $u = u + 1$ and acquire $\Omega(u)$ and $\Upsilon(u)$ by the formula (6.86). If $u < N$, execute Step 2, else execute Step 4.
Step 4.	Stop.

Remark 6.8 *Through Theorem 6.7, the distributed filter gains have been obtained by calculating a set of RLIMs (6.84). It is remarkable that the main innovations of this section are presented in those aspects: (1) a original distributed filter model has been established for sensor networks whose topologies change in a random way; (2) the RR protocol has been applied in sensor networks in order to save precious communication resources; and (3) the DFD algorithm which is built in a recursive form is suitable for online applications. Apparently, the result derived in this section is more valuable for practical application.*

6.4 Illustrative Examples

In this section, three simulation examples are given to demonstrate the approaches presented in this chapter.

6.4.1 Example 1

In this example, we present a simulation example to illustrate the effectiveness of the distributed finite-horizon state estimator design scheme proposed in Section 6.1 for sensor networks which contain switching topologies and redundant channels.

The sensor network is represented by a directed graph $\mathcal{G}^{(l)} = (\mathcal{V}^{(l)}, \mathcal{E}^{(l)}, \mathcal{A}^{(l)})$ with the set of nodes $\mathcal{V}^{(1)} = \{1,2,3,4,5\}$ and $\mathcal{V}^{(2)} = \{1,2,3,4,5\}$, set of edges $\mathcal{E}^{(1)} = \{(1,1),(1,5),(2,2),(2,5),(3,3),(3,4),(4,3),(4,4),(5,2),(5,5)\}$ and $\mathcal{E}^{(2)} = \{(1,1),(1,4),(2,2),(2,4),(3,1),(3,3),(4,2),(4,4),(5,1),(5,5)\}$, the respective adjacency matrices are shown as follows.

$$\mathcal{A}^{(1)} = \begin{bmatrix} 1 & 0 & 0 & 0 & 1 \\ 0 & 1 & 0 & 0 & 1 \\ 0 & 0 & 1 & 1 & 0 \\ 0 & 0 & 1 & 1 & 0 \\ 0 & 1 & 0 & 0 & 1 \end{bmatrix}, \quad \mathcal{A}^{(2)} = \begin{bmatrix} 1 & 0 & 0 & 1 & 0 \\ 0 & 1 & 0 & 1 & 0 \\ 1 & 0 & 1 & 0 & 0 \\ 0 & 1 & 0 & 1 & 0 \\ 1 & 0 & 0 & 0 & 1 \end{bmatrix}.$$

The parameters of considered time-varying nonlinear discrete system (6.1) are given as follows :

$$A(k) = \begin{bmatrix} 0.2\sin(2k) & -0.3 \\ 0.3 & -0.1 \end{bmatrix}, \quad G(k) = \begin{bmatrix} 2.4 & 2.3 \end{bmatrix}^T, \quad M(k) = \begin{bmatrix} 0.2 & 0.2 \end{bmatrix}$$

and the nonlinear functions $f(x(k))$

$$f(x(k)) = \begin{bmatrix} \tan(x_1(k)) & \tan(x_2(k)) \end{bmatrix}^T$$

where the lth element of the system (6.1) is presented as $x_l(k)$ $(l = 1,2)$, and we can easily obtain that the nonlinear functions $f(x(k))$ satisfies (6.4) with the parameters as follows:

$$L_1 = \begin{bmatrix} -0.30 & 0.29 \\ 0.32 & 0.40 \end{bmatrix}, \quad L_2 = \begin{bmatrix} -0.33 & 0.27 \\ 0.41 & 0.80 \end{bmatrix}.$$

We suppose that $m = 3$, and the parameters of measurements are defined as:

$$C_{11}(k) = \begin{bmatrix} 2.2\sin(2k) & 2.3\sin(2k) \end{bmatrix}, \; C_{12}(k) = \begin{bmatrix} 2.2\sin(2k) & 2.2\sin(2k) \end{bmatrix},$$
$$C_{13}(k) = \begin{bmatrix} 2.3\sin(2k) & 2.2\sin(2k) \end{bmatrix}, \; C_{21}(k) = \begin{bmatrix} 2.3\sin(2k) & 2.2\sin(2k) \end{bmatrix},$$
$$C_{22}(k) = \begin{bmatrix} 2.1\sin(2k) & 2.3\sin(2k) \end{bmatrix}, \; C_{23}(k) = \begin{bmatrix} 2.2\sin(2k) & 2.3\sin(3k) \end{bmatrix},$$
$$C_{31}(k) = \begin{bmatrix} 2.2 & 2.1 \end{bmatrix}, \; C_{32}(k) = \begin{bmatrix} 2.2 & 2.3 \end{bmatrix}, \; C_{33}(k) = \begin{bmatrix} 2.2 & 2.4 \end{bmatrix},$$
$$C_{41}(k) = \begin{bmatrix} 2.3 & 2.4 \end{bmatrix}, \; C_{42}(k) = \begin{bmatrix} 2.2 & 2.4 \end{bmatrix}, \; C_{43}(k) = \begin{bmatrix} 2.2 & 2.4 \end{bmatrix},$$
$$C_{51}(k) = \begin{bmatrix} 2.2\sin(4k) & 2.1\sin(2k) \end{bmatrix}, \; C_{52}(k) = \begin{bmatrix} 2.3\sin(3k) & 2.2\sin(2k) \end{bmatrix},$$
$$C_{53}(k) = \begin{bmatrix} 2.2\sin(2k) & 2.4\sin(2k) \end{bmatrix}, \; D_1(k) = 2.9, \; D_2(k) = 2.5, \; D_3(k) = 2.6,$$
$$D_4(k) = 2.9, \; D_5(k) = 2.6, \; \bar{\alpha}_{11} = 0.5, \; \bar{\alpha}_{12} = 0.7, \; \bar{\alpha}_{13} = 0.7, \; \bar{\alpha}_{21} = 0.5,$$
$$\bar{\alpha}_{22} = 0.4, \; \bar{\alpha}_{23} = 0.7, \; \bar{\alpha}_{31} = 0.8, \; \bar{\alpha}_{32} = 0.8, \; \bar{\alpha}_{33} = 0.8, \; \bar{\alpha}_{41} = 0.7,$$
$$\bar{\alpha}_{42} = 0.7, \; \bar{\alpha}_{43} = 0.6, \; \bar{\alpha}_{51} = 0.5, \; \bar{\alpha}_{52} = 0.7, \; \bar{\alpha}_{53} = 0.8.$$

In this example, the H_∞ performance index γ is taken as 0.9, and the positive definite matrices are chosen as $S_i(m) = \text{diag}\{2, 2\}$ ($i = 1, 2, \ldots, 5$ and $m = 1, 2$). Using the designed algorithm to solve the RLMIs (6.28), we can achieve the parameters of designed state estimators in Section 6.1 and the partial parameters are shown in the following Table 6.1. We set the original state $x_0 = \begin{bmatrix} 1.1 & 1.1 \end{bmatrix}^T$ and $\hat{x}_i = \begin{bmatrix} 0 & 0 \end{bmatrix}^T$ ($i = 1, 2, \ldots, 5$). The noise $w(k)$ is selected as $\frac{\cos(2k+1)}{3k+1}$, and the external disturbance $v(k)$ is taken as $\frac{\sin(10k+1)}{3k+1}$. The results are shown in Figs. 6.1–6.4. Fig. 6.1 presents the changes of the system topology. Fig. 6.2 shows the estimation errors of $z(k)$ in each sensor. Fig. 6.3 depicts state estimation error x_1, and Fig. 6.4 depicts state estimation error x_2. Obviously, the simulation results indicate the effectiveness of designed state estimators in Section 6.1.

6.4.2 Example 2

In this example, the effectiveness of the non-fragile distributed fault estimation method proposed in Section 6.2 is demonstrated for nonlinear time-varying systems with randomly occurring gain variations in sensor networks.

The sensor network is represented by a directed graph $\mathcal{G} = (\mathcal{V}, \mathcal{E}, \mathcal{A})$ with the set of nodes $\mathcal{V} = \{1, 2, 3, 4, 5\}$, set of edges $\mathcal{E} = \{(1,1), (1,2), (1,5), (2,2), (2,3), (3,1), (3,3), (4,2), (4,4), (5,2), (5,5)\}$ and the following adjacency matrix:

$$\mathcal{A} = \begin{bmatrix} 1 & 1 & 0 & 0 & 1 \\ 0 & 1 & 1 & 0 & 0 \\ 1 & 0 & 1 & 0 & 0 \\ 0 & 1 & 0 & 1 & 0 \\ 0 & 1 & 0 & 0 & 1 \end{bmatrix}.$$

TABLE 6.1
The Partial Parameters of State Estimators

k	**0**	**1**	**2**	$\cdots$
$K_{11}^1(k)$	$\begin{bmatrix} 0.1671 & 0.2228 \\ 0.1671 & 0.2228 \end{bmatrix}$	$\begin{bmatrix} -0.1795 & -0.2393 \\ -0.1795 & -0.2393 \end{bmatrix}$	$\begin{bmatrix} -0.1415 & -0.1886 \\ -0.1415 & -0.1886 \end{bmatrix}$	$\cdots$
$K_{15}^1(k)$	$\begin{bmatrix} 0.0923 & 0.1230 \\ 0.0923 & 0.1230 \end{bmatrix}$	$\begin{bmatrix} -0.0694 & -0.0925 \\ -0.0694 & -0.0925 \end{bmatrix}$	$\begin{bmatrix} -0.0572 & -0.0763 \\ -0.0572 & -0.0763 \end{bmatrix}$	$\cdots$
$H_{11}^1(k)$	$\begin{bmatrix} 0.5789 & 0.5789 \end{bmatrix}^T$	$\begin{bmatrix} 0.6935 & 0.6935 \end{bmatrix}^T$	$\begin{bmatrix} 1.3465 & 1.3465 \end{bmatrix}^T$	$\cdots$
$H_{15}^1(k)$	$\begin{bmatrix} 0.9430 & 0.9430 \end{bmatrix}^T$	$\begin{bmatrix} 1.0154 & 1.0154 \end{bmatrix}^T$	$\begin{bmatrix} 1.0912 & 1.0912 \end{bmatrix}^T$	$\cdots$
$K_{11}^2(k)$	$\begin{bmatrix} 0.5065 & 0.5045 \\ 0.5065 & 0.5045 \end{bmatrix}$	$\begin{bmatrix} 0.2746 & 0.1828 \\ 0.2746 & 0.1828 \end{bmatrix}$	$\begin{bmatrix} 0.2479 & 0.1586 \\ 0.2479 & 0.1586 \end{bmatrix}$	$\cdots$
$K_{14}^2(k)$	$\begin{bmatrix} 0.3872 & 0.3440 \\ 0.3872 & 0.3440 \end{bmatrix}$	$\begin{bmatrix} 0.2582 & 0.1699 \\ 0.2582 & 0.1699 \end{bmatrix}$	$\begin{bmatrix} 0.4697 & 0.4523 \\ 0.4697 & 0.4523 \end{bmatrix}$	$\cdots$
$H_{11}^2(k)$	$\begin{bmatrix} 0.9476 & 0.9476 \end{bmatrix}^T$	$\begin{bmatrix} 0.6515 & 0.6515 \end{bmatrix}^T$	$\begin{bmatrix} 1.3142 & 1.3142 \end{bmatrix}^T$	$\cdots$
$H_{14}^2(k)$	$\begin{bmatrix} 1.6464 & 1.6464 \end{bmatrix}^T$	$\begin{bmatrix} 1.3736 & 1.3736 \end{bmatrix}^T$	$\begin{bmatrix} 1.2511 & 1.2511 \end{bmatrix}^T$	$\cdots$

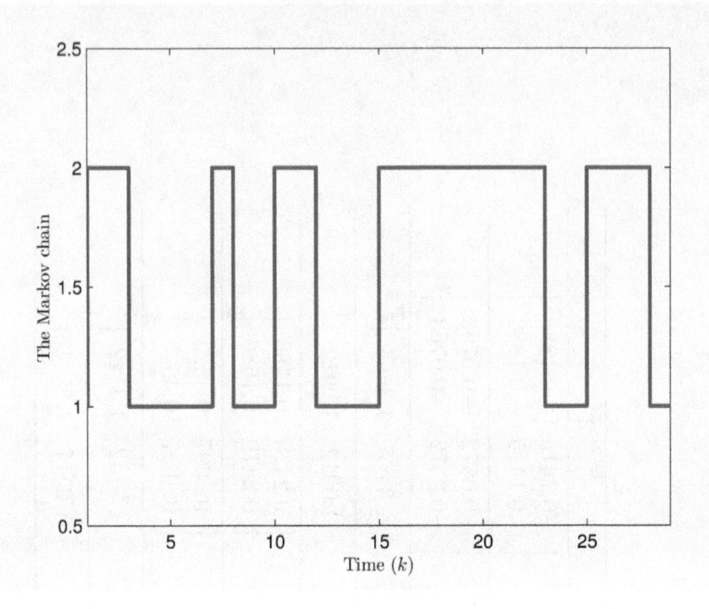

FIGURE 6.1
Modes evolution.

The parameters of the time-varying nonlinear discrete system are given by

$$A(k) = \begin{bmatrix} 0.6\sin(4k) & 0.5 \\ 0 & -0.3 \end{bmatrix}, \quad D(k) = \begin{bmatrix} 0.2 & 0.6 \end{bmatrix}^T, \quad G(k) = \begin{bmatrix} 0.5 & 0.8 \end{bmatrix}^T$$

and the nonlinear function is set to be

$$h(x(k)) = \begin{bmatrix} -0.5x_1 + 0.4x_2 & \frac{0.1x_1}{\sqrt{x_1^2 x_2^2 + 10}} \end{bmatrix}^T$$

where x_l $(l = 1, 2)$ represents the l-th element of the system state $x(k)$. It is easy to verify that the above nonlinear function $h(x(k))$ satisfies (6.34) with

$$\Psi = \begin{bmatrix} -0.29 & 0.29 \\ 0 & 0.6 \end{bmatrix}, \quad \Omega = \begin{bmatrix} -0.29 & 0.29 \\ 0 & 0.4 \end{bmatrix}.$$

The measurement output matrices and the multiplicative gain variations

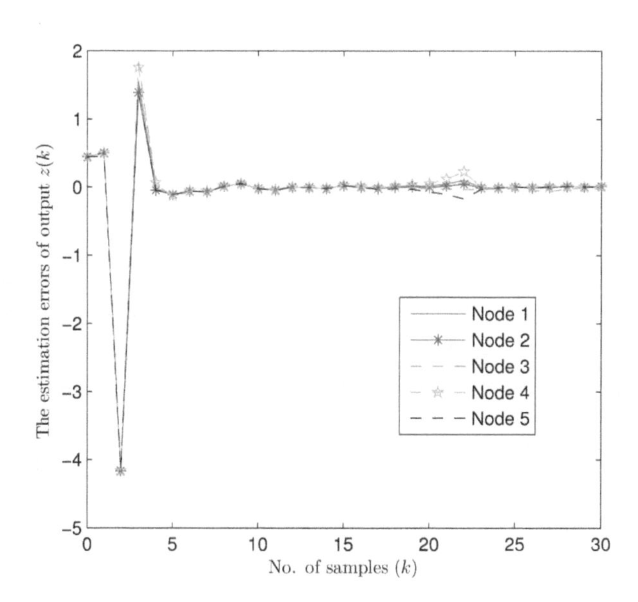

FIGURE 6.2
The estimation errors of the output $z(k)$.

are characterized as follows:

$$C_1(k) = \begin{bmatrix} 0.5 & 0.1\sin(2k) \end{bmatrix}, \quad C_2(k) = \begin{bmatrix} 0.4 & 0.2 \end{bmatrix}, \quad E_a = \begin{bmatrix} 0.3 & 0.3 \\ 0.3 & 0.3 \end{bmatrix},$$
$$C_3(k) = \begin{bmatrix} 0.6 & 0.4\sin(2k) \end{bmatrix}, \quad C_4(k) = \begin{bmatrix} 0.3\sin(4k) & 0 \end{bmatrix}, \quad H_b = 1,$$
$$C_5(k) = \begin{bmatrix} 0.2\sin(3k) & 0.1\sin(2k) \end{bmatrix}, \quad E_1(k) = 0.1, \quad E_b = 0.3,$$
$$E_2(k) = 0.31, \quad E_3(k) = 0.23, \quad E_4(k) = 0.2, \quad E_5(k) = 0.11, \quad H_1(k) = 0.6,$$
$$H_2(k) = 0.8, \quad H_3(k) = 0.7, \quad H_4(k) = 0.9, \quad H_5(k) = 0.4, \quad H_a = \begin{bmatrix} 1 & 0 \\ 0 & 1 \end{bmatrix}.$$

The exogenous disturbance sequences are chosen to be $\omega(k) = \exp(-k)$ and $v(k) = \frac{\sin(10k+1)}{3k+1}$, respectively. Moreover, we set $\bar{\alpha} = 0.8$, $S_i = \text{diag}\{2, 2\}$ $(i = 1, 2, \cdots, 5)$, $x(0) = \begin{bmatrix} 0.26 & -0.2 \end{bmatrix}^T$ and $\hat{x}_i(0) = 0$ $(i = 1, 2, \cdots, 5)$. According to the given $NFED$ algorithm, we solve the recursive linear matrix inequalities (6.52) according to Theorem 6.5 with the Matlab LMI toolbox. Some of the parameters of the desired estimators are given in Table 6.2. The fault to be estimated is $f(k) = \frac{\cos(10k)}{8k+1}$. Fig. 6.5 shows the fault estimation error. Fig. 6.6 plots the simulation result on the fault signal and its estimate. The simulation results have confirmed the effectiveness of the non-fragile distributed fault estimation technique presented in Section 6.2.

TABLE 6.2
Non-fragile Distributed Fault Estimator Parameters

k	**0**	**1**	**2**	$\cdots$
$K_{11}(k)$	$\begin{bmatrix} -0.0000 & 0.2957 \\ -0.0000 & 0.2957 \end{bmatrix}$	$\begin{bmatrix} -0.0000 & 0.2943 \\ -0.0000 & 0.2943 \end{bmatrix}$	$\begin{bmatrix} -0.0001 & 0.2984 \\ -0.0001 & 0.2984 \end{bmatrix}$	$\cdots$
$K_{12}(k)$	$\begin{bmatrix} 0.2956 & 0.2956 \\ 0.2956 & 0.2956 \end{bmatrix}$	$\begin{bmatrix} 0.2941 & 0.2942 \\ 0.2941 & 0.2942 \end{bmatrix}$	$\begin{bmatrix} 0.2982 & 0.2983 \\ 0.2982 & 0.2983 \end{bmatrix}$	$\cdots$
$K_{15}(k)$	$\begin{bmatrix} -0.2507 & -0.2508 \\ -0.2507 & -0.2508 \end{bmatrix}$	$\begin{bmatrix} -0.5874 & -0.5874 \\ -0.5874 & -0.5874 \end{bmatrix}$	$\begin{bmatrix} -0.2552 & -0.2552 \\ -0.2552 & -0.2552 \end{bmatrix}$	$\cdots$
$K_{22}(k)$	$\begin{bmatrix} -0.1369 & -0.1369 \\ -0.1369 & -0.1369 \end{bmatrix}$	$\begin{bmatrix} -0.1048 & -0.1049 \\ -0.1048 & -0.1049 \end{bmatrix}$	$\begin{bmatrix} -0.3540 & -0.3541 \\ -0.3540 & -0.3541 \end{bmatrix}$	$\cdots$
$K_{23}(k)$	$\begin{bmatrix} 0.1840 & 0.1841 \\ 0.1840 & 0.1841 \end{bmatrix}$	$\begin{bmatrix} 0.1823 & 0.1824 \\ 0.1823 & 0.1824 \end{bmatrix}$	$\begin{bmatrix} 0.1851 & 0.1853 \\ 0.1851 & 0.1853 \end{bmatrix}$	$\cdots$
$K_{31}(k)$	$\begin{bmatrix} 0.1137 & -0.0000 \\ 0.1137 & -0.0000 \end{bmatrix}$	$\begin{bmatrix} 0.1121 & -0.0000 \\ 0.1121 & -0.0000 \end{bmatrix}$	$\begin{bmatrix} 0.1159 & -0.0000 \\ 0.1159 & -0.0000 \end{bmatrix}$	$\cdots$
$K_{33}(k)$	$\begin{bmatrix} 0.1840 & 0.1841 \\ 0.1840 & 0.1841 \end{bmatrix}$	$\begin{bmatrix} 0.1823 & 0.1824 \\ 0.1823 & 0.1824 \end{bmatrix}$	$\begin{bmatrix} 0.1851 & 0.1853 \\ 0.1851 & 0.1853 \end{bmatrix}$	$\cdots$
$K_{42}(k)$	$\begin{bmatrix} -0.1566 & -0.1567 \\ -0.1566 & -0.1567 \end{bmatrix}$	$\begin{bmatrix} -0.3637 & -0.3638 \\ -0.3637 & -0.3638 \end{bmatrix}$	$\begin{bmatrix} -0.1599 & -0.1599 \\ -0.1599 & -0.1599 \end{bmatrix}$	$\cdots$
$K_{44}(k)$	$\begin{bmatrix} -0.0481 & 0.1841 \\ -0.0481 & 0.1841 \end{bmatrix}$	$\begin{bmatrix} -0.0485 & 0.1823 \\ -0.0485 & 0.1823 \end{bmatrix}$	$\begin{bmatrix} -0.0480 & 0.1853 \\ -0.0480 & 0.1853 \end{bmatrix}$	$\cdots$

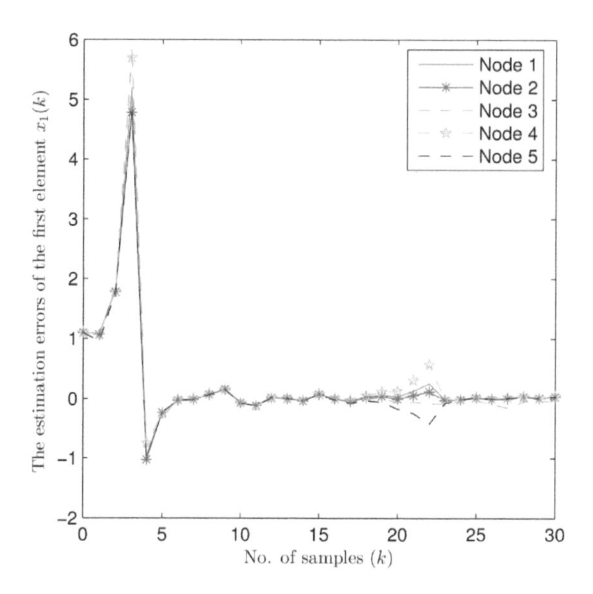

FIGURE 6.3
The estimation errors of the first element $x_1(k)$.

6.4.3 Example 3

We would reveal the effectiveness of the addressed finite-horizon distributed filter design scheme in Section 6.3 for SNs with switching topologies under the RR protocol.

We employ a directed graph $\mathbb{G}^{(r)} = (\mathbb{V}^{(r)}, \mathbb{E}^{(r)}, \mathbb{A}^{(r)})$ to describe the network framework. With the set of nodes $\mathbb{V}^{(1)} = \{1,2,3\}$ and $\mathbb{V}^{(2)} = \{1,2,3\}$, the set of edges $\mathbb{E}^{(1)} = \{(1,1),(1,3),(2,1),(2,2),(3,2),(3,3)\}$ and $\mathbb{E}^{(2)} = \{(1,1),(1,2),(2,2),(2,3),(3,1),(3,3)\}$. The respective adjacency matrices are shown as follows:

$$\mathbb{A}^{(1)} = \begin{bmatrix} 1 & 0 & 1 \\ 1 & 1 & 0 \\ 0 & 1 & 1 \end{bmatrix}, \quad \mathbb{A}^{(2)} = \begin{bmatrix} 1 & 1 & 0 \\ 0 & 1 & 1 \\ 1 & 0 & 1 \end{bmatrix}.$$

The parameters of the considered time-varying nonlinear discrete system (6.61) are given by

$$A(u) = \begin{bmatrix} 0.02\mathrm{sin(2u)} & -0.03 \\ 0.03 & -0.01 \end{bmatrix}, \quad B(u) = \begin{bmatrix} 0.5 & 0.5 \end{bmatrix}^T, \quad L(u) = \begin{bmatrix} 0.5 & 0.3 \end{bmatrix}.$$

Select the nonlinear function $h(x(u))$ as

$$h(x(u)) = \begin{bmatrix} \sin(x_1(u)) & \sin(x_2(u)) \end{bmatrix}^T$$

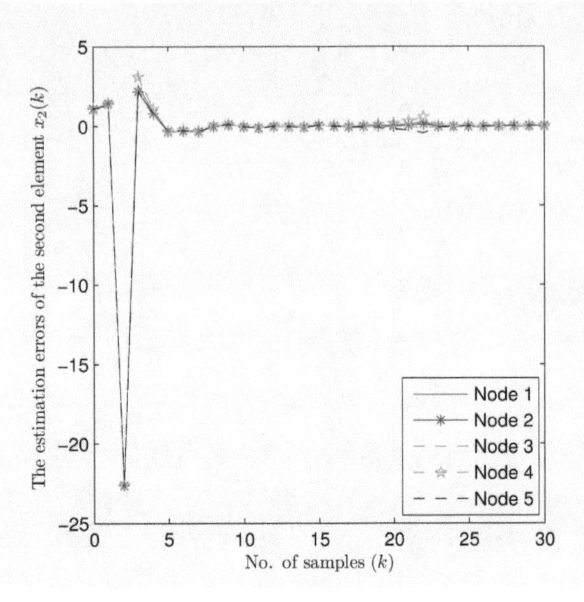

FIGURE 6.4
The estimation errors of the second element $x_2(k)$.

where $x_l(u)$ $(l = 1, 2)$ represents the lth element of the system (6.61). In order to satisfy the nonlinear function $h(x(u))$ presented in (6.62), the parameters are selected as follows:

$$M_1 = \begin{bmatrix} -10 & 12 \\ 15 & 16 \end{bmatrix}, \quad M_2 = \begin{bmatrix} -11 & 10 \\ 13 & 11 \end{bmatrix}.$$

Let $n = 3$ and the parameters of measurements be defined as

$$\begin{aligned} C_1(u) &= \begin{bmatrix} 0.02\sin(2u) & 0.03\sin(2u) \end{bmatrix}, \quad D_1(u) = 0.8, \\ C_2(u) &= \begin{bmatrix} 0.03\sin(2u) & 0.02\sin(2u) \end{bmatrix}, \quad D_2(u) = 0.9, \\ C_3(u) &= \begin{bmatrix} 0.02 & 0.01 \end{bmatrix}, \quad D_3(u) = 0.7, \quad \bar{\pi} = 0.8. \end{aligned}$$

The H_∞ performance index γ is chosen as 0.7.

The parameters of filters are acquired by employing the designed DFD algorithm to solve the RLMIs presented by (6.84). Set the original states $x(0) = \begin{bmatrix} 0.5 & 0.5 \end{bmatrix}^T$, $\hat{x}_1(0) = \begin{bmatrix} 0 & 0 \end{bmatrix}^T$, $\hat{x}_2(0) = \begin{bmatrix} 0 & 0.5 \end{bmatrix}^T$, $\hat{x}_3(0) = \begin{bmatrix} 0 & 0 \end{bmatrix}^T$ and $y_i(0) = 0$ $(i = 1, 2, 3)$. The noise $w(u)$ is selected as $\frac{\cos(2u+1)}{3u+1}$, and the disturbance $v(u)$ is taken as $\frac{\sin(10u+1)}{3u+1}$.

We would present the simulation results in Figs. 6.7–6.8. Fig. 6.7 shows the estimation errors of $z(u)$ and Fig. 6.8 presents the estimation of $z(u)$.

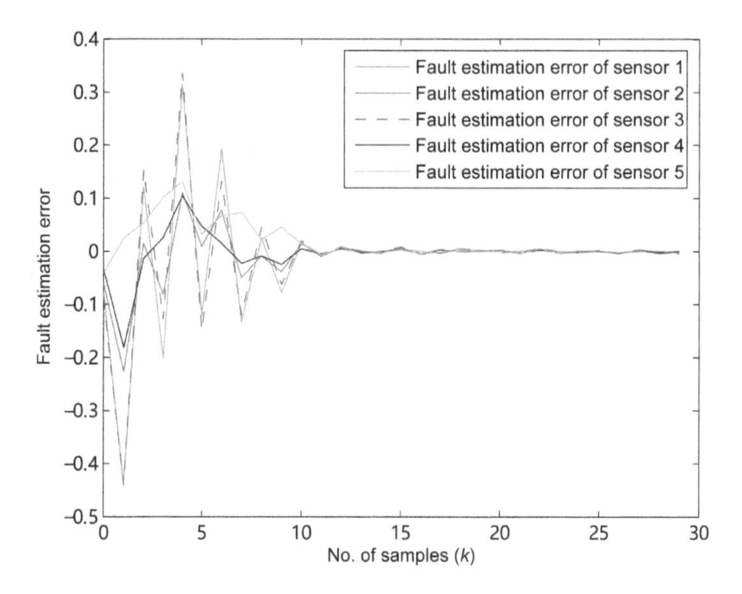

FIGURE 6.5
Fault estimation error.

Obviously, it is easy to confirm that the designed filtering method in Section 6.3 is useful for SNs.

6.5 Summary

In this chapter, first, we have made few attempts to design the distributed state estimators for a class of time-varying systems over sensor network with redundant channels and switching topologies. The redundant channels model govern by a set of Bernoulli distributed random variables is employed to increase the data transmission reliability for the reason that the packet dropouts often happen in sensor networks. With the aid of a stochastic Kronecker delta function, a Markovian chain has been introduced to describe the features of switching topologies. The distributed state estimators have been acquired by computing a set of recursive linear matrix inequalities such that the estimation error dynamics satisfies the given average H_∞ performance constraint. Second, the non-fragility problem of distributed fault estimators has also been considered. By employing a recursive matrix inequality approach, the non-fragile distributed fault estimators have been designed with

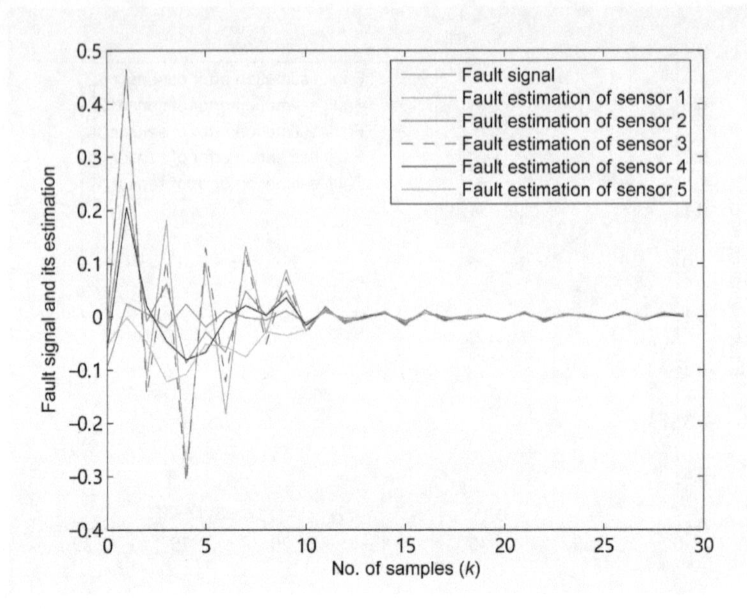

FIGURE 6.6
Fault signal and its estimate.

which the error dynamics of the fault estimation satisfies a prescribed average H_∞ performance constraint. Third, the time-varying distributed filter has been obtained for a set of nonlinear systems over sensor networks subject to switching topologies. The RR protocol, in which all the sensors would be given identical probability to use the communication channels based on a fixed cycle, has been employed to allocate the limited bandwidth reasonably. Through stochastic analysis technique, sufficient conditions have been established such that the dynamics of augmented error system satisfies a given performance constraint. Finally, the effectiveness of the developed distributed algorithms have been demonstrated by three simulation examples.

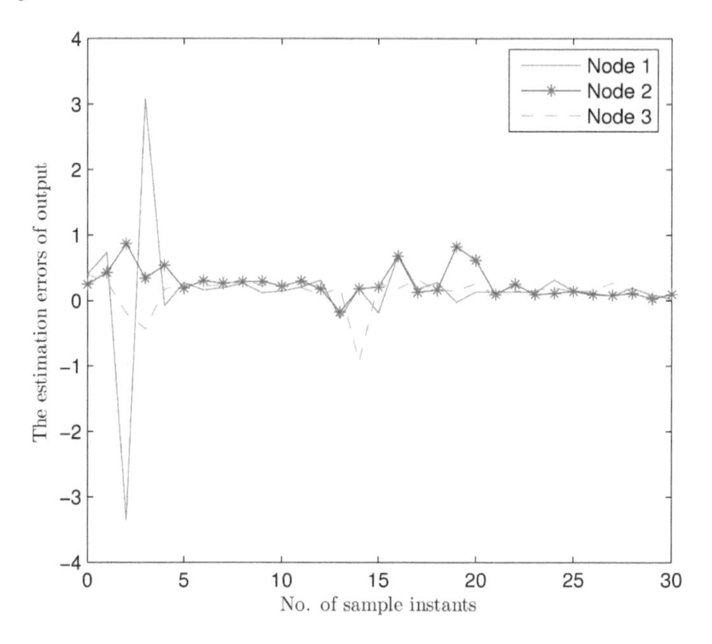

FIGURE 6.7
The estimation errors of output $z(u)$.

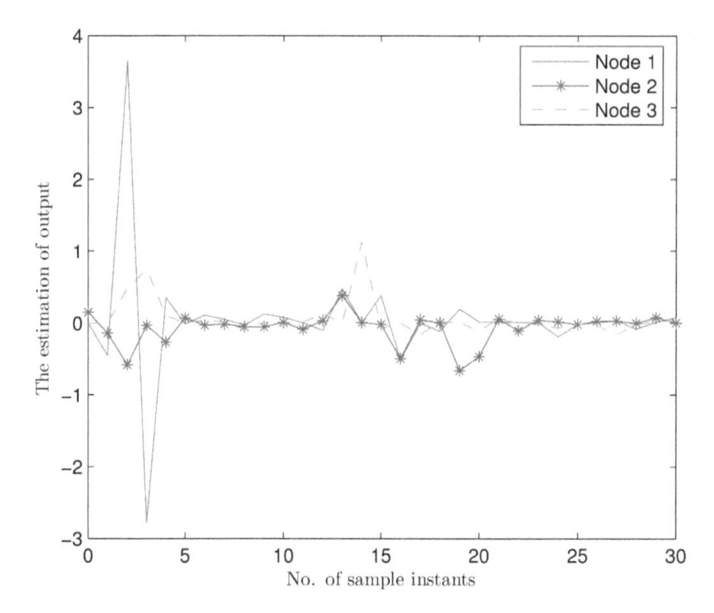

FIGURE 6.8
The estimation of output $z(u)$.

7

State Estimation for Complex Networks

This chapter deals with the variance-constrained H_∞ state estimation and the robust PNB state estimation problems for nonlinear time-varying complex networks. The phenomena of RVTs, stochastic inner coupling and measurement quantization are taken into account in the complex networks. A Kronecker delta function and Markovian jumping parameters are employed to characterize the random changes of network topologies. A Gaussian random variable is employed to model the stochastic disturbances in the inner coupling of complex networks. As a kind of incomplete measurements, measurement quantization is considered so as to account for the signal distortion phenomenon in the transmission process. Stochastic nonlinearities with known statistical characteristics are introduced to reflect the stochastic evolution of the complex networks. We aim to put forward a finite-horizon estimation method such that, in the simultaneous presence of quantized measurements and stochastic inner coupling, the prescribed variance constraints on the estimation error and the expected H_∞ performance requirements are ensured over a finite horizon. Sufficient conditions are constructed through a series of RLMIs and, subsequently, the estimator gain parameters are obtained. Besides, the robust finite-horizon H_∞ state estimator is designed for a class of time-varying complex networks with randomly occurring uncertainties, randomly occurring multiple delays as well as sensor saturations under random access protocol and available measurements from only a part of network nodes. The random access protocol is utilized to orchestrate the data transmission at each time step based on a Markov chain. Finally, simulation examples are demonstrated to illustrate the effectiveness and applicability of the developed estimator design algorithm.

DOI: 10.1201/9781003189497-7

7.1 State Estimation with Randomly Varying Topologies

7.1.1 Problem Formulation

Take account of the following array of stochastic discrete time-varying complex networks containing U coupled nodes:

$$\begin{cases} x_i(k+1) = h(k, x_i(k)) + \sum_{\varphi=1}^{s}\sum_{j=1}^{U} \delta(\chi(k), \varphi) w_{ij}^{(\varphi)} \Big(\Gamma + \omega(k)A(k)\Big) x_j(k) \\ \qquad + B_{v_i}(k)v_1(k) \\ z_i(k) = M_i(k)x_i(k), \quad i = 1, 2, \ldots, U \end{cases} \tag{7.1}$$

where $x_i(k) \in \mathbb{R}^n$ is the state vector of the ith node and $z_i(k) \in \mathbb{R}^r$ is the output of the ith node. The disturbance input $v_1(k) \in \mathbb{R}^{n_w}$ is a zero-mean Gaussian white noise sequence on the probability space $(\Omega, \mathscr{F}, \text{Prob})$ with covariance $V_1 > 0$. $A(k)$, $B_{v_i}(k)$ and $M_i(k)$ $(i = 1, 2, \ldots, U)$ are known matrices with suitable dimensions. The outer-coupling configuration matrix of the network $W^{(\varphi)} = [w_{ij}^{(\varphi)}]_{U \times U}$ $(\varphi = 1, 2, \ldots, s)$ is a non-zero matrix with $w_{ij}^{(\varphi)} \geq 0$ $(i \neq j)$. Furthermore, such coupling configuration matrices are symmetric (i.e., $W^{(\varphi)} = W^{(\varphi)T}$) and satisfy

$$\sum_{j=1}^{U} w_{ij}^{(\varphi)} = \sum_{j=1}^{U} w_{ji}^{(\varphi)} = 0, \quad i = 1, 2, \ldots, U, \ \varphi = 1, 2, \ldots, s. \tag{7.2}$$

$\Gamma = \text{diag}\{r_1 I, r_2 I, \ldots, r_n I\}$ is an inner-coupling matrix. $\omega(k) \in \mathbb{R}$ is a Gaussian white noise sequence on the probability space $(\Omega, \mathscr{F}, \text{Prob})$ with $\mathbb{E}\{\omega^2(k)\} = 1$. $\delta(\cdot, \cdot)$ is a Kronecker delta function, i.e.,

$$\delta(m, n) = \begin{cases} 0, & \text{if } m \neq n \\ 1, & \text{if } m = n. \end{cases}$$

$\chi(k)$ is a random variable which characterizes the randomly varying topologies of the complex network. In the current section, we consider the sequence $\{\chi(k)\}$ conforming to a discrete-time homogeneous Markov chain taking values in the following finite state space:

$$S = \{1, 2, \ldots, s\} \tag{7.3}$$

and $\beth = [\lambda_{\alpha\beta}]_{s \times s}$ is a transition probability matrix with its entities defined as

$$\lambda_{\alpha\beta} = \text{Prob}\{\chi(k+1) = \beta | \chi(k) = \alpha\}. \tag{7.4}$$

Remark 7.1 *As analyzed previously in the Introduction, the topologies of complex networks described by $W^{(\varphi)}$ ($\varphi = 1, 2, \ldots, s$) might vary in a probabilistic way. In fact, the transition probability matrix regulating the switches of network topologies can be determined a priori through statistical tests. It is worth pointing out that, by employing a Kronecker delta function, we can adopt the Markov process $\{\chi(k)\}$ to reflect the phenomenon of RVTs that is newly introduced to the research community of complex networks.*

Remark 7.2 *In numerous practical complex networks, it is often the situation that the inner-coupling matrix cannot be exactly known or identified owing to the uncertainties in the interconnections among subsystems. In model (7.1), we attempt to describe the stochastic coupling disturbances by using the scalar Gaussian process $\omega(k)$. Such a representation would promote the subsequent dynamics analysis while retaining the engineering insights of the random changes of the inner-coupling strengths.*

The nonlinear function $h(k, x_i(k))$ with $h(k, 0) = 0$ is a stochastic nonlinear function with the statistical characteristics as follows:

$$\begin{cases} \mathbb{E}\left\{h(k, x_i(k))\Big|x_i(k)\right\} = 0, \\ \mathbb{E}\left\{h(k, x_i(k))h^T(j, x_i(j))\big|x_i(k)\right\} = 0, \quad k \neq j, \\ \mathbb{E}\left\{h(k, x_i(k))h^T(k, x_i(k))\big|x_i(k)\right\} \\ = \sum_{r=1}^{q} \pi_r(k)\pi_r^T(k)\mathbb{E}\left\{x_i^T(k)\bar{\Gamma}_r(k)x_i(k)\right\} \\ := \sum_{r=1}^{q} \Xi_r(k)\mathbb{E}\left\{x_i^T(k)\bar{\Gamma}_r(k)x_i(k)\right\} \end{cases} \tag{7.5}$$

where $i = 1, 2, \ldots, U$ and q is a given nonnegative integer, $\Xi_r(k)$ and $\bar{\Gamma}_r(k)$ ($r = 1, 2, \ldots, q$) are given matrices with suitable dimensions.

In this section, the quantized measurement of the ith sensor is represented by

$$y_i(k) = \varrho\big(C_i(k)x_i(k)\big) + D_{v_i}(k)v_2(k), \qquad i = 1, 2, \ldots, U \tag{7.6}$$

where $y_i(k) \in \mathbb{R}^m$ means the measurement output of the ith node; and $v_2(k) \in \mathbb{R}^{n_v}$ stands for the disturbance input of the ith node, which is a zero-mean Gaussian white noise sequence on the probability space $(\Omega, \mathscr{F}, \text{Prob})$ with covariance $V_2 > 0$. It is supposed that $\chi(k)$, $\omega(k)$, $v_1(k)$, $v_2(k)$ and $h(k, x_i(k))$ ($i = 1, 2, \ldots, U$; $\varphi = 1, 2, \ldots, s$) are mutually independent. $C_i(k)$ and $D_{v_i}(k)$ ($i = 1, 2, \ldots, U$) are given matrices with suitable dimensions.

The quantizer $\varrho(\cdot)$ is represented by

$$\varrho(\vartheta) \triangleq \begin{bmatrix} \varrho_1(\vartheta_1) & \varrho_2(\vartheta_2) & \cdots & \varrho_m(\vartheta_m) \end{bmatrix}^T, \quad \forall \vartheta \in \mathbb{R}^m$$

where the quantizer $\varrho(\cdot)$ belongs to the logarithmic type, that is, for each $\varrho_j(\cdot)$ $(1 \leq j \leq m)$, the set of quantization levels is indicated by

$$\Sigma_j = \left\{ \pm u_i^{(j)}, u_i^{(j)} = \rho_j^i u_0^{(j)}, i = 0, \ \pm 1, \ \pm 2, \ \cdots \right\} \cup \{0\},$$
$$0 \leq \rho_j \leq 1, \quad u_0^{(j)} > 0,$$

and the quantizer separates the total segment in accordance with the quantization levels. Thus, the logarithmic quantizer $\varrho_j(\cdot)$ can be reflected by

$$\varrho_j(\vartheta_j) = \begin{cases} u_i^{(j)}, & \frac{1}{1+\kappa_j} u_i^{(j)} \leq \vartheta_j \leq \frac{1}{1-\kappa_j} u_i^{(j)}, \\ 0, & \vartheta_j = 0, \\ -\varrho_j(-\vartheta_j), & \vartheta_j \leq 0 \end{cases}$$

with $\kappa_j = (1-\rho_j)/(1+\rho_j)$. Based on [146], we have

$$\varrho_j(\vartheta_j) = (1 + \Delta_k^j)\vartheta_j \quad \text{with} \quad |\Delta_k^j| \leq \kappa_j. \tag{7.7}$$

Defining $\Delta_k \triangleq \text{diag}\{\Delta_k^1, \Delta_k^2, \ldots, \Delta_k^m\}$, the measurement (7.6) can be described as follows:

$$y_i(k) = \big(I + \Delta_k\big) C_i(k) x_i(k) + D_{v_i}(k) v_2(k), \qquad i = 1, 2, \ldots, U. \tag{7.8}$$

In view of the measurements $y_i(k)$ $(i = 1, 2, \ldots, U)$, we set up state estimator for complex network (7.1) as follows:

$$\begin{cases} \hat{x}_i(k+1) = L_i(k, \chi(k)) \hat{x}_i(k) + Z_i(k, \chi(k)) y_i(k), \\ \hat{z}_i(k) = M_i(k) \hat{x}_i(k), \qquad i = 1, 2, \ldots, U \end{cases} \tag{7.9}$$

where $\hat{x}_i(k) \in \mathbb{R}^n$ denotes the estimate of the state $x_i(k)$, $\hat{z}_i(k) \in \mathbb{R}^r$ means the output of estimator on the ith node, and $L_i(k, \chi(k))$ and $Z_i(k, \chi(k))$ are the estimator gains to be decided.

In order to simplify representation, we define the following notation:

$$\begin{aligned}
& x(k) \triangleq \left[x_1^T(k) \quad x_2^T(k) \quad \cdots \quad x_U^T(k) \right]^T, \\
& \hat{x}(k) \triangleq \left[\hat{x}_1^T(k) \quad \hat{x}_2^T(k) \quad \cdots \quad \hat{x}_U^T(k) \right]^T, \\
& z(k) \triangleq \left[z_1^T(k) \quad z_2^T(k) \quad \cdots \quad z_U^T(k) \right]^T, \\
& \hat{z}(k) \triangleq \left[\hat{z}_1^T(k) \quad \hat{z}_2^T(k) \quad \cdots \quad \hat{z}_U^T(k) \right]^T, \\
& \mathcal{H}(k, x(k)) \triangleq \left[h^T(k, x_1(k)) \quad h^T(k, x_2(k)) \quad \cdots \quad h^T(k, x_U(k)) \right]^T, \qquad (7.10) \\
& M(k) \triangleq \text{diag}\{M_1(k), M_2(k), \ldots, M_U(k)\}, \\
& B_v(k) \triangleq \left[B_{v_1}^T(k) \quad B_{v_2}^T(k) \quad \cdots \quad B_{v_U}^T(k) \right]^T, \\
& C(k) \triangleq \text{diag}\{C_1(k), C_2(k), \ldots, C_U(k)\}, \\
& D_v(k) \triangleq \left[D_{v_1}^T(k) \quad D_{v_2}^T(k) \quad \ldots \quad D_{v_U}^T(k) \right]^T,
\end{aligned}$$

$$L(k,\chi(k)) \triangleq \text{diag}\{L_1(k,\chi(k)), L_2(k,\chi(k)), \ldots, L_U(k,\chi(k))\},$$
$$Z(k,\chi(k)) \triangleq \text{diag}\{Z_1(k,\chi(k)), Z_2(k,\chi(k)), \ldots, Z_U(k,\chi(k))\}.$$

Let the state estimation error of the ith node be $e_i(k) \triangleq x_i(k) - \hat{x}_i(k)$ and the output estimation error of the ith node be $\tilde{z}_i(k) \triangleq z_i(k) - \hat{z}_i(k)$ $(i = 1, 2, \ldots, U)$. Then, the estimation error dynamics for the complex network is acquired from (7.1), (7.8) and (7.9) as follows:

$$\begin{aligned} e(k+1) &= \Bigg[\Big(\sum_{\varphi=1}^{s} \delta(\chi(k),\varphi) W^{(\varphi)} \otimes \Gamma\Big) - L(k,\chi(k)) - Z(k,\chi(k)) \\ &\quad \times (I + \Delta_k) C(k)\Bigg] x(k) + L(k,\chi(k)) e(k) \\ &\quad + \Big(\sum_{\varphi=1}^{s} \delta(\chi(k),\varphi) W^{(\varphi)} \otimes A(k)\Big) \omega(k) x(k) + B_v(k) v_1(k) \\ &\quad - Z(k,\chi(k)) D_v(k) v_2(k) + \mathcal{H}(k, x(k)) \\ \tilde{z}(k) &= M(k) e(k) \end{aligned} \tag{7.11}$$

where

$$\begin{aligned} e(k) &\triangleq \begin{bmatrix} e_1^T(k) & e_2^T(k) & \cdots & e_U^T(k) \end{bmatrix}^T, \\ \tilde{z}(k) &\triangleq \begin{bmatrix} \tilde{z}_1^T(k) & \tilde{z}_2^T(k) & \cdots & \tilde{z}_U^T(k) \end{bmatrix}^T. \end{aligned}$$

In consideration of (7.5), the statistical characteristics of $\mathcal{H}(k, x(k))$ is expressed by

$$\mathbb{E}\left\{\mathcal{H}(k, x(k)) \middle| x(k)\right\} = 0, \tag{7.12}$$

$$\mathbb{E}\left\{\mathcal{H}(k, x(k)) \mathcal{H}^T(j, x(k)) \middle| x(k)\right\} = 0, \quad k \neq j, \tag{7.13}$$

$$\begin{aligned} &\mathbb{E}\Big\{\mathcal{H}(k, x(k)) \mathcal{H}^T(k, x(k)) \mid x(k)\Big\} \\ &= \sum_{r=1}^{q} \left(\mathbf{1}_U \otimes \pi_r(k)\right) \left(\mathbf{1}_U \otimes \pi_r(k)\right)^T \mathbb{E}\left\{x^T(k) \left(I \otimes \bar{\Gamma}_r(k)\right) x(k)\right\} \\ &= \sum_{r=1}^{q} \left(\mathbf{1}_U \mathbf{1}_U^T\right) \otimes \Xi_r(k) \mathbb{E}\left\{x^T(k) \left(I \otimes \bar{\Gamma}_r(k)\right) x(k)\right\}. \end{aligned} \tag{7.14}$$

Letting $\bar{\Delta} \triangleq \text{diag}\{\kappa_1, \kappa_2, \ldots, \kappa_m\}$ and $F(k) \triangleq \Delta_k \bar{\Delta}^{-1}$, we know that $F(k)$ is an uncertain real-valued time-varying matrix satisfying $F^T(k) F(k) \leq I$. Besides, by setting $\eta(k) \triangleq \begin{bmatrix} x^T(k) & e^T(k) \end{bmatrix}^T$ and $v(k) \triangleq \begin{bmatrix} v_1^T(k) & v_2^T(k) \end{bmatrix}^T$, the following augmented system is obtained:

$$\left\{ \begin{aligned} \eta(k+1) =& (\mathcal{C}_w(k,\chi(k)) + \Delta\mathcal{C}_w(k,\chi(k))) \eta(k) + \mathcal{A}_w(k,\chi(k)) \\ & \times \omega(k) \eta(k) + \mathcal{I}\mathcal{H}(k, x(k)) + \mathcal{B}_k(k,\chi(k)) v(k) \\ \tilde{z}(k) =& \mathcal{M}(k) \eta(k) \end{aligned} \right. \tag{7.15}$$

where

$$
\begin{aligned}
&\mathcal{C}_w(k,\chi(k)) \triangleq \begin{bmatrix} \mathcal{T}_w(k,\chi(k)) & 0 \\ \mathcal{L}_w(k,\chi(k)) & L(k,\chi(k)) \end{bmatrix}, \quad \mathcal{M}(k) \triangleq \begin{bmatrix} 0 & M(k) \end{bmatrix}, \\
&\mathcal{T}_w(k,\chi(k)) \triangleq \sum_{\varphi=1}^{s} \delta(\chi(k),\varphi) W^{(\varphi)} \otimes \Gamma, \\
&\bar{\mathcal{A}}_w(k,\chi(k)) \triangleq \sum_{\varphi=1}^{s} \delta(\chi(k),\varphi) W^{(\varphi)} \otimes A(k), \\
&\mathcal{L}_w(k,\chi(k)) \triangleq \mathcal{T}_w(k) - L(k,\chi(k)) - Z(k,\chi(k))C(k), \\
&\bar{E}(k) \triangleq \begin{bmatrix} (I \otimes \bar{\Delta})C(k) & 0 \end{bmatrix}, \bar{H}(k,\chi(k)) \triangleq \begin{bmatrix} 0 & -Z^T(k,\chi(k)) \end{bmatrix}^T, \qquad (7.16) \\
&\mathcal{A}_w(k,\chi(k)) \triangleq \begin{bmatrix} \bar{\mathcal{A}}_w(k,\chi(k)) & 0 \\ \bar{\mathcal{A}}_w(k,\chi(k)) & 0 \end{bmatrix}, \quad \mathcal{I} \triangleq \begin{bmatrix} I \\ I \end{bmatrix}, \\
&\mathcal{B}_k(k,\chi(k)) \triangleq \begin{bmatrix} B_v(k) & 0 \\ B_v(k) & -Z(k,\chi(k))D_v(k) \end{bmatrix}, \\
&\Delta\mathcal{C}_w(k,\chi(k)) \triangleq \bar{H}(k,\chi(k))(I \otimes F(k))\bar{E}(k).
\end{aligned}
$$

The state covariance matrix of the dynamical system (7.15) is defined as

$$
\mathbb{X}(k) := \mathbb{E}\left\{\eta(k)\eta^T(k)\right\} = \mathbb{E}\left\{ \begin{bmatrix} x(k) \\ e(k) \end{bmatrix} \begin{bmatrix} x(k) \\ e(k) \end{bmatrix}^T \right\}. \qquad (7.17)
$$

The aim of this section is to design the time-varying state estimators of form (7.9) for the stochastic discrete time-varying complex network (7.1) under the incomplete measurements (7.6). Particularly, we are devoted to finding the state estimator parameters $L_i(k,\chi(k))$ and $Z_i(k,\chi(k))$ ($i = 1,2,\dots,U$, $k = 0,1,\dots,N-1$) such that the following two requirements are met simultaneously:

- (Q1): For the known disturbance attenuation level $\gamma > 0$, the positive definite matrices $\Sigma_v(\chi(k))$, $\Sigma_\eta(\chi(k))$ and the initial state $\eta(0)$, the filtering error $\tilde{z}(k)$ meets the following H_∞ performance constraint:

$$
\begin{aligned}
J_1 &\triangleq \mathbb{E}\left\{ \sum_{k=0}^{N-1} \left(\|\tilde{z}(k)\|^2 - \gamma^2 \|v(k)\|^2_{\Sigma_v(\chi(k))} \right) \right\} \\
&\quad -\gamma^2 \mathbb{E}\left\{\eta^T(0)\Sigma_\eta(\chi(k))\eta(0)\right\} \\
&< 0 \ (\forall \{\omega(k), v(k)\} \neq 0) \qquad (7.18)
\end{aligned}
$$

  where $\|v(k)\|^2_{\Sigma_v(\chi(k))} = v^T(k)\Sigma_v(\chi(k))v(k)$.

- (Q2): The estimation error covariance satisfies the following requirements:

$$
J_2 \triangleq \mathbb{E}\left\{e(k)e^T(k)\right\} \leq \Upsilon(k) \qquad (7.19)
$$

where $\Upsilon(k)$ $(0 \leq k < N)$ is a set of known matrices determining the acceptable estimation accuracy in accordance with the practical requirements.

Remark 7.3 *In comparison with the classical optimal estimator which minimizes the error covariance, the developed estimator of this section focuses on guaranteeing that the estimation error covariance is upper bounded by a specified value. Simultaneously, the performance requirement of certain disturbance attenuation level can be achieved. In other words, the desired estimator should satisfy both the covariance and H_∞ performance constraints.*

7.1.2 Analysis of H_∞ and Covariance Performances

Now we proceed to analyze the performances of the complex networks (7.1).

A. H_∞ Performance

Let us begin with the analysis of the H_∞ performance, i.e., the establishment of sufficient conditions that ensure the H_∞ performance constraint for a given estimator.

Theorem 7.1 *Consider the complex network (7.1) and let the state estimator gains $L_i(k,\alpha)$ and $Z_i(k,\alpha)$ in (7.9) be known. For a positive scalar $\gamma > 0$, positive definite matrices $\Sigma_v(\alpha) > 0$ and $\Sigma_\eta(\alpha) > 0$, the H_∞ performance constraint defined in (7.18) is fulfilled for all nonzero $v(k)$ if, with the initial condition*

$$P(0,\alpha) \leq \gamma^2 \Sigma_\eta(\alpha), \ \alpha \in S,$$

there is a set of positive definite matrices $\{P(k,\alpha)\}_{1\leq k\leq N+1,\ \alpha\in S}$ satisfying the following recursive matrix inequalities:

$$\Phi(k,\alpha) = \begin{bmatrix} \Phi_{11}(k,\alpha) & * \\ 0 & \mathcal{B}_k^T(k,\alpha)\bar{P}(k+1,\alpha)\mathcal{B}_k(k,\alpha) - \gamma^2\Sigma_v(\alpha) \end{bmatrix} < 0 \quad (7.20)$$

where

$$\begin{aligned} \Phi_{11}(k,\alpha) \triangleq & (\mathcal{C}_w(k,\alpha) + \Delta\mathcal{C}_w(k,\alpha))^T \bar{P}(k+1,\alpha)(\mathcal{C}_w(k,\alpha) + \Delta\mathcal{C}_w(k,\alpha)) \\ & + \mathcal{A}_w^T(k,\alpha)\bar{P}(k+1,\alpha)\mathcal{A}_w(k,\alpha) + \sum_{r=1}^{q} \hat{\Gamma}_r(k)\cdot \mathrm{tr}\big[\mathcal{I}^T \\ & \times \bar{P}(k+1,\alpha)\mathcal{I}\big(\mathbf{1}_U\mathbf{1}_U^T\big)\otimes \Xi_r(k)\big] - P(k,\alpha) + \mathcal{M}^T(k)\mathcal{M}(k), \end{aligned}$$

$$\bar{P}(k,\alpha) \triangleq \sum_{\beta=1}^{s} \lambda_{\alpha\beta}P(k,\beta), \ \hat{\Gamma}_r(k) \triangleq \mathrm{diag}\{I\otimes \bar{\Gamma}_r(k), 0\}.$$

Proof *Define*

$$J(k,\chi(k)=\alpha) \triangleq \eta^T(k+1)\bar{P}(k+1,\alpha)\eta(k+1) - \eta^T(k)P(k,\alpha)\eta(k). \quad (7.21)$$

In view of (7.12) and the estimator (7.9), we acquire that

$$\mathbb{E}\left\{\mathcal{H}(k,x(k))|\eta(k)\right\}=0. \tag{7.22}$$

It is known from (7.15) that

$$\begin{aligned}
&\mathbb{E}\{J(k,\chi(k)=\alpha)\}\\
=&\mathbb{E}\left\{\eta^T(k)(\mathcal{C}_w(k,\alpha)+\Delta\mathcal{C}_w(k,\alpha))^T\bar{P}(k+1,\alpha)(\mathcal{C}_w(k,\alpha)\right.\\
&+\Delta\mathcal{C}_w(k,\alpha))\eta(k)+\eta^T(k)\mathcal{A}_w^T(k,\alpha)\bar{P}(k+1,\alpha)\mathcal{A}_w(k,\alpha)\eta(k)\\
&+v^T(k)\mathcal{B}_k^T(k,\alpha)\bar{P}(k+1,\alpha)\mathcal{B}_k(k,\alpha)v(k)+\mathcal{H}^T(k,x(k))\mathcal{I}^T\\
&\left.\times\bar{P}(k+1,\alpha)\mathcal{I}\mathcal{H}(k,x(k))-\eta^T(k)P(k,\alpha)\eta(k)\right\}.
\end{aligned} \tag{7.23}$$

Taking account of (7.14), we have

$$\begin{aligned}
&\mathbb{E}\left\{\mathcal{H}^T(k,x(k))\mathcal{I}^T\bar{P}(k+1,\alpha)\mathcal{I}\mathcal{H}(k,x(k))\right\}\\
=&\mathbb{E}\left\{\mathrm{tr}\left[\mathcal{I}^T\bar{P}(k+1,\alpha)\mathcal{I}\mathcal{H}(k,x(k))\mathcal{H}^T(k,x(k))\right]\right\}\\
=&\mathbb{E}\left\{\eta^T(k)\sum_{r=1}^{q}\hat{\Gamma}_r(k)\cdot\mathrm{tr}\left[\mathcal{I}^T\bar{P}(k+1,\alpha)\mathcal{I}(\mathbf{1}_U\mathbf{1}_U^T)\otimes\Xi_r(k)\right]\eta(k)\right\}.
\end{aligned} \tag{7.24}$$

Adding the zero term $\tilde{z}^T(k)\tilde{z}(k)-\gamma^2v^T(k)\Sigma_v(\alpha)v(k)-\tilde{z}^T(k)\tilde{z}(k)+\gamma^2v^T(k)\Sigma_v(\alpha)\,v(k)$ *to* $\mathbb{E}\{J(k,\chi(k)=\alpha)\}$ *leads to*

$$\begin{aligned}
&\mathbb{E}\{J(k,\chi(k)=\alpha)\}\\
=\;&\mathbb{E}\Bigg\{\begin{bmatrix}\eta^T(k) & v^T(k)\end{bmatrix}\Phi(k,\alpha)\begin{bmatrix}\eta(k)\\ v(k)\end{bmatrix}-\tilde{z}^T(k)\tilde{z}(k)\\
&+\gamma^2v^T(k)\Sigma_v(\alpha)v(k)\Bigg\}.
\end{aligned} \tag{7.25}$$

Summing up (7.25) on both sides from 0 *to* $N-1$ *concerning* k *yields*

$$\begin{aligned}
&\sum_{k=0}^{N-1}\mathbb{E}\{J(k,\chi(k)=\alpha)\}\\
=\;&\mathbb{E}\left\{\eta^T(N)\bar{P}(N-1,\alpha)\eta(N)-\eta^T(0)P(0,\alpha)\eta(0)\right\}\\
=\;&\mathbb{E}\Bigg\{\sum_{k=0}^{N-1}\begin{bmatrix}\eta^T(k) & v^T(k)\end{bmatrix}\Phi(k,\alpha)\begin{bmatrix}\eta(k)\\ v(k)\end{bmatrix}\Bigg\}\\
&-\mathbb{E}\Bigg\{\sum_{k=0}^{N-1}\left(\tilde{z}^T(k)\tilde{z}(k)-\gamma^2v^T(k)\Sigma_v(\alpha)v(k)\right)\Bigg\}.
\end{aligned} \tag{7.26}$$

Thus, the performance index in (7.18) can be rewritten in the following form:

$$J_1=\mathbb{E}\Bigg\{\sum_{k=0}^{N-1}\begin{bmatrix}\eta^T(k) & v^T(k)\end{bmatrix}\Phi(k,\alpha)\begin{bmatrix}\eta(k)\\ v(k)\end{bmatrix}\Bigg\}$$

$$\begin{aligned}&-\mathbb{E}\{\eta^T(N)\bar{P}(N-1,\alpha)\eta(N)\} \\ &+\eta^T(0)(P(0,\alpha)-\gamma^2\Sigma_\eta(\alpha))\eta(0).\end{aligned} \tag{7.27}$$

From the conditions $\Phi(k,\alpha) < 0$ and $\bar{P}(N-1,\alpha) > 0$ and the initial condition $P(0,\alpha) \leq \gamma^2\Sigma_\eta(\alpha)$, it is seen that $J_1 < 0$. The proof is now complete.

B. Variance Analysis

Now, we are in the stage to analyze the variance-constrained estimator design issue for the addressed nonlinear stochastic complex networks.

Theorem 7.2 *Consider the complex network (7.1) and let the state estimator gains $L_i(k,\alpha)$ and $Z_i(k,\alpha)$ be known. We have $Q(k,\alpha) \geq \mathbb{X}(k)$ ($\forall k \in \{1,2,\dots,N+1\}$) if, with the initial condition $Q(0,\alpha) = \mathbb{X}(0)$ ($\alpha \in S$), there is a set of positive definite matrices $\{Q(k,\alpha)\}_{1\leq k\leq N+1,\ \alpha\in S}$ satisfying the recursive matrix inequalities as follows:*

$$\bar{Q}(k+1,\alpha) \geq \Psi(Q(k,\alpha)) \tag{7.28}$$

where

$$\begin{aligned}\bar{Q}(k,\alpha) &\triangleq \sum_{\beta=1}^{s}\lambda_{\alpha\beta}Q(k,\beta),\ V \triangleq \mathrm{diag}\{V_1,V_2\}, \\ \Psi(Q(k,\alpha)) &\triangleq (\mathcal{C}_w(k,\alpha)+\Delta\mathcal{C}_w(k,\alpha))Q(k,\alpha)(\mathcal{C}_w(k,\alpha)+\Delta\mathcal{C}_w(k,\alpha))^T \\ &\quad +\mathcal{A}_w(k,\alpha)Q(k,\alpha)\mathcal{A}_w^T(k,\alpha)+\sum_{r=1}^{q}\mathcal{I}\left(\mathbf{1}_U\mathbf{1}_U^T\right)\otimes\Xi_r(k)\mathcal{I}^T \\ &\quad \times\mathrm{tr}\left[\hat{\Gamma}_r(k)Q(k,\alpha)\right]+\mathcal{B}_k(k,\alpha)V\mathcal{B}_k^T(k,\alpha).\end{aligned} \tag{7.29}$$

Proof *On the basis of (7.19), the Lyapunov-type equation regulating the evolution of covariance $\mathbb{X}(k)$ is obtained as follows:*

$$\begin{aligned}&\mathbb{X}(k+1) \\ =\ &\mathbb{E}\left\{\eta(k+1)\eta^T(k+1)\right\} \\ =\ &(\mathcal{C}_w(k,\alpha)+\Delta\mathcal{C}_w(k,\alpha))\eta(k)\eta^T(k)(\mathcal{C}_w(k,\alpha)+\Delta\mathcal{C}_w(k,\alpha))^T \\ &+\mathcal{A}_w(k,\alpha)\eta(k)\eta^T(k)\mathcal{A}_w^T(k,\alpha)+\mathcal{I}\mathcal{H}(k,x(k))\mathcal{H}^T(k,x(k))\mathcal{I}^T \\ &+\mathcal{B}_k(k,\alpha)v(k)v^T(k)\mathcal{B}_k^T(k,\alpha).\end{aligned} \tag{7.30}$$

Owing to the fact that

$$\begin{aligned}&\mathbb{E}\{\mathcal{I}\mathcal{H}(k,x(k))\mathcal{H}^T(k,x(k))\mathcal{I}^T\} \\ =\ &\mathcal{I}\sum_{r=1}^{q}\left(\mathbf{1}_U\mathbf{1}_U^T\right)\otimes\Xi_r(k)\mathbb{E}\left\{x^T(k)\left(I\otimes\bar{\Gamma}_r(k)\right)x(k)\right\}\mathcal{I}^T\end{aligned} \tag{7.31}$$

$$= \sum_{r=1}^{q} \mathcal{I}(\mathbf{1}_U \mathbf{1}_U^T) \otimes \Xi_r(k)\mathcal{I}^T \cdot \mathrm{tr}\left[\hat{\Gamma}_r(k)\mathbb{X}(k)\right],$$

one has

$$\begin{aligned}
\mathbb{X}(k+1) &= (\mathcal{C}_w(k,\alpha) + \Delta\mathcal{C}_w(k,\alpha))\mathbb{X}(k)(\mathcal{C}_w(k,\alpha) + \Delta\mathcal{C}_w(k,\alpha))^T \\
&\quad + \mathcal{A}_w(k,\alpha)\mathbb{X}(k)\mathcal{A}_w^T(k,\alpha) + \sum_{r=1}^{q} \mathcal{I}(\mathbf{1}_U \mathbf{1}_U^T) \\
&\quad \otimes \Xi_r(k)\mathcal{I}^T \cdot \mathrm{tr}\left[\hat{\Gamma}_r(k)\mathbb{X}(k)\right] + \mathcal{B}_k(k,\alpha)V\mathcal{B}_k^T(k,\alpha) \\
&= \Psi(\mathbb{X}(k)).
\end{aligned}$$

It is clear that $Q(0,\alpha) \geq \mathbb{X}(0)$. *Letting* $Q(k,\alpha) \geq \mathbb{X}(k)$, *one acquires the following inequalities:*

$$\bar{Q}(k+1,\alpha) \geq \Psi(Q(k,\alpha)) \geq \Psi(\mathbb{X}(k)) = \mathbb{X}(k+1), \tag{7.32}$$

which accomplishes the proof.

The following corollary is easily accessible from Theorem 7.2.

Corollary 7.1 *The inequality holds*

$$\begin{aligned}
\mathbb{E}\{e(k)e^T(k)\} &= \begin{bmatrix} 0 & I \end{bmatrix} \mathbb{X}(k) \begin{bmatrix} 0 & I \end{bmatrix}^T \\
&\leq \begin{bmatrix} 0 & I \end{bmatrix} Q(k,\alpha) \begin{bmatrix} 0 & I \end{bmatrix}^T, \alpha \in S, \forall k.
\end{aligned}$$

To summarize the aforementioned analysis, we present the following theorem which contains both the H_∞ performance index and the covariance constraint simultaneously by utilizing the RLMI method.

Theorem 7.3 *Take account of the complex network (7.1) and suppose that the gains* $L_i(k,\alpha)$ *and* $Z_i(k,\alpha)$ *in (7.9) are known. For a positive scalar* $\gamma > 0$, *positive definite matrices* $\Sigma_v(\alpha) > 0$ *and* $\Sigma_\eta(\alpha) > 0$, *if there are families of positive definite matrices* $\{P(k,\alpha)\}_{1\leq k\leq N+1,\ \alpha\in S}$, $\{Q(k,\alpha)\}_{1\leq k\leq N+1,\ \alpha\in S}$ *and* $\{\eta_r(k)\}_{0\leq k\leq N}$ $(r = 1,2,\ldots,q)$ *satisfying the following recursive matrix inequalities:*

$$\begin{bmatrix} -\eta_r(k) & * \\ \mathcal{I}(\mathbf{1}_U \otimes \pi_r(k)) & -\bar{P}^{-1}(k+1,\alpha) \end{bmatrix} < 0, \tag{7.33}$$

$$\begin{bmatrix}
\bar{F}_{11}(k,\alpha) & * & * \\
0 & -\gamma^2\Sigma_v(\alpha) & * \\
\mathcal{C}_w(k,\alpha) + \Delta\mathcal{C}_w(k,\alpha) & 0 & -\bar{P}^{-1}(k+1,\alpha) \\
\mathcal{A}_w(k,\alpha) & 0 & 0 \\
\mathcal{M}(k) & 0 & 0 \\
0 & \mathcal{B}_k(k,\alpha) & 0
\end{bmatrix}$$

$$\begin{matrix} * & * & * \\ * & * & * \\ * & * & * \\ -\bar{P}^{-1}(k+1,\alpha) & * & * \\ 0 & -I & * \\ 0 & 0 & -\bar{P}^{-1}(k+1,\alpha) \end{matrix}\Bigg] < 0, \tag{7.34}$$

$$\begin{bmatrix} -\bar{Q}(k+1,\alpha) & * & * & * & * \\ Q(k,\alpha)(\mathcal{C}_w(k,\alpha)+\Delta\mathcal{C}_w(k,\alpha))^T & -Q(k,\alpha) & * & * & * \\ Q(k,\alpha)\mathcal{A}_w^T(k,\alpha) & 0 & -Q(k,\alpha) & * & * \\ \mathcal{B}_k^T(k,\alpha) & 0 & 0 & -V^{-1} & * \\ \Omega_{31}(k) & 0 & 0 & 0 & \Omega_{33}(k,\alpha) \end{bmatrix} < 0 \tag{7.35}$$

with the initial conditions

$$\begin{cases} P(0,\alpha) \leq \gamma^2 \Sigma_\eta(\alpha) \\ Q(0) = \mathbb{X}(0) \end{cases} \tag{7.36}$$

where

$$\begin{aligned} \bar{F}_{11}(k,\alpha) &\triangleq \sum_{r=1}^{q} \hat{\Gamma}_r(k)\eta_r(k) - P(k,\alpha), \\ \Omega_{33}(k,\alpha) &\triangleq \mathrm{diag}\{-\sigma_1(k,\alpha)I, -\sigma_2(k,\alpha)I, \ldots, -\sigma_q(k,\alpha)I, \}, \\ \Omega_{31}(k) &\triangleq [\mathcal{I}(\mathbf{1}_U \otimes \pi_1(k)), \cdots, \mathcal{I}(\mathbf{1}_U \otimes \pi_q(k))]^T, \\ \sigma_i(k,\alpha) &\triangleq (\mathrm{tr}[\hat{\Gamma}_i(k)Q(k,\alpha)])^{-1}, i = 1, 2, \ldots, q. \end{aligned}$$

Then, for the estimation error system (7.15), we obtain $J_1 < 0$ *and* $\mathbb{E}\{e(k)e^T(k)\} \leq \begin{bmatrix} 0 & I \end{bmatrix} Q(k,\alpha) \begin{bmatrix} 0 & I \end{bmatrix}^T$ $(\forall k \in \{0, 1, \ldots, N+1\},\ \alpha \in S)$.

Proof *Based on the aforementioned analysis on the* H_∞ *performance and the estimation error covariance, we merely need to indicate that, under initial conditions (7.36), inequalities (7.33) and (7.34) imply (7.20), and (7.35) derives (7.28).*

From Lemma 2.1, we know that

$$\big(\mathbf{1}_U \otimes \pi_r(k)\big)^T \mathcal{I}^T \bar{P}(k+1,\alpha)\mathcal{I}\big(\mathbf{1}_U \otimes \pi_r(k)\big) < \eta_r(k), (r = 1, 2, \ldots, q) \tag{7.37}$$

holds if and only if (7.33) holds. Besides, with the property of matrix trace, (7.37) can be rewritten as

$$\text{tr}\left[\mathcal{I}^T\bar{P}(k+1,\alpha)\mathcal{I}(\mathbf{1}_U\mathbf{1}_U^T)\otimes\Xi_r(k)\right] < \eta_r(k), (r=1,2,\ldots,q) \tag{7.38}$$

and (7.34) is acquired by

$$\digamma(k,\alpha) = \begin{bmatrix} \digamma_{11}(k,\alpha) & * \\ 0 & \mathcal{B}_k^T(k,\alpha)\bar{P}(k+1,\alpha)\mathcal{B}_k(k,\alpha) - \gamma^2\Sigma_v(\alpha) \end{bmatrix} < 0 \tag{7.39}$$

where

$$\begin{aligned} \digamma_{11}(k,\alpha) \triangleq& (\mathcal{C}_w(k,\alpha)+\Delta\mathcal{C}_w(k,\alpha))^T\bar{P}(k+1,\alpha)(\mathcal{C}_w(k,\alpha)+\Delta\mathcal{C}_w(k,\alpha)) \\ &+ \mathcal{A}_w^T(k,\alpha)\bar{P}(k+1,\alpha)\mathcal{A}_w(k,\alpha) + \sum_{r=1}^{q}\hat{\Gamma}_r(k)\eta_r(k) - P(k,\alpha) \\ &+ \mathcal{M}^T(k)\mathcal{M}(k). \end{aligned}$$

Thus, it is obvious that (7.20) is acquired by (7.33) and (7.34) under the same initial condition.

As a similar case, we acquire easily that (7.28) holds if and only if (7.35) holds. Hence, considering Theorems 7.1–7.2 and Corollary 7.1, the H_∞ index defined in (7.18) meets $J_1 < 0$ and the error covariance of complex networks (7.1) reaches $\mathbb{E}\{e(k)e^T(k)\} \le \begin{bmatrix}0 & I\end{bmatrix} Q(k,\alpha) \begin{bmatrix}0 & I\end{bmatrix}^T$ ($\forall k \in \{0,1,\ldots,N+1\}$, $\alpha \in S$), simultaneously. The proof is accomplished.

So far, the variance-constrained H_∞ state estimation issue has been analyzed for a kind of nonlinear time-varying complex networks with RVTs, stochastic inner coupling and measurement quantization. In the following subsection, we are to design the estimator gain matrices by using the RLMI method.

7.1.3 Design of Finite-Horizon State Estimators

In this part, an algorithm is put forward to settle the estimator design issue. It is indicated that the estimator gain matrices can be acquired via solving a set of RLMIs. That is to say, at sampling instant k ($k > 0$), by virtue of solving a series of LMIs, we can calculate the desired estimator gains and the matrices that are required for solving the LMIs at the next time step.

Theorem 7.4 *For a known disturbance attenuation level $\gamma > 0$, positive definite matrices $\Sigma_v(\alpha) > 0$ and $\Sigma_\eta(\alpha) = \begin{bmatrix}\Sigma_{\eta1}(\alpha) & \Sigma_{\eta2}(\alpha) \\ \Sigma_{\eta3}(\alpha) & \Sigma_{\eta4}(\alpha)\end{bmatrix} > 0$ and a sequence of pre-specified variance upper bounds $\{\Upsilon(k)\}_{0\le k\le N+1}$, the variance-constrained H_∞ estimator design issue is settled for the stochastic complex* *networks (7.1) with randomly varying topologies if, with the initial conditions*

$$\begin{cases} \begin{bmatrix} S(0,\alpha)-\gamma^2\Sigma_{\eta 1}(\alpha) & -\gamma^2\Sigma_{\eta 2}(\alpha) \\ -\gamma^2\Sigma_{\eta 3}(\alpha) & Y(0,\alpha)-\gamma^2\Sigma_{\eta 4}(\alpha) \end{bmatrix} \leq 0 \\ \mathbb{E}\{e(0)e^T(0)\} = Q_2(0,\alpha) \leq \Upsilon(0), \end{cases} \tag{7.40}$$

there are groups of positive definite matrices

$$\{\check{S}(k,\alpha)\}_{0\leq k\leq N+1,\ \alpha\in S},\quad \{\check{Y}(k,\alpha)\}_{0\leq k\leq N+1,\ \alpha\in S},$$
$$\{Q_1(k,\alpha)\}_{0\leq k\leq N+1,\ \alpha\in S},\quad \{Q_2(k,\alpha)\}_{0\leq k\leq N+1,\ \alpha\in S},$$

positive scalars

$$\{\rho_1(k,\alpha)\}_{0\leq k\leq N,\ \alpha\in S},\quad \{\rho_2(k,\alpha)\}_{0\leq k\leq N,\ \alpha\in S},$$
$$\{\eta_r(k)\}_{0\leq k\leq N}\ (r=1,2,\ldots,q)$$

and groups of real-valued matrices

$$\{Q_3(k,\alpha)\}_{0\leq k\leq N+1,\ \alpha\in S},\quad \{L_i(k,\alpha)\}_{0\leq k\leq N,\ \alpha\in S},\quad \{Z_i(k,\alpha)\}_{0\leq k\leq N,\ \alpha\in S}$$

satisfying the RLMIs as follows:

$$\begin{bmatrix} -\eta_r(k) & * & * \\ \mathbf{1}_U\otimes\pi_r(k) & -\check{S}(k+1,\alpha) & * \\ \mathbf{1}_U\otimes\pi_r(k) & 0 & -\check{Y}(k+1,\alpha) \end{bmatrix} < 0, \tag{7.41}$$

$$\begin{bmatrix} \daleth_{11}(k,\alpha) & * \\ \daleth_{21}(k,\alpha) & \daleth_{22}(k,\alpha) \end{bmatrix} < 0, \tag{7.42}$$

$$\begin{bmatrix} \Theta_{11}(k,\alpha) & * \\ \Theta_{21}(k,\alpha) & \Theta_{22}(k,\alpha) \end{bmatrix} < 0, \tag{7.43}$$

$$\bar{Q}_2(k+1,\alpha)-\Upsilon(k+1)\leq 0 \tag{7.44}$$

with the parameters updated by

$$\sum_{\beta=1}^{s}\lambda_{\alpha\beta}S(k+1,\beta)=\check{S}^{-1}(k+1,\alpha),\quad \sum_{\beta=1}^{s}\lambda_{\alpha\beta}Y(k+1,\beta)=\check{Y}^{-1}(k+1,\alpha) \tag{7.45}$$

where

$$\beth_{11}(k,\alpha) \triangleq \begin{bmatrix} \beth_{1111}(k,\alpha) & * & * & * & * \\ 0 & -Y(k,\alpha) & * & * & * \\ 0 & 0 & -\gamma^2\Sigma_v(\alpha) & * & * \\ \sum_{\varphi=1}^{s}\delta(\alpha,\varphi)W^{(\varphi)}\otimes\Gamma & 0 & 0 & -\check{S}(k+1,\alpha) & * \\ \beth_{1151}(k,\alpha) & L(k,\alpha) & 0 & 0 & -\check{Y}(k+1,\alpha) \end{bmatrix},$$

$$\beth_{1111}(k,\alpha) \triangleq \sum_{r=1}^{q} I\otimes(\bar{\Gamma}_r(k)\eta_r(k)) - S(k,\alpha) + \rho_1(k,\alpha)C^T(k)\left(I\otimes\bar{\Delta}^T\bar{\Delta}\right)C(k),$$

$$\beth_{1151}(k,\alpha) \triangleq \sum_{\varphi=1}^{s}\delta(\alpha,\varphi)W^{(\varphi)}\otimes\Gamma - L(k,\alpha) - Z(k,\alpha)C(k),$$

$$\beth_{21}(k,\alpha) \triangleq \begin{bmatrix} \sum_{\varphi=1}^{s}\delta(\alpha,\varphi)W^{(\varphi)}\otimes A(k) & 0 & 0 & 0 & 0 & 0 \\ \sum_{\varphi=1}^{s}\delta(\alpha,\varphi)W^{(\varphi)}\otimes A(k) & 0 & 0 & 0 & 0 & 0 \\ 0 & M(k) & 0 & 0 & 0 & 0 \\ 0 & 0 & B_v(k) & 0 & 0 & 0 \\ 0 & 0 & B_v(k) & -Z(k,\alpha)D_v(k) & 0 & 0 \\ 0 & 0 & 0 & 0 & 0 & -Z^T(k,\alpha) \end{bmatrix},$$

$$\beth_{22}(k,\alpha) \triangleq \mathrm{diag}\{-\check{S}(k+1,\alpha), -\check{Y}(k+1,\alpha), -I, -\check{S}(k+1,\alpha), -\check{Y}(k+1,\alpha), -\rho_1(k,\alpha)I\},$$

$$\Theta_{11}(k,\alpha) \triangleq \begin{bmatrix} -\bar{Q}_1(k+1,\alpha) & * & * & * & * & * \\ -\bar{Q}_3(k+1,\alpha) & -\bar{Q}_2(k+1,\alpha) & * & * & * & * \\ \Theta_{1131}(k,\alpha) & \Theta_{1132}(k,\alpha) & -Q_1(k,\alpha) & * & * & * \\ \Theta_{1141}(k,\alpha) & \Theta_{1142}(k,\alpha) & -Q_3(k,\alpha) & -Q_2(k,\alpha) & * & * \\ \Theta_{1151}(k,\alpha) & \Theta_{1152}(k,\alpha) & 0 & 0 & -Q_1(k,\alpha) & * \\ \Theta_{1161}(k,\alpha) & \Theta_{1162}(k,\alpha) & 0 & 0 & -Q_3(k,\alpha) & -Q_2(k,\alpha) \end{bmatrix},$$

$$\Theta_{1132}(k,\alpha) \triangleq Q_1(k,\alpha)\left(\sum_{\varphi=1}^{s}\delta(\alpha,\varphi)W^{(\varphi)}\otimes\Gamma - L(k,\alpha) - Z(k,\alpha)C(k)\right)^T + Q_3^T(k)L^T(k,\alpha),$$

$$\Theta_{1142}(k,\alpha) \triangleq Q_3(k,\alpha)\left(\sum_{\varphi=1}^{s}\delta(\alpha,\varphi)W^{(\varphi)}\otimes\Gamma - L(k,\alpha) - Z(k,\alpha)C(k)\right)^T + Q_2(k)L^T(k,\alpha),$$

$$\Theta_{1131}(k,\alpha) \triangleq Q_1(k,\alpha)\left(\sum_{\varphi=1}^{s}\delta(\alpha,\varphi)W^{(\varphi)}\otimes\Gamma\right)^T,$$

$$\Theta_{1141}(k,\alpha) \triangleq Q_3(k,\alpha)\left(\sum_{\varphi=1}^{s}\delta(\alpha,\varphi)W^{(\varphi)}\otimes\Gamma\right)^T,$$

$$\Theta_{1151}(k,\alpha) = \Theta_{1152}(k,\alpha) \triangleq Q_1(k,\alpha)\left(\sum_{\varphi=1}^{s}\delta(\alpha,\varphi)W^{(\varphi)}\otimes A(k)\right)^T,$$

$$\bar{Q}_1(k,\alpha) \triangleq \sum_{\beta=1}^{s}\lambda_{\alpha\beta}Q_1(k,\beta),\ \bar{Q}_2(k,\alpha) \triangleq \sum_{\beta=1}^{s}\lambda_{\alpha\beta}Q_2(k,\beta),$$

$$\Theta_{1161}(k,\alpha) = \Theta_{1162}(k,\alpha) \triangleq Q_3(k,\alpha)\left(\sum_{\varphi=1}^{s}\delta(\alpha,\varphi)W^{(\varphi)}\otimes A(k)\right)^T,$$

$$\Theta_{21}(k,\alpha) \triangleq \begin{bmatrix} \hat{F}(k) & \hat{F}(k) & 0 \\ B_v^T(k) & B_v^T(k) & 0 \\ 0 & -D_v^T(k)Z^T(k,\alpha) & 0 \\ 0 & -Z^T(k,\alpha) & 0 \\ 0 & 0 & \Theta_{210}(k,\alpha) \end{bmatrix}$$

$$\begin{bmatrix} 0 & 0 & 0 \\ 0 & 0 & 0 \\ 0 & 0 & 0 \\ 0 & 0 & 0 \\ \rho_2(k,\alpha)(I \otimes \bar{\Delta})C(k)Q_3^T(k,\alpha) & 0 & 0 \end{bmatrix},$$

$$\Theta_{22}(k,\alpha) \triangleq \text{diag}\{\tilde{\digamma}(k,\alpha), -V_1^{-1}, -V_2^{-1}, -\rho_2(k,\alpha)I, -\rho_2(k,\alpha)I\},$$

$$\bar{Q}_3(k,\alpha) \triangleq \sum_{\beta=1}^{s} \lambda_{\alpha\beta} Q_3(k,\beta),$$

$$\tilde{\digamma}(k,\alpha) \triangleq \text{diag}\{-(\text{tr}\tilde{Q}_r(k,\alpha))^{-1}I, \ldots, -(\text{tr}\tilde{Q}_r(k,\alpha))^{-1}I\},$$

$$\Theta_{210}(k,\alpha) \triangleq \rho_2(k,\alpha)(I \otimes \bar{\Delta})C(k)Q_1^T(k,\alpha),$$

$$\tilde{Q}_r(k,\alpha) \triangleq \begin{bmatrix} (I \otimes \bar{\Gamma}_r(k))Q_1(k,\alpha) & (I \otimes \bar{\Gamma}_r(k))Q_3^T(k,\alpha) \\ 0 & 0 \end{bmatrix},$$

$$\hat{\digamma}(k) \triangleq \begin{bmatrix} \mathbf{1}_U \otimes \pi_1(k) & \mathbf{1}_U \otimes \pi_r(k) & \cdots & \mathbf{1}_U \otimes \pi_q(k) \end{bmatrix}^T.$$

Proof *The proof is based on Theorem 7.3. Decomposing the variables $P(k,\alpha)$ and $Q(k,\alpha)$ as follows:*

$$P(k,\alpha) = \begin{bmatrix} S(k,\alpha) & 0 \\ 0 & Y(k,\alpha) \end{bmatrix}, \; P^{-1}(k,\alpha) = \begin{bmatrix} \hat{S}(k,\alpha) & 0 \\ 0 & \hat{Y}(k,\alpha) \end{bmatrix},$$

$$\bar{P}(k,\alpha) = \begin{bmatrix} \bar{S}(k,\alpha) & 0 \\ 0 & \bar{Y}(k,\alpha) \end{bmatrix}, \; \bar{S}(k,\alpha) = \sum_{\beta=1}^{s} \lambda_{\alpha\beta} S(k,\beta),$$

$$\bar{Y}(k,\alpha) = \sum_{\beta=1}^{s} \lambda_{\alpha\beta} Y(k,\beta), \; \bar{P}^{-1}(k,\alpha) = \begin{bmatrix} \check{S}(k,\alpha) & 0 \\ 0 & \check{Y}(k,\alpha) \end{bmatrix},$$

$$Q(k,\alpha) = \begin{bmatrix} Q_1(k,\alpha) & * \\ Q_3(k,\alpha) & Q_2(k,\alpha) \end{bmatrix}.$$

It is convenient to see that (7.41) and (7.33) are equivalent.

For dealing with the uncertainty $F(k)$, we rewrite (7.34) as follows:

$$\begin{bmatrix} \digamma_{11}(k,\alpha) & * & * \\ 0 & -\gamma^2\Sigma_v(\alpha) & * \\ \mathcal{C}_w(k,\alpha) & 0 & -\bar{P}^{-1}(k+1,\alpha) \\ \mathcal{A}_w(k,\alpha) & 0 & 0 \\ \mathcal{M}(k) & 0 & 0 \\ 0 & \mathcal{B}_k(k,\alpha) & 0 \end{bmatrix}$$

$$\begin{array}{ccc} * & * & * \\ * & * & * \\ * & * & * \\ -\bar{P}^{-1}(k+1,\alpha) & * & * \\ 0 & -I & * \\ 0 & 0 & -\bar{P}^{-1}(k+1,\alpha) \end{array}\Bigg] \tag{7.46}$$

$$+\hat{H}(k,\alpha)(I \otimes F(k))\hat{E}(k) + \hat{E}^T(k)(I \otimes F(k))^T \hat{H}^T(k,\alpha) < 0$$

where

$$\hat{H}^T(k,\alpha) \triangleq \begin{bmatrix} 0 & 0 & \bar{H}^T(k,\alpha) & 0 & 0 & 0 \end{bmatrix},\ \hat{E}(k) \triangleq \begin{bmatrix} \bar{E}(k) & 0 & 0 & 0 & 0 & 0 \end{bmatrix}.$$

From Lemmas 2.1 and 3.1, it is seen that (7.34) holds if and only if (7.42) holds. Simultaneously, we recognize that (7.35) holds if and only if (7.43) holds. Therefore, based on Theorem 7.3, we have $J < 0$ *and* $\mathbb{E}\{e(k)e^T(k)\} \leq \begin{bmatrix} 0 & I \end{bmatrix} Q(k,\alpha) \begin{bmatrix} 0 & I \end{bmatrix}^T$ $(\forall k \in \{0,1,\ldots,N+1\})$. *From (7.44), it is convenient to see that*

$$\mathbb{E}\{e(k)e^T(k)\} \leq \begin{bmatrix} 0 & I \end{bmatrix} Q(k,\alpha) \begin{bmatrix} 0 & I \end{bmatrix}^T \leq \Upsilon(k), \quad (\forall k \in \{0,1,\ldots,N\}).$$

It is now concluded that the performance indices (Q1) and (Q2) are both achieved, which completes the proof.

In accordance with Theorem 7.4, we summarize the Variance-constrained Estimator Design (*VED*) algorithm as follows.

Algorithm *VED*	
Step 1.	Given the H_∞ performance index γ, the positive definite matrices $\Sigma_v(\alpha)$ and $\Sigma_\eta(\alpha)$, and the initial conditions $x_i(0)$ and $\hat{x}_i(0)$, select the matrices $\{Q_1(0,\alpha), S(0,\alpha), Y(0,\alpha)\}$ which meet the initial condition (7.40).
Step 2.	Acquire the values of positive definite matrices $\{\check{S}(k+1,\alpha), \check{Y}(k+1,\alpha), \bar{Q}_1(k+1,\alpha), \bar{Q}_2(k+1,\alpha)\}$, matrix $\{\bar{Q}_3(k+1,\alpha)\}$ and estimator gains $\{L_i(k,\alpha), Z_i(k,\alpha)\}$ at the sampling time step k by solving the LMIs (7.41)-(7.44).
Step 3.	Set $k = k+1$ and derive $\{S(k+1,\alpha), Y(k+1,\alpha)\}$ by adopting the parameter update formula (7.45).
Step 4.	If $k < N$, then go to Step 2, else go to Step 4.
Step 5.	Stop.

Remark 7.4 *In Theorem 7.4, the gain parameters of the variance-constrained* H_∞ *estimator under measurement quantization are acquired over*

a finite horizon via solving recursive linear matrix inequalities (7.41)–(7.44). Such an estimator is obtained recursively and is therefore fit for online implementation. It is worth mentioning that for the stochastic discrete time-varying complex network (7.1) considered in this section, there are several major aspects which make the design of the state estimator complex, i.e. RVTs, stochastic inner coupling, measurement quantization and nonlinearities. In Theorem 7.4, sufficient conditions, which contain all of the information on these aspects, are set up for a finite-horizon state estimator to meet the prescribed H_∞ performance requirement and error variance constraints.

7.2 Partial-Nodes-Based State Estimation under Random Access Protocol

7.2.1 Problem Formulation

Upon a finite horizon $[0, \mathrm{N}]$, we take account of the following class of time-varying CNs with Λ coupled nodes:

$$\begin{aligned} x_\imath(\hbar+1) &= (A_\imath(\hbar)+\phi(\hbar)\Delta A_\imath(\hbar))x_\imath(\hbar)+F_\imath(\hbar)\varsigma(x_\imath(\hbar)) \\ &\quad +A_{m\imath}(\hbar)\sum_{l=1}^{d}\theta_{\imath l}(\hbar)x_\imath(\hbar-m_l(\hbar)) \\ &\quad +\sum_{\jmath=1}^{\Lambda}w_{\imath\jmath}\Gamma x_\jmath(\hbar)+E_\imath(\hbar)\varpi(\hbar), \\ y_\imath(\hbar) &= \sigma(C_\imath(\hbar)x_\imath(\hbar))+D_\imath(\hbar)\varpi(\hbar), \\ z_\imath(\hbar) &= H_\imath(\hbar)x_\imath(\hbar), \\ x_\imath(\jmath) &= \rho_\imath(\jmath), \quad \imath=1,2,\ldots,\Lambda;\ \jmath\in\mathbb{Z} \end{aligned} \tag{7.47}$$

where $x_\imath(\hbar)=\begin{bmatrix}x_{\imath1}(\hbar) & x_{\imath2}(\hbar) & \cdots & x_{\imath r_x}(\hbar)\end{bmatrix}^T \in \mathbb{R}^{r_x}$ denotes the state vector of the $\imath$th node, $y_\imath(\hbar)\in\mathbb{R}^{r_y}$ $(1\le r_y\le r_x)$ represents the measurement signal, and $z_\imath(\hbar)\in\mathbb{R}^{r_z}$ stands for the output to be estimated. The nonlinear vector-valued function $\varsigma(\cdot)$ $(\varsigma(0)=0)$ is constrained by

$$\|\varsigma(x)-\varsigma(y)\|\le\|\Omega(x-y)\|, \quad \forall x,y\in\mathbb{R}^{r_x} \tag{7.48}$$

where $\Omega = \mathrm{diag}\{\alpha_1,\alpha_2,\ldots,\alpha_{r_x}\} > 0$ is a given matrix. $\Gamma = \mathrm{diag}\{\gamma_1,\gamma_2,\ldots,\gamma_{r_x}\}\ge 0$ describes an inner-coupling matrix which connects the $\jmath$th $(\jmath=1,2,\ldots,r_x)$ state variable if $\gamma_\jmath\neq 0$. $\varpi(\hbar)\in l_2([0,\mathrm{N}];\mathbb{R})$ means the disturbance. $A_\imath(\hbar)$, $A_{m\imath}(\hbar)$, $F_\imath(\hbar)$, $E_\imath(\hbar)$, $C_\imath(\hbar)$, $D_\imath(\hbar)$ and $H_\imath(\hbar)$ are given real constant matrices with suitable dimensions. $\rho_\imath(\jmath)$ $(\imath=1,2,\ldots,\Lambda,\ \jmath\in\mathbb{Z})$ denote the initial values of nodes. $W=[w_{\imath\jmath}]\in\mathbb{R}^{\Lambda\times\Lambda}$ is the outer-coupling configuration matrix of the network with $w_{ij} \geq 0$ $(i \neq j)$ but not all zero. As usual, $W = [w_{ij}]$ is symmetric and satisfies

$$\sum_{j=1}^{\Lambda} w_{ij} = \sum_{j=1}^{\Lambda} w_{ji} = 0, \; i = 1, 2, \ldots, \Lambda.$$

$\phi(\hbar)$ is a random variable obeying the Gaussian distribution whose expectation is $\bar{\phi}$ and variance is $\tilde{\phi}^2$.

In the term $\sum_{l=1}^{d} \theta_{il}(\hbar) x(\hbar - m_l(\hbar))$, $m_l(\hbar)$ $(l = 1, 2, \ldots, d)$ describes the stochastic time-varying communication delays satisfying $\underline{m} \leq m_l(\hbar) \leq \bar{m}$, where $\underline{m}$ and $\bar{m}$ are known positive integers representing the lower and upper bounds of the communication delays. The stochastic variable $\theta_{il}(\hbar)$ is a Bernoulli-distributed sequence taking values of 1 or 0 with

$$\text{Prob}\{\theta_{il}(\hbar) = 1\} = \bar{\theta}_{il}, \; \text{Prob}\{\theta_{il}(\hbar) = 0\} = 1 - \bar{\theta}_{il}.$$

The parameter uncertainty $\Delta A_i(\hbar)$ satisfies the following norm-bounded form:

$$\Delta A_i(\hbar) = A_{1i}(\hbar) A_{2i}(\hbar) A_{3i}(\hbar) \tag{7.49}$$

where $A_{1i}(\hbar)$ and $A_{3i}(\hbar)$ are known constant matrices with proper dimensions, and $A_{2i}(\hbar)$ is the uncertain matrix with $A_{2i}^T(\hbar) A_{2i}(\hbar) \leq I$.

The term $\sigma(C_i(\hbar) x_i(\hbar))$ represents the sensor saturations where the saturation function $\sigma(\cdot)$: $\mathbb{R}^{r_y} \mapsto \mathbb{R}^{r_y}$ is defined as follows:

$$\sigma(a) \triangleq \begin{bmatrix} \sigma_1(a_1) & \sigma_2(a_2) & \cdots & \sigma_{r_y}(a_{r_y}) \end{bmatrix}^T.$$

Here, $\sigma_\ell(a_\ell) = \text{sign}(a_\ell) \min\{a_{\ell,\max}, |a_\ell|\}$ $(\ell = 1, 2, \ldots, r_y)$ and $a_{\ell,\max}$ is the ℓth element of the saturation level $a_{\max}$.

For the saturation function $\sigma(C_i(\hbar) x_i(\hbar))$ in (7.47), there exist two diagonal matrices L_1 and L_2 satisfying $0 \leq L_1 < I \leq L_2$ such that the following decomposition holds:

$$\sigma(C_i(\hbar) x_i(\hbar)) = L_1 C_i(\hbar) x_i(\hbar) + \varrho(C_i(\hbar) x_i(\hbar)) \tag{7.50}$$

where $\varrho(C_i(\hbar) x_i(\hbar))$ is the nonlinear vector-valued function satisfying the following sector condition:

$$\varrho^T(C_i(\hbar) x_i(\hbar)) \Big(\varrho(C_i(\hbar) x_i(\hbar)) - L C_i(\hbar) x_i(\hbar) \Big) \leq 0 \tag{7.51}$$

with $L = L_2 - L_1$.

Denoting $o(\hbar) \triangleq \begin{bmatrix} o_1^T(\hbar) & o_2^T(\hbar) & \cdots & o_\Lambda^T(\hbar) \end{bmatrix}^T$ $(o = x, z)$, a compact form of $x_i(\hbar + 1)$ and $z_i(\hbar)$ $(i = 1, 2, \ldots, \Lambda)$ is obtained from (7.47) as follows:

$$x(\hbar + 1) \;\; = \;\; \big(A(\hbar) + \bar{\phi} \Delta A(\hbar)\big) x(\hbar) + \tilde{\phi}(\hbar) \Delta A(\hbar) x(\hbar)$$

$$\begin{aligned}
&+\sum_{l=1}^{d}\bar{\Theta}_l A_m(\hbar)x(\hbar-m_l(\hbar)) \\
&+\sum_{l=1}^{d}\tilde{\Theta}_l(\hbar)A_m(\hbar)x(\hbar-m_l(\hbar))+(W\otimes\Gamma)x(\hbar) \\
&+F(\hbar)\zeta(x(\hbar))+E(\hbar)\varpi(\hbar), \\
z(\hbar) &= H(\hbar)x(\hbar) \qquad (7.52)
\end{aligned}$$

where

$$\begin{aligned}
\digamma(\hbar) &\triangleq \mathrm{diag}\{\digamma_1(\hbar),\digamma_2(\hbar),\ldots,\digamma_\Lambda(\hbar)\}\ (\digamma=A,\Delta A,A_m,F,H), \\
\tilde{\theta}_{\imath l}(\hbar) &\triangleq \theta_{\imath l}(\hbar)-\bar{\theta}_{\imath l},\bar{\Theta}_l\triangleq \mathrm{diag}\{\bar{\theta}_{1l}I,\bar{\theta}_{2l}I,\ldots,\bar{\theta}_{\Lambda l}I\}, \\
\tilde{\Theta}_l(\hbar) &\triangleq \mathrm{diag}\{\tilde{\theta}_{1l}(\hbar)I,\tilde{\theta}_{2l}(\hbar)I,\ldots,\tilde{\theta}_{\Lambda l}(\hbar)I\}, \\
E(\hbar) &\triangleq \begin{bmatrix} E_1^T(\hbar) & E_2^T(\hbar) & \cdots & E_\Lambda^T(\hbar)\end{bmatrix}^T, \\
\zeta(x(\hbar)) &\triangleq \begin{bmatrix}\zeta^T(x_1(\hbar)) & \zeta^T(x_2(\hbar)) & \cdots & \zeta^T(x_\Lambda(\hbar))\end{bmatrix}^T.
\end{aligned}$$

Without loss of generality, we assume that only the outputs of the first μ $(1\le\mu<\Lambda)$ nodes can be accessed. The compact form of $y_\imath(\hbar)$ $(\imath=1,2,\ldots,\mu)$ is acquired as follows:

$$y(\hbar)=\bar{C}(\hbar)x(\hbar)+\varrho(\bar{C}(\hbar)x(\hbar))+D(\hbar)\varpi(\hbar) \qquad (7.53)$$

where

$$\begin{aligned}
&y(\hbar)\triangleq\begin{bmatrix}y_1^T(\hbar) & y_2^T(\hbar) & \cdots & y_\mu^T(\hbar)\end{bmatrix}^T,\bar{C}(\hbar)\triangleq\begin{bmatrix}C(\hbar) & 0\end{bmatrix}, \\
&\varrho(\bar{C}(\hbar)x(\hbar))\triangleq\big[\ \varrho^T(C_1(\hbar)x_1(\hbar)) \quad \varrho^T(C_2(\hbar)x_2(\hbar)) \\
&\qquad\qquad \cdots \quad \varrho^T(C_\mu(\hbar)x_\mu(\hbar))\ \big]^T, \\
&D(\hbar)\triangleq\begin{bmatrix}D_1^T(\hbar) & D_2^T(\hbar) & \cdots & D_\mu^T(\hbar)\end{bmatrix}^T, \\
&C(\hbar)\triangleq\mathrm{diag}\{L_1C_1(\hbar),L_1C_2(\hbar),\ldots,L_1C_\mu(\hbar)\}.
\end{aligned}$$

For the CN (7.47), the available output signals are only from the first μ nodes. Furthermore, in order to save the transmission cost and alleviate the communication burden, the RAP is adopted to determine which node has the privilege to transmit the signal to the estimator (to be discussed later) at one time step. Denote $r(\hbar)$ $(\hbar\in[0,\mathrm{N}],\ r(\hbar)\in\{1,2,\ldots,\mu\})$ as the randomly selected node which obtains access to the communication network at instant $\hbar$. The evolution of $r(\hbar)$ can be modelled by a Markov chain, whose transition probability matrix is $\Pi=[\pi_{\epsilon\kappa}]_{\mu\times\mu}$ with

$$\pi_{\epsilon\kappa}=\mathrm{Prob}\{r(\hbar+1)=\kappa|r(\hbar)=\epsilon\},\quad \forall\epsilon,\kappa\in\{1,2,\ldots,\mu\}$$

where $\pi_{\epsilon\kappa}\ge0$ is the transition probability from mode ϵ to κ at time instant $\hbar$ and $\sum_{\kappa=1}^{r_y}\pi_{\epsilon\kappa}=1$.

Let $\tilde{y}_\imath(\hbar)$ $(\imath = 1, 2, \ldots, \mu)$ be the output signal of node $\imath$ after transmitting through the network under the RAP. The updating rule for $\tilde{y}_\imath(\hbar)$ can be expressed by

$$\tilde{y}_\imath(\hbar) = \begin{cases} y_\imath(\hbar), & \text{if} \quad \imath = r(\hbar), \\ \tilde{y}_\imath(\hbar - 1), & \text{otherwise} \end{cases} \tag{7.54}$$

with $\tilde{y}_\imath(\jmath - 1) = \psi_\imath(\jmath)$ $(\imath = 1, 2, \ldots, \mu;\ \jmath \in \mathbb{Z})$, which can be further written by

$$\tilde{y}_\imath(\hbar) = \delta(r(\hbar) - \imath)y_\imath(\hbar) + (1 - \delta(r(\hbar) - \imath))\tilde{y}_\imath(\hbar - 1) \tag{7.55}$$

where $\delta(\cdot) \in \{0, 1\}$ is the Kronecker delta function. Then, one has

$$\tilde{y}(\hbar) \;=\; \Pi_{r(\hbar)}y(\hbar) + (I - \Pi_{r(\hbar)})\tilde{y}(\hbar - 1) \tag{7.56}$$

where

$$\begin{aligned} \tilde{y}(\hbar) &\triangleq \begin{bmatrix} \tilde{y}_1^T(\hbar) & \tilde{y}_2^T(\hbar) & \cdots & \tilde{y}_\mu^T(\hbar) \end{bmatrix}^T, \\ \Pi_{r(\hbar)} &\triangleq \mathrm{diag}\{\delta(r(\hbar) - 1)I, \delta(r(\hbar) - 2)I, \ldots, \delta(r(\hbar) - \mu)I\}. \end{aligned}$$

Remark 7.5 *In the traditional data transmission mechanism, all signals are transmitted through the shared network simultaneously, which may result in network congestion, data corruption and waste of network resource. Recently, much research has been carried out to reduce the communication frequency and the data transmission amount. For example, by utilizing the event-triggered mechanism, signals would be transmitted into the network only when a certain event-triggering condition is satisfied [155]. Besides, by introducing the protocols into the network, only one node is allowed to access the network to transmit signals at each time step, and the determination of transmission node may be random/cyclic or may conform to a specified quadratic selection principle. Here, we adopt the random access protocol to regulate the data transmission, that is, one node is randomly selected at every transmission instant to send measurement signals while the other nodes keep the output values of the last instant.*

Denoting $\tilde{x}(\hbar) \triangleq \begin{bmatrix} x^T(\hbar) & \tilde{y}^T(\hbar - 1) \end{bmatrix}^T$ and $r(\hbar) \triangleq \epsilon$, one has

$$\begin{aligned} \tilde{x}(\hbar + 1) &= \big(\bar{A}_\epsilon(\hbar) + \bar{\phi}\Delta\bar{A}(\hbar)\big)\tilde{x}(\hbar) + \bar{\Pi}_\epsilon\varrho(\bar{C}(\hbar)x(\hbar)) \\ &\quad + \tilde{\phi}(\hbar)\Delta\bar{A}(\hbar)\tilde{x}(\hbar) + \bar{F}(\hbar)\varsigma(I_\varsigma\tilde{x}(\hbar)) \\ &\quad + \sum_{l=1}^{d}\bar{A}_{ml}(\hbar)\tilde{x}(\hbar - m_l(\hbar)) + \sum_{l=1}^{d}\tilde{A}_{ml}(\hbar) \\ &\quad \times\tilde{x}(\hbar - m_l(\hbar)) + \bar{E}_\epsilon(\hbar)\varpi(\hbar), \\ \tilde{y}(\hbar) &= \bar{C}_{1\epsilon}(\hbar)\tilde{x}(\hbar) + \Pi_\epsilon\varrho(\bar{C}(\hbar)x(\hbar)) + \Pi_\epsilon D(\hbar)\varpi(\hbar), \\ z(\hbar) &= H(\hbar)I_\varsigma\tilde{x}(\hbar) \end{aligned} \tag{7.57}$$

where

$$\bar{A}_\epsilon(\hbar) \triangleq \begin{bmatrix} A(\hbar)+W\otimes\Gamma & 0 \\ \Pi_\epsilon\bar{C}(\hbar) & I-\Pi_\epsilon \end{bmatrix}, \bar{F}(\hbar) \triangleq \begin{bmatrix} F(\hbar) \\ 0 \end{bmatrix},$$
$$\bar{E}_\epsilon(\hbar) \triangleq \begin{bmatrix} E^T(\hbar) & (\Pi_\epsilon D(\hbar))^T \end{bmatrix}^T, \bar{\Pi}_\epsilon \triangleq \begin{bmatrix} 0 & \Pi_\epsilon^T \end{bmatrix}^T,$$
$$\Delta\bar{A}(\hbar) \triangleq \mathrm{diag}\{\Delta A(\hbar), 0\}, \bar{C}_{1\epsilon}(\hbar) \triangleq \begin{bmatrix} \Pi_\epsilon\bar{C}(\hbar) & I-\Pi_\epsilon \end{bmatrix},$$
$$\tilde{A}_{ml}(\hbar) \triangleq \mathrm{diag}\{\tilde{\Theta}_l(\hbar)A_m(\hbar), 0\}, \bar{A}_{ml}(\hbar) \triangleq \mathrm{diag}\{\bar{\Theta}_l A_m(\hbar), 0\},$$
$$I_\zeta \triangleq \begin{bmatrix} I_{(Sr_x)\times(Sr_x)} & 0_{(Sr_x)\times(\mu r_y)} \end{bmatrix}.$$

Taking into account of the harsh working conditions and changeable network environments, measurements from partial nodes may be inaccessible. Thus, one needs to estimate the states of the whole network via outputs from only partial network nodes. In order to estimate the states of network (7.47) based on the measurements of the first μ nodes, the following state estimators are constructed:

$$\begin{aligned}
\hat{x}_{1,\imath}(\hbar+1) =& G_{1,\imath\epsilon}(\hbar)\hat{x}_{1,\imath}(\hbar) + K_{1,\imath\epsilon}(\hbar)\tilde{y}_\imath(\hbar), \quad \imath = 1,2,\ldots,\mu \\
\hat{x}_{2,\imath}(\hbar+1) =& G_{2,\imath\epsilon}(\hbar)\hat{x}_{2,\imath}(\hbar) + K_{2,\imath\epsilon}(\hbar)\tilde{y}_\imath(\hbar), \quad \imath = 1,2,\ldots,\mu \\
\hat{x}_{1,\imath}(\hbar+1) =& G_{1,\imath\epsilon}(\hbar)\hat{x}_{1,\imath}(\hbar), \quad \imath = \mu+1,\mu+2,\ldots,\Lambda \\
\hat{z}_\imath(\hbar) =& M_{\imath\epsilon}(\hbar)\hat{x}_{1,\imath}(\hbar), \quad \imath = 1,2,\ldots,\Lambda \\
\hat{x}_{1,\imath}(\jmath) =& \varphi_{1,\imath}(\jmath), \quad \imath = 1,2,\ldots,\Lambda; \ \jmath \in \mathbb{Z} \\
\hat{x}_{2,\imath}(\jmath) =& \varphi_{2,\imath}(\jmath), \quad \imath = 1,2,\ldots,\mu; \ \jmath \in \mathbb{Z}
\end{aligned} \tag{7.58}$$

where $\hat{x}_{1,\imath}(\hbar)$ is the estimate of $x_\imath(\hbar)$, $\hat{x}_{2,\imath}(\hbar)$ is the estimate of $\tilde{y}_\imath(\hbar-1)$, $\hat{z}_\imath(\hbar)$ is the estimate of $z_\imath(\hbar)$, and $G_{1,\imath\epsilon}(\hbar)$, $G_{2,\imath\epsilon}(\hbar)$, $K_{1,\imath\epsilon}(\hbar)$, $K_{2,\imath\epsilon}(\hbar)$ and $M_{\imath\epsilon}(\hbar)$ are the gains of the estimators to be determined.

By setting $\hat{x}(\hbar) \triangleq \begin{bmatrix} \hat{x}_{1,1}^T(\hbar) & \cdots & \hat{x}_{1,\Lambda}^T(\hbar) & \hat{x}_{2,1}^T(\hbar) & \cdots & \hat{x}_{2,\mu}^T(\hbar) \end{bmatrix}^T$ and $\hat{z}(\hbar) \triangleq \begin{bmatrix} \hat{z}_1^T(\hbar) & \cdots & \hat{z}_\Lambda^T(\hbar) \end{bmatrix}^T$, the estimators for the CN (7.47) are rewritten in the following compact form:

$$\begin{aligned}
\hat{x}(\hbar+1) &= G_\epsilon(\hbar)\hat{x}(\hbar) + K_\epsilon(\hbar)\tilde{y}(\hbar), \\
\hat{z}(\hbar) &= M_\epsilon(\hbar)I_\zeta\hat{x}(\hbar)
\end{aligned} \tag{7.59}$$

where

$$G_\epsilon(\hbar) \triangleq \mathrm{diag}\{G_{1,1\epsilon}(\hbar), \ldots, G_{1,\Lambda\epsilon}(\hbar), G_{2,1\epsilon}(\hbar), \ldots, G_{2,\mu\epsilon}(\hbar)\},$$
$$K_\epsilon(\hbar) \triangleq \begin{bmatrix} K_{x\epsilon}^T(\hbar) & 0^T_{((\Lambda-\mu)r_x)\times(\mu r_y)} & K_{y\epsilon}^T(\hbar) \end{bmatrix}^T,$$
$$M_\epsilon(\hbar) \triangleq \mathrm{diag}\{M_{1\epsilon}(\hbar), M_{2\epsilon}(\hbar), \ldots, M_{\Lambda\epsilon}(\hbar)\},$$
$$K_{x\epsilon}(\hbar) \triangleq \mathrm{diag}\{K_{1,1\epsilon}(\hbar), \ldots, K_{1,\mu\epsilon}(\hbar)\},$$
$$K_{y\epsilon}(\hbar) \triangleq \mathrm{diag}\{K_{2,1\epsilon}(\hbar), \ldots, K_{2,\mu\epsilon}(\hbar)\}.$$

Let $e(\hbar) \triangleq \tilde{x}(\hbar) - \hat{x}(\hbar)$ and $z_e(\hbar) \triangleq z(\hbar) - \hat{z}(\hbar)$ be the estimation errors

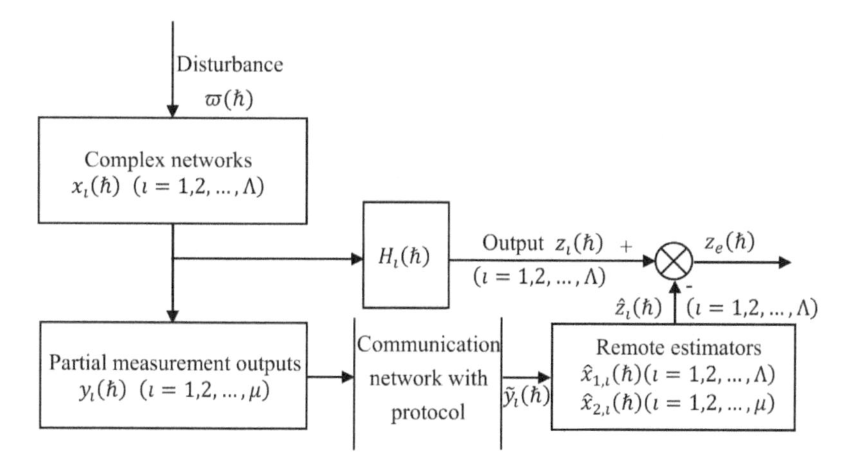

FIGURE 7.1
State estimation problem for CN (7.47).

of the state and the output, respectively. The schematic diagram of the state estimation problem is shown in Fig. 7.1 for CN (7.47) as follows:

Denoting $\xi(\hbar) \triangleq \left[\tilde{x}^T(\hbar) \quad e^T(\hbar)\right]^T$, the following augmented system is obtained:

$$\begin{aligned}
\xi(\hbar+1) &= (\mathcal{A}_\epsilon(\hbar) + \bar{\phi}\Delta\bar{\mathcal{A}}(\hbar))\xi(\hbar) + \mathcal{C}_\epsilon(\hbar)\varrho(\bar{C}(\hbar)x(\hbar)) \\
&\quad +\tilde{\phi}(\hbar)\Delta\bar{\mathcal{A}}(\hbar)\xi(\hbar) + \sum_{l=1}^{d}\bar{\mathcal{A}}_{ml}(\hbar)\xi(\hbar - m_l(\hbar)) \\
&\quad +\sum_{l=1}^{d}\tilde{\mathcal{A}}_{ml}(\hbar)\xi(\hbar - m_l(\hbar)) \\
&\quad +\bar{\mathcal{F}}(\hbar)\zeta(I_\varsigma\mathcal{I}_\varsigma\xi(\hbar)) + \bar{\mathcal{E}}_\epsilon(\hbar)\varpi(\hbar), \\
z_e(\hbar) &= \mathcal{H}_\epsilon(\hbar)\xi(\hbar)
\end{aligned} \tag{7.60}$$

where

$$\begin{aligned}
\mathcal{A}_\epsilon(\hbar) &\triangleq \begin{bmatrix} \bar{A}_\epsilon(\hbar) & 0 \\ \bar{A}_\epsilon(\hbar) - G_\epsilon(\hbar) - K_\epsilon(\hbar)\bar{C}_{1\epsilon}(\hbar) & G_\epsilon(\hbar) \end{bmatrix}, \\
\Delta\bar{\mathcal{A}}(\hbar) &\triangleq \begin{bmatrix} \Delta\bar{A}(\hbar) & 0 \\ \Delta\bar{A}(\hbar) & 0 \end{bmatrix}, \ \tilde{\mathcal{A}}_{ml}(\hbar) \triangleq \begin{bmatrix} \tilde{A}_{ml}(\hbar) & 0 \\ \tilde{A}_{ml}(\hbar) & 0 \end{bmatrix}, \\
\mathcal{C}_\epsilon(\hbar) &\triangleq \begin{bmatrix} \bar{\Pi}_\epsilon \\ \bar{\Pi}_\epsilon - K_\epsilon(\hbar)\Pi_\epsilon \end{bmatrix}, \bar{\mathcal{A}}_{ml}(\hbar) \triangleq \begin{bmatrix} \bar{A}_{ml}(\hbar) & 0 \\ \bar{A}_{ml}(\hbar) & 0 \end{bmatrix}, \\
\bar{\mathcal{F}}(\hbar) &\triangleq \begin{bmatrix} \bar{F}(\hbar) \\ \bar{F}(\hbar) \end{bmatrix}, \ \bar{\mathcal{E}}_\epsilon(\hbar) \triangleq \begin{bmatrix} \bar{E}_\epsilon(\hbar) \\ \bar{E}_\epsilon(\hbar) - K_\epsilon(\hbar)\Pi_\epsilon D(\hbar) \end{bmatrix},
\end{aligned}$$

$$\begin{aligned}\mathcal{H}_\epsilon(\hbar) &\triangleq \left[(H(\hbar)-M_\epsilon(\hbar))I_\zeta \quad M_\epsilon(\hbar)I_\zeta\right],\\ \mathcal{I}_\zeta &\triangleq \left[I_{(Sr_x+\mu r_y)\times(Sr_x+\mu r_y)} \quad 0_{(Sr_x+\mu r_y)\times(Sr_x+\mu r_y)}\right].\end{aligned}$$

We are now ready to formulate the problem to be investigated. The main objective is to compute the gain parameters of time-varying PNB state estimators (7.58) such that the augmented dynamics satisfies the following H_∞ performance requirement:

$$\begin{aligned}\mathbb{J}(\hbar) \triangleq & \sum_{\hbar=0}^{\mathbb{N}-1} \mathbb{E}\{\|z_e(\hbar)\|^2\} - \gamma^2\Bigg(\sum_{\hbar=0}^{\mathbb{N}-1}\|\varpi(\hbar)\|^2 \\ & + \sum_{t=-\bar{m}}^{0}\xi^T(t)Y(t)\xi(t)\Bigg) < 0 \end{aligned} \tag{7.61}$$

where γ is a prescribed disturbance rejection level and $Y(t)$ $(t=-\bar{m},-\bar{m}+1,\ldots,0)$ are known matrices.

7.2.2 Main Results

In this subsection, the H_∞ performance of dynamics (7.60) is analyzed, and sufficient conditions are established for the existence of the state estimators (7.58) whose gain matrices can be computed by solving the recursive linear matrix inequalities.

Theorem 7.5 *Let the disturbance rejection level $\gamma>0$ and the estimator gains $G_{1,\imath\epsilon}(\hbar)$, $M_{\imath\epsilon}(\hbar)$ $(\imath=1,2,\ldots,\Lambda)$, $G_{2,\imath\epsilon}(\hbar)$, $K_{1,\imath\epsilon}(\hbar)$ and $K_{2,\imath\epsilon}(\hbar)$ $(\imath=1,2,\ldots,\mu)$ be given. The estimation error dynamics (7.60) achieves the H_∞ performance constraint (7.61) if there exist positive definite matrices $Q_\epsilon(\hbar)>0$, $R_v(\hbar)>0$ $(v=1,\ldots,d)$ and positive scalars χ_1 and χ_2 satisfying the following inequalities:*

$$\Psi_\epsilon(\hbar) = \begin{bmatrix} \Psi_{11\epsilon}(\hbar) & \Psi_{12\epsilon}(\hbar) & \Psi_{13\epsilon}(\hbar) & \Psi_{14\epsilon}(\hbar) & \Psi_{15\epsilon}(\hbar) \\ * & \Psi_{22\epsilon}(\hbar) & \Psi_{23\epsilon}(\hbar) & \Psi_{24\epsilon}(\hbar) & \Psi_{25\epsilon}(\hbar) \\ * & * & \Psi_{33\epsilon}(\hbar) & \Psi_{34\epsilon}(\hbar) & \Psi_{35\epsilon}(\hbar) \\ * & * & * & \Psi_{44\epsilon}(\hbar) & \Psi_{45\epsilon}(\hbar) \\ * & * & * & * & \Psi_{55\epsilon}(\hbar) \end{bmatrix} < 0 \tag{7.62}$$

with the initial condition:

$$\begin{aligned}& Q_\epsilon(0)-\gamma^2 Y(0)<0,\\ & (\bar{m}-\underline{m}+1)\sum_{v=1}^{d} R_v(u)-\gamma^2 Y(u)<0, u=-\bar{m},-\bar{m}+1,\ldots,-1\end{aligned} \tag{7.63}$$

where

$$\Psi_{11\epsilon}(\hbar) \triangleq (\mathcal{A}_\epsilon(\hbar)+\bar{\phi}\Delta\bar{A}(\hbar))^T\bar{Q}_\epsilon(\hbar+1)(\mathcal{A}_\epsilon(\hbar)$$

$$
\begin{aligned}
& +\bar{\phi}\Delta\bar{\mathcal{A}}(\hbar)) + \tilde{\phi}^2 \Delta\bar{\mathcal{A}}^T(\hbar)\bar{Q}_\epsilon(\hbar+1)\Delta\bar{\mathcal{A}}(\hbar) \\
& -Q_\epsilon(\hbar) + (\bar{m} - \underline{m} + 1)\sum_{v\triangleq 1}^{d} R_v(\hbar) \\
& +\chi_1 \mathcal{I}_\zeta^T I_\zeta^T(-I \otimes \Omega^T\Omega) I_\zeta \mathcal{I}_\zeta + \mathcal{H}_\epsilon^T(\hbar)\mathcal{H}_\epsilon(\hbar), \\
\Psi_{12\epsilon}(\hbar) \triangleq\ & (\mathcal{A}_\epsilon(\hbar) + \bar{\phi}\Delta\bar{\mathcal{A}}(\hbar))^T \bar{Q}_\epsilon(\hbar+1)\bar{\mathcal{A}}_m(\hbar), \\
\Psi_{13\epsilon}(\hbar) \triangleq\ & (\mathcal{A}_\epsilon(\hbar) + \bar{\phi}\Delta\bar{\mathcal{A}}(\hbar))^T \bar{Q}_\epsilon(\hbar+1)\bar{\mathcal{F}}(\hbar), \\
\Psi_{14\epsilon}(\hbar) \triangleq\ & (\mathcal{A}_\epsilon(\hbar) + \bar{\phi}\Delta\bar{\mathcal{A}}(\hbar))^T \bar{Q}_\epsilon(\hbar+1)\bar{\mathcal{E}}_\epsilon(\hbar), \\
\Psi_{55\epsilon}(\hbar) \triangleq\ & \mathcal{C}_\epsilon^T(\hbar)\bar{Q}_\epsilon(\hbar+1)\mathcal{C}_\epsilon(\hbar) - \chi_2 I, \\
\Psi_{15\epsilon}(\hbar) \triangleq\ & (\mathcal{A}_\epsilon(\hbar) + \bar{\phi}\Delta\bar{\mathcal{A}}(\hbar))^T \bar{Q}_\epsilon(\hbar+1)\mathcal{C}_\epsilon(\hbar) \\
& +\chi_2 \mathcal{I}_\zeta^T I_\zeta^T \bar{L}^T, \quad \bar{L} \triangleq \begin{bmatrix} \check{L} & 0 \end{bmatrix}, \\
\Psi_{22\epsilon}(\hbar) \triangleq\ & \bar{\mathcal{A}}_m^T(\hbar)\bar{Q}_\epsilon(\hbar+1)\bar{\mathcal{A}}_m(\hbar) + \hat{\mathcal{A}}_m^T(\hbar) \\
& \times(I \otimes \bar{Q}_\epsilon(\hbar+1))\hat{\mathcal{A}}_m(\hbar) - \bar{R}(\hbar), \\
\Psi_{23\epsilon}(\hbar) \triangleq\ & \bar{\mathcal{A}}_m^T(\hbar)\bar{Q}_\epsilon(\hbar+1)\bar{\mathcal{F}}(\hbar), \\
\bar{R}(\hbar) \triangleq\ & \mathrm{diag}\{R_1(\hbar - m_1(\hbar)), R_2(\hbar - m_2(\hbar)), \ldots, \\
& R_d(\hbar - m_d(\hbar))\}, \\
\Psi_{24\epsilon}(\hbar) \triangleq\ & \bar{\mathcal{A}}_m^T(\hbar)\bar{Q}_\epsilon(\hbar+1)\bar{\mathcal{E}}_\epsilon(\hbar), \\
\Psi_{25\epsilon}(\hbar) \triangleq\ & \bar{\mathcal{A}}_m^T(\hbar)\bar{Q}_\epsilon(\hbar+1)\mathcal{C}_\epsilon(\hbar), \\
\Psi_{33\epsilon}(\hbar) \triangleq\ & \bar{\mathcal{F}}^T(\hbar)\bar{Q}_\epsilon(\hbar+1)\bar{\mathcal{F}}(\hbar) - \chi_1 I, \\
\check{L} \triangleq\ & \mathrm{diag}\{LC_1(\hbar), LC_2(\hbar), \ldots, LC_\mu(\hbar)\}, \\
\Psi_{34\epsilon}(\hbar) \triangleq\ & \bar{\mathcal{F}}^T(\hbar)\bar{Q}_\epsilon(\hbar+1)\bar{\mathcal{E}}_\epsilon(\hbar), \\
\Psi_{35\epsilon}(\hbar) \triangleq\ & \bar{\mathcal{F}}^T(\hbar)\bar{Q}_\epsilon(\hbar+1)\mathcal{C}_\epsilon(\hbar), \\
\Psi_{44\epsilon}(\hbar) \triangleq\ & \bar{\mathcal{E}}_\epsilon^T(\hbar)\bar{Q}_\epsilon(\hbar+1)\bar{\mathcal{E}}_\epsilon(\hbar) - \gamma^2 I, \\
\Psi_{45\epsilon}(\hbar) \triangleq\ & \bar{\mathcal{E}}_\epsilon^T(\hbar)\bar{Q}_\epsilon(\hbar+1)\mathcal{C}_\epsilon(\hbar), \\
\bar{\mathcal{A}}_m(\hbar) \triangleq\ & \begin{bmatrix} \bar{\mathcal{A}}_{m1}(\hbar) & \bar{\mathcal{A}}_{m2}(\hbar) & \ldots & \bar{\mathcal{A}}_{md}(\hbar) \end{bmatrix}, \\
\hat{\mathcal{A}}_m(\hbar) \triangleq\ & \mathrm{diag}\{\hat{\mathcal{A}}_{m1}(\hbar), \hat{\mathcal{A}}_{m2}(\hbar), \ldots, \hat{\mathcal{A}}_{md}(\hbar)\}, \\
\hat{\mathcal{A}}_{ml}(\hbar) \triangleq\ & \begin{bmatrix} \hat{A}_{ml}(\hbar) & 0 \\ \hat{A}_{ml}(\hbar) & 0 \end{bmatrix}, \quad \bar{Q}_\epsilon(\hbar) \triangleq \sum_{\kappa=1}^{\mu} \pi_{\epsilon\kappa} Q_\kappa(\hbar), \\
\hat{\mathcal{A}}_{ml}(\hbar) \triangleq\ & \mathrm{diag}\{\hat{\Theta}_l A_m(\hbar), 0\}, \\
\hat{\Theta}_l \triangleq\ & \mathrm{diag}\{\sqrt{\bar{\theta}_{1l}(1-\bar{\theta}_{1l})}I, \sqrt{\bar{\theta}_{2l}(1-\bar{\theta}_{2l})}I, \ldots, \\
& \sqrt{\bar{\theta}_{\Lambda l}(1-\bar{\theta}_{\Lambda l})}I\}.
\end{aligned}
$$

Proof *Define Lyapunov functions as follows:*

$$
J(\hbar, r(\hbar)) \triangleq J_1(\hbar, r(\hbar)) + J_2(\hbar) + J_3(\hbar) \tag{7.64}
$$

where

$$
\begin{aligned}
J_1(\hbar, r(\hbar)) &\triangleq \xi^T(\hbar)Q_\epsilon(\hbar)\xi(\hbar),\\
J_2(\hbar) &\triangleq \sum_{v=1}^{d}\sum_{u=\hbar-m_v(\hbar)}^{\hbar-1} \xi^T(u)R_v(u)\xi(u),\\
J_3(\hbar) &\triangleq \sum_{v=1}^{d}\sum_{n=-\bar{m}+1}^{-\underline{m}}\sum_{u=\hbar+n}^{\hbar-1} \xi^T(u)R_v(u)\xi(u).
\end{aligned}
$$

In terms of (7.60), the mathematical expectation of the difference of $J(\hbar, r(\hbar))$ is computed as

$$
\begin{aligned}
&\mathbb{E}\{\Delta J_1(\hbar)\}\\
= \ &\mathbb{E}\{\xi^T(\hbar+1)\bar{Q}_\epsilon(\hbar+1)\xi(\hbar+1) - \xi^T(\hbar)Q_\epsilon(\hbar)\xi(\hbar)\}\\
= \ &\xi^T(\hbar)(\mathcal{A}_\epsilon(\hbar) + \bar{\phi}\Delta\bar{\mathcal{A}}(\hbar))^T\bar{Q}_\epsilon(\hbar+1)(\mathcal{A}_\epsilon(\hbar)\\
&+\bar{\phi}\Delta\bar{\mathcal{A}}(\hbar))\xi(\hbar) + 2\xi^T(\hbar)(\mathcal{A}_\epsilon(\hbar) + \bar{\phi}\Delta\bar{\mathcal{A}}(\hbar))^T\\
&\times\bar{Q}_\epsilon(\hbar+1)\bar{\mathcal{A}}_m(\hbar)\xi(\hbar - m(\hbar)) + 2\xi^T(\hbar)(\mathcal{A}_\epsilon(\hbar)\\
&+\bar{\phi}\Delta\bar{\mathcal{A}}(\hbar))^T\bar{Q}_\epsilon(\hbar+1)\bar{\mathcal{F}}(\hbar)\zeta(I_\zeta\mathcal{I}_\zeta\xi(\hbar))\\
&+2\xi^T(\hbar)(\mathcal{A}_\epsilon(\hbar) + \bar{\phi}\Delta\bar{\mathcal{A}}(\hbar))^T\bar{Q}_\epsilon(\hbar+1)\bar{\mathcal{E}}_\epsilon(\hbar)\varpi(\hbar)\\
&+2\xi^T(\hbar)(\mathcal{A}_\epsilon(\hbar) + \bar{\phi}\Delta\bar{\mathcal{A}}(\hbar))^T\bar{Q}_\epsilon(\hbar+1)\mathcal{C}_\epsilon(\hbar)\\
&\times\varrho(\bar{C}(\hbar)x(\hbar)) + \xi^T(\hbar - m(\hbar))\bar{\mathcal{A}}_m^T(\hbar)\bar{Q}_\epsilon(\hbar+1)\\
&\times\Big(\bar{\mathcal{A}}_m(\hbar)\xi(\hbar - m(\hbar))\Big) + 2\xi^T(\hbar - m(\hbar))\bar{\mathcal{A}}_m^T(\hbar)\\
&\times\bar{Q}_\epsilon(\hbar+1)\bar{\mathcal{F}}(\hbar)\zeta(I_\zeta\mathcal{I}_\zeta\xi(\hbar)) + 2\xi^T(\hbar - m(\hbar))\\
&\times\bar{\mathcal{A}}_m^T(\hbar)\bar{Q}_\epsilon(\hbar+1)\bar{\mathcal{E}}_\epsilon(\hbar)\varpi(\hbar) + 2\xi^T(\hbar - m(\hbar))\\
&\times\bar{\mathcal{A}}_m^T(\hbar)\bar{Q}_\epsilon(\hbar+1)\mathcal{C}_\epsilon(\hbar)\varrho(\bar{C}(\hbar)x(\hbar))\\
&+\zeta^T(I_\zeta\mathcal{I}_\zeta\xi(\hbar))\bar{\mathcal{F}}^T(\hbar)\bar{Q}_\epsilon(\hbar+1)\bar{\mathcal{F}}(\hbar)\zeta(I_\zeta\mathcal{I}_\zeta\xi(\hbar))\\
&+2\zeta^T(I_\zeta\mathcal{I}_\zeta\xi(\hbar))\bar{\mathcal{F}}^T(\hbar)\bar{Q}_\epsilon(\hbar+1)\bar{\mathcal{E}}_\epsilon(\hbar)\varpi(\hbar)\\
&+2\zeta^T(I_\zeta\mathcal{I}_\zeta\xi(\hbar))\bar{\mathcal{F}}^T(\hbar)\bar{Q}_\epsilon(\hbar+1)\mathcal{C}_\epsilon(\hbar)\varrho(\bar{C}(\hbar)x(\hbar))\\
&+\varpi^T(\hbar)\bar{\mathcal{E}}_\epsilon^T(\hbar)\bar{Q}_\epsilon(\hbar+1)\bar{\mathcal{E}}_\epsilon(\hbar)\varpi(\hbar) - \xi^T(\hbar)Q_\epsilon(\hbar)\\
&\times\xi(\hbar) + 2\varpi^T(\hbar)\bar{\mathcal{E}}_\epsilon^T(\hbar)\bar{Q}_\epsilon(\hbar+1)\mathcal{C}_\epsilon(\hbar)\varrho(\bar{C}(\hbar)x(\hbar))\\
&+\tilde{\phi}^2\xi^T(\hbar)\Delta\bar{\mathcal{A}}^T(\hbar)\bar{Q}_\epsilon(\hbar+1)\Delta\bar{\mathcal{A}}(\hbar)\xi(\hbar)\\
&+\xi^T(\hbar - m(\hbar))\hat{\mathcal{A}}_m^T(\hbar)(I\otimes\bar{Q}_\epsilon(\hbar+1))\\
&\times\hat{\mathcal{A}}_m(\hbar)\xi(\hbar - m(\hbar)) + \varrho^T(\bar{C}(\hbar)x(\hbar))\mathcal{C}_\epsilon^T(\hbar)\\
&\times\bar{Q}_\epsilon(\hbar+1)\mathcal{C}_\epsilon(\hbar)\varrho(\bar{C}(\hbar)x(\hbar)),
\end{aligned}
\tag{7.65}
$$

where

$$
\xi(\hbar - m(\hbar)) \triangleq \begin{bmatrix} \xi^T(\hbar - m_1(\hbar)) & \xi^T(\hbar - m_2(\hbar)) & \cdots & \xi^T(\hbar - m_d(\hbar)) \end{bmatrix}^T,
$$

$$\begin{aligned}\mathbb{E}\{\Delta J_2(\hbar)\} \leq& \sum_{v=1}^{d}\xi^T(\hbar)R_v(\hbar)\xi(\hbar)-\xi^T(\hbar-m(\hbar))\bar{R}(\hbar)\xi(\hbar-m(\hbar))\\ &+\sum_{v=1}^{d}\sum_{u=\hbar+1-\bar{m}}^{\hbar-\underline{m}}\xi^T(u)R_v(u)\xi(u) \end{aligned}\quad (7.66)$$

and

$$\begin{aligned}\mathbb{E}\{\Delta J_3(\hbar)\} =& \sum_{v=1}^{d}(\bar{m}-\underline{m})\xi^T(\hbar)R_v(\hbar)\xi(\hbar)\\ &-\sum_{v=1}^{d}\sum_{n=\hbar-\bar{m}+1}^{\hbar-\underline{m}}\xi^T(n)R_v(n)\xi(n). \end{aligned}\quad (7.67)$$

On the basis of the nonlinear constraints (7.48) and (7.51), it can be found that

$$\begin{bmatrix} x_\imath(\hbar)\\ \zeta(x_\imath(\hbar))\end{bmatrix}^T \begin{bmatrix} -\Omega^T\Omega & 0\\ * & I\end{bmatrix}\begin{bmatrix} x_\imath(\hbar)\\ \zeta(x_\imath(\hbar))\end{bmatrix} \leq 0, \quad (7.68)$$

$$\begin{bmatrix} x_\imath(\hbar)\\ \varrho(C_\imath(\hbar)x_\imath(\hbar))\end{bmatrix}^T \begin{bmatrix} 0 & \frac{-(LC_\imath(\hbar))^T}{2}\\ * & I\end{bmatrix}\begin{bmatrix} x_\imath(\hbar)\\ \varrho(C_\imath(\hbar)x_\imath(\hbar))\end{bmatrix} \leq 0. \quad (7.69)$$

According to (7.65)–(7.67) and the corresponding compact forms of (7.68)–(7.69), one derives that

$$\begin{aligned}&\mathbb{E}\{\Delta J(\hbar, r(\hbar))\}\\ \leq\ & \mathbb{E}\{\Delta J_1(\hbar, r(\hbar))\}+\mathbb{E}\{\Delta J_2(\hbar)\}+\mathbb{E}\{\Delta J_3(\hbar)\}\\ &-\chi_1\Big(\zeta^T(I_\zeta\mathcal{I}_\zeta\xi(\hbar))\zeta(I_\zeta\mathcal{I}_\zeta\xi(\hbar))-\xi^T(\hbar)\mathcal{I}_\zeta^T I_\zeta^T\\ &\times(-I\otimes\Omega^T\Omega)I_\zeta\mathcal{I}_\zeta\xi(\hbar)\Big)-\chi_2\Big(-\xi^T(\hbar)\mathcal{I}_\zeta^T I_\zeta^T\bar{L}^T\\ &\times\varrho(\bar{C}(\hbar)x(\hbar))+\varrho^T(\bar{C}(\hbar)x(\hbar))\varrho(\bar{C}(\hbar)x(\hbar))\Big). \end{aligned}\quad (7.70)$$

On the basis of (7.70), it is clear that

$$\begin{aligned}\mathbb{J}(\hbar) =& \mathbb{J}(\hbar)+\sum_{\hbar=0}^{\mathbb{N}-1}\mathbb{E}\{\Delta J(\hbar, r(\hbar))-\Delta J(\hbar, r(\hbar))\}\\ \leq& \sum_{\hbar=0}^{\mathbb{N}-1}\eta^T(\hbar)\Psi_\epsilon(\hbar)\eta(\hbar)-\gamma^2\sum_{t=-\bar{m}}^{0}\xi^T(t)Y(t)\xi(t)\\ &-J(\mathbb{N}, r(\mathbb{N}))+J(0, r(0)) \end{aligned}\quad (7.71)$$

where

$$\eta(\hbar) \triangleq \begin{bmatrix} \xi^T(\hbar) & \xi^T(\hbar-m(\hbar)) & \zeta^T(\mathcal{I}_\zeta\xi(\hbar)) & \varpi^T(\hbar) & \varrho^T(\bar{C}(\hbar)x(\hbar))\end{bmatrix}^T.$$

By employing (7.64), one obtains

$$
\begin{aligned}
&-\gamma^2 \sum_{t=-\bar{m}}^{0} \xi^T(t)Y(t)\xi(t) + J(0, r(0)) \\
\leq\ & \xi^T(0)(Q_\epsilon(0) - \gamma^2 Y(0))\xi(0) + \sum_{u=-\bar{m}}^{-1} \xi^T(u) \\
&\times \left((\bar{m} - \underline{m} + 1) \sum_{v=1}^{d} R_v(u) - \gamma^2 Y(u) \right) \xi(u).
\end{aligned}
$$

Taking (7.62)–(7.64) into account, it is known that $\eta^T(\hbar)\Psi_\epsilon(\hbar)\eta(\hbar) < 0$, $-\gamma^2 \sum_{t=-\bar{m}}^{0} \xi^T(t)Y(t)\xi(t) + J(0, r(0)) < 0$ *and* $J(\mathbb{N}, r(\mathbb{N})) > 0$, *then it is obtained from (7.71) that* $\mathbb{J}(\hbar) < 0$, *which completes the proof.*

Next, we are to design the parameter matrices of the time-varying estimators (7.58).

Theorem 7.6 *Let the disturbance rejection level* $\gamma > 0$ *be given. The estimation error dynamics (7.60) achieves the* H_∞ *performance constraint (7.61) if there exist positive definite matrices* $Q_\epsilon(\hbar) = \text{diag}\{Q_{1\epsilon}(\hbar), Q_{2\epsilon}(\hbar)\} > 0$ $(Q_{1\epsilon}(\hbar) = \text{diag}\{Q_{11\epsilon}(\hbar), Q_{12\epsilon}(\hbar), \ldots Q_{1(\Lambda+\mu)\epsilon}(\hbar)\}$, $Q_{2\epsilon}(\hbar) = \text{diag}\{Q_{21\epsilon}(\hbar), Q_{22\epsilon}(\hbar), \ldots, Q_{2(\Lambda+\mu)\epsilon}(\hbar)\})$, $R_v(\hbar) > 0$ $(v = 1, \ldots, d)$, *matrices* $Z_{G\epsilon}(\hbar) = \text{diag}\{Z_{G1\epsilon}(\hbar), Z_{G2\epsilon}(\hbar)\}$ $(Z_{G1\epsilon}(\hbar) = \text{diag}\{Z_{G11\epsilon}(\hbar), Z_{G12\epsilon}(\hbar), \ldots, Z_{G1S\epsilon}(\hbar)\}$, $Z_{G2\epsilon}(\hbar) = \text{diag}\{Z_{G21\epsilon}(\hbar), Z_{G22\epsilon}(\hbar), \ldots, Z_{G2\mu\epsilon}(\hbar)\})$, $\bar{Z}_{K\epsilon}(\hbar) = \begin{bmatrix} Z^T_{Kx\epsilon}(\hbar) & 0 & Z^T_{Ky\epsilon}(\hbar) \end{bmatrix}^T$ $(Z_{Kx\epsilon}(\hbar) = \text{diag}\{Z_{K11\epsilon}(\hbar), Z_{K12\epsilon}(\hbar), \ldots, Z_{K1\mu\epsilon}(\hbar)\}$, $Z_{Ky\epsilon}(\hbar) = \text{diag}\{Z_{K21\epsilon}(\hbar), Z_{K22\epsilon}(\hbar), \ldots, Z_{K2\mu\epsilon}(\hbar)\})$ *and positive scalars* χ_1, χ_2 *and* χ_3 *satisfying the following inequalities:*

$$
\begin{bmatrix}
\Psi_{0\epsilon}(\hbar) & * & * & * \\
L_\epsilon(\hbar) & -\hat{Q}_\epsilon(\hbar+1) & * & * \\
0 & \Phi^T_{1\epsilon}(\hbar) & -\chi_3 I & * \\
\chi_3\Phi_2(\hbar) & 0 & 0 & -\chi_3 I
\end{bmatrix} < 0 \tag{7.72}
$$

with the initial condition (7.63), where

$$
\begin{aligned}
L_\epsilon(\hbar) &\triangleq \begin{bmatrix} \tilde{L}_{1\epsilon}(\hbar) & \tilde{L}_{2\epsilon}(\hbar) \end{bmatrix}, \\
\tilde{L}_{1\epsilon}(\hbar) &\triangleq \begin{bmatrix}
\mathcal{A}_{Q\epsilon}(\hbar) & \bar{Q}_\epsilon(\hbar+1)\bar{\mathcal{A}}_m(\hbar) \\
0 & 0 \\
0 & (I \otimes \bar{Q}_\epsilon(\hbar+1))\hat{\mathcal{A}}_m(\hbar) \\
\mathcal{H}_\epsilon(\hbar) & 0
\end{bmatrix},
\end{aligned}
$$

$$
\begin{aligned}
\tilde{L}_{2\epsilon}(\hbar) &\triangleq \begin{bmatrix} \bar{Q}_\epsilon(\hbar+1)\bar{\mathcal{F}}(\hbar) & \mathcal{E}_{Q\epsilon}(\hbar) & \mathcal{C}_{Q\epsilon}(\hbar) \\ 0 & 0 & 0 \\ 0 & 0 & 0 \\ 0 & 0 & 0 \end{bmatrix}, \\
\mathcal{A}_{Q\epsilon}(\hbar) &\triangleq \bar{Q}_\epsilon(\hbar+1)\mathcal{A}_{0\epsilon}(\hbar) + Z_{1\epsilon}(\hbar), \\
\mathcal{C}_{Q\epsilon}(\hbar) &\triangleq \bar{Q}_\epsilon(\hbar+1)\mathcal{C}_{0\epsilon}(\hbar) + Z_{2\epsilon}(\hbar), \\
\mathcal{E}_{Q\epsilon}(\hbar) &\triangleq \bar{Q}_\epsilon(\hbar+1)\mathcal{E}_{0\epsilon}(\hbar) + Z_{3\epsilon}(\hbar), \\
\mathcal{A}_{0\epsilon}(\hbar) &\triangleq \begin{bmatrix} \bar{A}_\epsilon(\hbar) & 0 \\ \bar{A}_\epsilon(\hbar) & 0 \end{bmatrix}, \; Z_{2\epsilon}(\hbar) \triangleq \begin{bmatrix} 0 \\ -\bar{Z}_{K\epsilon}(\hbar)\Pi_\epsilon \end{bmatrix}, \\
Z_{1\epsilon}(\hbar) &\triangleq \begin{bmatrix} 0 & 0 \\ -Z_{G\epsilon}(\hbar) - \bar{Z}_{K\epsilon}(\hbar)\bar{C}_{1\epsilon}(\hbar) & Z_{G\epsilon}(\hbar) \end{bmatrix}, \\
\mathcal{C}_{0\epsilon}(\hbar) &\triangleq \begin{bmatrix} \bar{\Pi}_\epsilon^T & \bar{\Pi}_\epsilon^T \end{bmatrix}^T, \mathcal{E}_{0\epsilon}(\hbar) \triangleq \begin{bmatrix} \bar{E}_\epsilon^T(\hbar) & \bar{E}_\epsilon^T(\hbar) \end{bmatrix}^T, \\
Z_{3\epsilon}(\hbar) &\triangleq \begin{bmatrix} 0 & (-\bar{Z}_{K\epsilon}(\hbar)\Pi_\epsilon D(\hbar))^T \end{bmatrix}^T, \\
\Psi_{0\epsilon}(\hbar) &\triangleq \begin{bmatrix} \Psi_{011\epsilon}(\hbar) & * \\ \Psi_{021} & \Psi_{022}(\hbar) \end{bmatrix}, \\
\Psi_{021} &\triangleq \begin{bmatrix} 0 & 0 & 0 & (\chi_2 \bar{L} I_\varsigma \mathcal{I}_\varsigma)^T \end{bmatrix}^T, \\
\Psi_{022}(\hbar) &\triangleq \mathrm{diag}\{-\bar{R}(\hbar), -\chi_1 I, -\gamma^2 I, -\chi_2 I\}, \\
\Psi_{011\epsilon}(\hbar) &\triangleq -Q_\epsilon(\hbar) + (\bar{m} - \underline{m} + 1)\sum_{v=1}^{d} R_v(\hbar) \\
&\quad + \chi_1 \mathcal{I}_\varsigma^T I_\varsigma^T (-I \otimes \Omega^T \Omega) I_\varsigma \mathcal{I}_\varsigma, \\
\hat{Q}_\epsilon(\hbar+1) &\triangleq \mathrm{diag}\{\bar{Q}_\epsilon(\hbar+1), \bar{Q}_\epsilon(\hbar+1), \\
&\quad I \otimes \bar{Q}_\epsilon(\hbar+1), I\}, \\
\Phi_{1\epsilon}(\hbar) &\triangleq [\; (\bar{Q}_\epsilon(\hbar+1)\bar{\phi}\bar{\mathcal{A}}_1(\hbar))^T \\
&\quad (\bar{Q}_\epsilon(\hbar+1)\tilde{\phi}\bar{\mathcal{A}}_1(\hbar))^T \quad 0 \quad 0 \;]^T, \\
\Phi_2(\hbar) &\triangleq \begin{bmatrix} \bar{\mathcal{A}}_3(\hbar) & 0 & 0 & 0 & 0 \end{bmatrix}, \\
\bar{\mathcal{A}}_1(\hbar) &\triangleq \mathrm{diag}\{\breve{A}_1(\hbar), \breve{A}_1(\hbar)\}, \bar{\mathcal{A}}_3(\hbar) \triangleq \begin{bmatrix} \breve{A}_3(\hbar) & 0 \\ \breve{A}_3(\hbar) & 0 \end{bmatrix}, \\
\breve{A}_1(\hbar) &\triangleq \mathrm{diag}\{\hat{A}_1(\hbar), 0\}, \; \breve{A}_3(\hbar) \triangleq \mathrm{diag}\{\hat{A}_3(\hbar), 0\}, \\
\hat{A}_1(\hbar) &\triangleq \mathrm{diag}\{A_{11}(\hbar), A_{12}(\hbar), \ldots, A_{1\Lambda}(\hbar)\}, \\
\hat{A}_3(\hbar) &\triangleq \mathrm{diag}\{A_{31}(\hbar), A_{32}(\hbar), \ldots, A_{3\Lambda}(\hbar)\}.
\end{aligned}
$$

Furthermore, the gain matrices of the state estimators are given by

$$
\begin{aligned}
G_{1,\imath\epsilon}(\hbar) &= \bar{Q}_{2\imath\epsilon}^{-1}(\hbar+1) Z_{G1\imath\epsilon}(\hbar), \; \imath = 1, 2, \ldots, \Lambda \\
G_{2,\imath\epsilon}(\hbar) &= \bar{Q}_{2(\Lambda+\imath)\epsilon}^{-1}(\hbar+1) Z_{G2\imath\epsilon}(\hbar), \; \imath = 1, 2, \ldots, \mu
\end{aligned}
$$

$$
\begin{aligned}
K_{1,\imath\epsilon}(\hbar) &= \bar{Q}_{2\imath\epsilon}^{-1}(\hbar+1)Z_{K1\imath\epsilon}(\hbar),\ \imath=1,2,\ldots,\mu \\
K_{2,\imath\epsilon}(\hbar) &= \bar{Q}_{2(\Lambda+\imath)\epsilon}^{-1}(\hbar+1)Z_{K2\imath\epsilon}(\hbar), \imath=1,2,\ldots,\mu.
\end{aligned} \tag{7.73}
$$

Proof *First, (7.62) can be rewritten as follows:*

$$
\Psi_\epsilon(\hbar) = \Psi_{0\epsilon}(\hbar) + \breve{\Psi}_\epsilon^T(\hbar)\hat{Q}_\epsilon(\hbar+1)\breve{\Psi}_\epsilon(\hbar) < 0 \tag{7.74}
$$

where

$$
\begin{aligned}
\breve{\Psi}_\epsilon(\hbar) &\triangleq \begin{bmatrix}\Psi_{1\epsilon}^T(\hbar) & \Psi_2^T(\hbar) & \Psi_3^T(\hbar) & \Psi_{4\epsilon}^T(\hbar)\end{bmatrix}^T, \\
\Psi_{1\epsilon}(\hbar) &\triangleq \begin{bmatrix}\mathcal{A}_\epsilon(\hbar)+\bar{\phi}\Delta\bar{\mathcal{A}}(\hbar) & \bar{\mathcal{A}}_m(\hbar) & \bar{\mathcal{F}}(\hbar) & \bar{\mathcal{E}}_\epsilon(\hbar) & \mathcal{C}_\epsilon(\hbar)\end{bmatrix}, \\
\Psi_2(\hbar) &\triangleq \begin{bmatrix}\tilde{\phi}\Delta\bar{\mathcal{A}}(\hbar) & 0 & 0 & 0 & 0\end{bmatrix}, \\
\Psi_3(\hbar) &\triangleq \begin{bmatrix}0 & \hat{\mathcal{A}}_m(\hbar) & 0 & 0 & 0\end{bmatrix}, \\
\Psi_{4\epsilon}(\hbar) &\triangleq \begin{bmatrix}\mathcal{H}_\epsilon(\hbar) & 0 & 0 & 0 & 0\end{bmatrix}.
\end{aligned}
$$

By applying Lemma 2.1 (Schur Complement Lemma) to (7.74), the following inequality is acquired:

$$
\Upsilon_\epsilon(\hbar) = \begin{bmatrix}\Psi_{0\epsilon}(\hbar) & * \\ \hat{\Psi}_\epsilon(\hbar) & -\hat{Q}_\epsilon(\hbar+1)\end{bmatrix} < 0
$$

where

$$
\begin{aligned}
\hat{\Psi}_\epsilon(\hbar) \triangleq \Big[& \Psi_{1\epsilon}^T(\hbar)\bar{Q}_\epsilon(\hbar+1) \quad \Psi_2^T(\hbar)\bar{Q}_\epsilon(\hbar+1) \\
& \Psi_3^T(\hbar)(I\otimes\bar{Q}_\epsilon(\hbar+1)) \quad \Psi_{4\epsilon}^T(\hbar) \Big]^T.
\end{aligned}
$$

According to whether it contains the uncertain terms or not, $\Upsilon_\epsilon(\hbar)$ can be split into two parts as follows:

$$
\Upsilon_\epsilon(\hbar) = \begin{bmatrix}\Psi_{0\epsilon}(\hbar) & * \\ \bar{\Psi}_\epsilon(\hbar) & -\hat{Q}_\epsilon(\hbar+1)\end{bmatrix} + \begin{bmatrix}0 & * \\ \Delta\bar{\Psi}_\epsilon(\hbar) & 0\end{bmatrix} < 0 \tag{7.75}
$$

where

$$
\bar{\Psi}_\epsilon(\hbar) \triangleq \begin{bmatrix}\bar{Q}_\epsilon(\hbar+1)\bar{\Psi}_{1\epsilon}(\hbar) \\ 0 \\ (I\otimes\bar{Q}_\epsilon(\hbar+1))\Psi_3(\hbar) \\ \Psi_{4\epsilon}(\hbar)\end{bmatrix},
$$

$$
\Delta\bar{\Psi}_\epsilon(\hbar) \triangleq \begin{bmatrix}\bar{Q}_\epsilon(\hbar+1)\Delta\Psi_{1\epsilon}(\hbar) \\ \bar{Q}_\epsilon(\hbar+1)\Psi_2(\hbar) \\ 0 \\ 0\end{bmatrix},
$$

$$\bar{\Psi}_{1\epsilon}(\hbar) \triangleq \begin{bmatrix} \mathcal{A}_\epsilon(\hbar) & \bar{\mathcal{A}}_m(\hbar) & \bar{\mathcal{F}}(\hbar) & \bar{\mathcal{E}}_\epsilon(\hbar) & \mathcal{C}_\epsilon(\hbar) \end{bmatrix},$$
$$\Delta\Psi_{1\epsilon}(\hbar) \triangleq \begin{bmatrix} \bar{\phi}\Delta\bar{\mathcal{A}}(\hbar) & 0 & 0 & 0 & 0 \end{bmatrix}.$$

Furthermore, it is easy to see that $\Delta\bar{\Psi}_\epsilon(\hbar) = \Phi_{1\epsilon}(\hbar)\bar{\mathcal{A}}_2(\hbar)\Phi_2(\hbar)$, *where*

$$\begin{aligned} \Delta\bar{\mathcal{A}}(\hbar) &\triangleq \bar{\mathcal{A}}_1(\hbar)\bar{\mathcal{A}}_2(\hbar)\bar{\mathcal{A}}_3(\hbar), \\ \bar{\mathcal{A}}_2(\hbar) &\triangleq \text{diag}\{\breve{A}_2(\hbar), \breve{A}_2(\hbar)\}, \ \breve{A}_2(\hbar) \triangleq \text{diag}\{\hat{A}_2(\hbar), 0\}, \\ \hat{A}_2(\hbar) &\triangleq \text{diag}\{A_{21}(\hbar), A_{22}(\hbar), \ldots, A_{2\Lambda}(\hbar)\}. \end{aligned}$$

Then, by employing Lemma 3.1 (S-procedure Lemma), one has that (7.75) holds if and only if the following is true:

$$\begin{bmatrix} \Psi_{0\epsilon}(\hbar) & * & * & * \\ \bar{\Psi}_\epsilon(\hbar) & -\hat{Q}_\epsilon(\hbar+1) & * & * \\ 0 & \Phi_{1\epsilon}^T(\hbar) & -\chi_3 I & * \\ \chi_3\Phi_2(\hbar) & 0 & 0 & -\chi_3 I \end{bmatrix} < 0. \tag{7.76}$$

Letting $Z_{G\epsilon}(\hbar) \triangleq \bar{Q}_{2\epsilon}(\hbar+1)G_\epsilon(\hbar)$ *and* $\bar{Z}_{K\epsilon}(\hbar) \triangleq \bar{Q}_{2\epsilon}(\hbar+1)K_\epsilon(\hbar)$, *it is obvious that (7.76) is equivalent to (7.72). Therefore, the proof is complete.*

In terms of Theorem 7.6, a Finite-Horizon Robust Partial-Nodes-Based Estimator Design (*FHRPNBED*) algorithm is put forward as follows.

Algorithm *FHRPNBED*

Step 1. Set values for the H_∞ performance index γ and the initial states $\rho_\imath(\jmath)$, $\varphi_{1,\imath}(\jmath)$ ($\imath = 1, \ldots, \Lambda$, $\jmath \in \mathbb{Z}$), $\psi_\imath(\jmath)$ and $\varphi_{2,\imath}(\jmath)$ ($\imath = 1, \ldots, \mu$, $\jmath \in \mathbb{Z}$). Select the matrices $Q_\epsilon(0)$, $Y(0)$, $Y(u)$ and $R_v(u)$ ($u = -\bar{m}, \ldots, -1$, $v = 1, \ldots, d$) which satisfy the initial condition (7.63) and let $\hbar = 0$.

Step 2. Acquire the values of matrices $\bar{Q}_{2\imath\epsilon}(\hbar+1), Z_{G1\imath\epsilon}(\hbar), M_{\imath\epsilon}(\hbar)$, ($\imath = 1, \ldots, \Lambda$), $\bar{Q}_{2(\Lambda+\imath)\epsilon}(\hbar+1), Z_{G2\imath\epsilon}(\hbar), Z_{K1\imath\epsilon}(\hbar), Z_{K2\imath\epsilon}(\hbar)$ ($\imath = 1, \ldots, \mu$) by solving the LMI (7.72) for the $\hbar$th sampling instant with $Q_\epsilon(\hbar)$ and compute the estimator gain matrices $G_{1,\imath\epsilon}(\hbar)$ ($\imath = 1, \ldots, \Lambda$), $G_{2,\imath\epsilon}(\hbar)$, $K_{1,\imath\epsilon}(\hbar)$, $K_{2,\imath\epsilon}(\hbar)$ ($\imath = 1, \ldots, \mu$) via (7.73).

Step 3. Let $\hbar = \hbar + 1$ and update the parameter $Q_\epsilon(\hbar)$.

Step 4. If $\hbar < \mathbb{N}$, then turn to Step 2, else quit.

Remark 7.6 *In Theorem 7.5, by virtue of the stochastic analysis methods and matrix computation techniques, the* H_∞ *performance of the estimators (7.58) is analyzed, which means that the influence from the noises on the* *error dynamics of the finite-horizon state estimation would be attenuated at a predetermined level. On the basis of the analysis results in Theorem 7.5 and by employing inequality techniques, sufficient conditions are presented in Theorem 7.6 to ensure the existence of the finite-horizon PNB robust state estimators which can satisfy the H_∞ performance requirement. Then, the time-varying gain matrices of the state estimators are calculated via solving the recursive linear matrix inequalities, which can be helpful for online implementation. Note that all the network parameters are included in the recursive linear matrix inequalities (7.72), such as the time-varying parameter matrices, probability characteristics of randomly occurring uncertainties and delays, as well as the selection probability of transmission node.*

7.3 Illustrative Examples

7.3.1 Example 1

This simulation example is given to indicate the effectiveness of the developed variance-constrained state estimation approach in Section 7.1 for nonlinear complex networks with RVTs, stochastic inner-coupling and the quantized measurements.

Take account of a nonlinear complex network (7.1) consisting of 3 nodes with the following outer-coupling configuration matrices:

$$W^{(1)} = \begin{bmatrix} -0.4 & 0.2 & 0.2 \\ 0.2 & -0.4 & 0.2 \\ 0.2 & 0.2 & -0.4 \end{bmatrix}, \quad W^{(2)} = \begin{bmatrix} -0.5 & 0.3 & 0.2 \\ 0.3 & -0.6 & 0.3 \\ 0.2 & 0.3 & -0.5 \end{bmatrix}.$$

The nonlinear function $h(k, x_i(k))$ is selected to be:

$$h(k, x_i(k)) = \begin{bmatrix} 0.1 \\ 0.3 \end{bmatrix} (0.2x_{i1}(k)\xi_1(k) + 0.3x_{i2}(k)\xi_2(k))$$

where $x_{ir}(k)$ $(r = 1, 2)$ represents the rth element of $x_i(k)$, and $\xi_r(k)$ $(r = 1, 2)$ denote uncorrelated Gaussian white noise sequences with zero mean and unit variances. Moreover, $\xi_r(k)$ are also uncorrelated with $\omega(k)$. We notice that the considered stochastic nonlinearities satisfy

$$\begin{aligned} & \mathbb{E}\{h(k, x_i(k))|x_i(k)\} = 0, \\ & \mathbb{E}\left\{h(k, x_i(k))h^T(k, x_i(k))|x_i(k)\right\} \\ = & \begin{bmatrix} 0.1 \\ 0.3 \end{bmatrix} \begin{bmatrix} 0.1 \\ 0.3 \end{bmatrix}^T \mathbb{E}\left\{x_i^T(k) \begin{bmatrix} 0.04 & 0 \\ 0 & 0.09 \end{bmatrix} x_i(k)\right\}. \end{aligned}$$

Other parameters of complex networks (7.1) are shown as

$$\begin{aligned}
A(k) &= \begin{bmatrix} 0.2 & -0.2 \\ 0.1+0.3\sin(3k) & -0.5 \end{bmatrix}, \; \beth = \begin{bmatrix} 0.3 & 0.7 \\ 0.6 & 0.4 \end{bmatrix}, \\
\Gamma &= \mathrm{diag}\{0.3, 0.4\}, \; M_3(k) = \begin{bmatrix} 0.5 & -0.2\sin(2k) \end{bmatrix}, \\
B_{v_1}(k) &= \begin{bmatrix} \sin(2k) \\ -0.2 \end{bmatrix}, \; B_{v_2}(k) = \begin{bmatrix} \sin(4k) \\ -0.3 \end{bmatrix}, \; B_{v_3}(k) = \begin{bmatrix} \sin(2k) \\ -0.5 \end{bmatrix}, \\
M_1(k) &= \begin{bmatrix} 0.2 & -0.4\sin(2k) \end{bmatrix}, \; M_2(k) = \begin{bmatrix} 0.3 & 0.2\sin(4k) \end{bmatrix}, \\
C_1(k) &= \begin{bmatrix} 0.6 & -\sin(2k) \end{bmatrix}, \; C_2(k) = \begin{bmatrix} 0.4 & \sin(3k) \end{bmatrix}, \\
C_3(k) &= \begin{bmatrix} 0.5 & -\sin(4k) \end{bmatrix}, \; D_{v_1}(k) = \sin(3k), \\
D_{v_2}(k) &= \sin(5k), \; D_{v_3}(k) = 3\sin(k).
\end{aligned}$$

Choose the quantizer parameters as $\kappa_1 = \kappa_2 = \kappa_3 = 0.25$ and $u_0^{(1)} = u_0^{(1)} = u_0^{(1)} = 3$. Let the H_∞ performance index be $\gamma = 1$, $\Sigma_v(1) = \Sigma_v(2) = \mathrm{diag}\{1, 1\}$, $\{\Upsilon(k)\}_{1\le k\le N} = I_3 \otimes \mathrm{diag}\{0.3, 0.3\}$, and $V_1 = V_2 = 1$, and select the initial matrices in terms of (7.40). On the basis of the *VED* algorithm, the solutions to RLMIs (7.41)–(7.44) are acquired, as shown in Table 7.1.

Assume that the initial states are $x_1(0) = \begin{bmatrix} 0.5 & 0.3 \end{bmatrix}^T$, $x_2(0) = \begin{bmatrix} 0.2 & -0.4 \end{bmatrix}^T$, $x_3(0) = \begin{bmatrix} -0.1 & 0.2 \end{bmatrix}^T$, $\hat{x}_1(0) = \begin{bmatrix} 1 & 1 \end{bmatrix}^T$, $\hat{x}_2(0) = \begin{bmatrix} 0.3 & -1.5 \end{bmatrix}^T$ and $\hat{x}_3(0) = \begin{bmatrix} -2.2 & 1 \end{bmatrix}^T$. Simulation results are indicated in Figs. 7.2–7.9, where Fig. 7.2 describes the current mode of the varying topology. Figs. 7.3–7.5 describe the output $z_i(k)$ $(i = 1, 2, 3)$ and the corresponding estimate $\hat{z}_i(k)$. Fig. 7.6 depicts the output estimation error $\tilde{z}_i(k)$ $(i = 1, 2, 3)$. Figs. 7.7–7.9 reflect the upper bounds of the estimation error covariance and the actual covariance of $e(k)$ acquired from Monte Carlo experiments. According to the above simulation results, we affirm that the developed *VED* algorithm in Section 7.1 indeed works effectively.

7.3.2 Example 2

This simulation example is presented to show the effectiveness of the proposed scheme in Section 7.2. For the time horizon $[0, 30]$, referring to reference [149], we select the parameters of CN (7.47) ($\Lambda = 5$ and $\mu = 2$) and other relevant parameters as follows:

$$\begin{aligned}
A_1(\hbar) &= \begin{bmatrix} 0.4\sin(2\hbar) & -0.4 \\ 0.3+0.1\sin(4\hbar) & -0.2 \end{bmatrix}, \; d = 2, \\
A_2(\hbar) &= \begin{bmatrix} 0.2 & -0.1 \\ -0.4 & 0.1+0.1\sin(2\hbar) \end{bmatrix}, \; \gamma = 0.9, \\
A_3(\hbar) &= \begin{bmatrix} 0.2 & -0.3 \\ 0.2 & -0.2 \end{bmatrix}, \; \Pi = \begin{bmatrix} 0.6 & 0.4 \\ 0.46 & 0.54 \end{bmatrix}, \\
A_4(\hbar) &= \begin{bmatrix} 0.3 & -0.1 \\ -0.2 & 0.3+0.1\sin(4\hbar) \end{bmatrix}, \; L = 0.2,
\end{aligned}$$

TABLE 7.1
Variance-constrained State Estimator Parameters

k	**0**	**1**	**2**	**3**	$\cdots$
$L_1(k,1)$	$\begin{bmatrix} -0.0517 & -0.0269 \\ -0.0303 & -0.0147 \end{bmatrix}$	$\begin{bmatrix} -0.0517 & 0.0410 \\ 0.0553 & -0.0438 \end{bmatrix}$	$\begin{bmatrix} -0.0263 & 0.0567 \\ 0.0504 & -0.1080 \end{bmatrix}$	$\begin{bmatrix} -0.0846 & -0.0458 \\ -0.0253 & 0.0085 \end{bmatrix}$	$\cdots$
$L_2(k,1)$	$\begin{bmatrix} -0.0992 & 0.0225 \\ -0.0028 & -0.0346 \end{bmatrix}$	$\begin{bmatrix} -0.0000 & 0.0139 \\ -0.0000 & -0.0758 \end{bmatrix}$	$\begin{bmatrix} -0.0003 & 0.2049 \\ -0.0002 & -0.0266 \end{bmatrix}$	$\begin{bmatrix} -0.0063 & -0.1838 \\ -0.0355 & -0.0960 \end{bmatrix}$	$\cdots$
$L_3(k,1)$	$\begin{bmatrix} -0.0768 & -0.0233 \\ -0.0060 & -0.0422 \end{bmatrix}$	$\begin{bmatrix} -0.0528 & -0.0267 \\ -0.0362 & -0.0183 \end{bmatrix}$	$\begin{bmatrix} -0.0181 & 0.0168 \\ 0.0586 & -0.0545 \end{bmatrix}$	$\begin{bmatrix} -0.0159 & 0.0081 \\ 0.0438 & -0.0779 \end{bmatrix}$	$\cdots$
$Z_1(k,1)$	$\begin{bmatrix} -0.0409 \\ 0.0114 \end{bmatrix}$	$\begin{bmatrix} -0.0414 \\ -0.1128 \end{bmatrix}$	$\begin{bmatrix} -0.1556 \\ -0.0712 \end{bmatrix}$	$\begin{bmatrix} -0.0652 \\ 0.1586 \end{bmatrix}$	$\cdots$
$Z_2(k,1)$	$\begin{bmatrix} -0.1257 \\ -0.0771 \end{bmatrix}$	$\begin{bmatrix} -0.4516 \\ -0.0979 \end{bmatrix}$	$\begin{bmatrix} -0.4266 \\ -0.1213 \end{bmatrix}$	$\begin{bmatrix} -0.5181 \\ 0.0518 \end{bmatrix}$	$\cdots$
$Z_3(k,1)$	$\begin{bmatrix} 0.0368 \\ -0.0189 \end{bmatrix}$	$\begin{bmatrix} -0.0398 \\ 0.1476 \end{bmatrix}$	$\begin{bmatrix} -0.1484 \\ -0.1340 \end{bmatrix}$	$\begin{bmatrix} -0.1935 \\ -0.0752 \end{bmatrix}$	$\cdots$

k	**0**	**1**	**2**	**3**	$\cdots$
$L_1(k,2)$	$\begin{bmatrix} -0.0881 - 0.0540 \\ -0.0720 - 0.0717 \end{bmatrix}$	$\begin{bmatrix} -0.04600.0364 \\ 0.0740 - 0.0587 \end{bmatrix}$	$\begin{bmatrix} -0.01820.0390 \\ 0.0549 - 0.1178 \end{bmatrix}$	$\begin{bmatrix} -0.0513 - 0.0385 \\ -0.0682 - 0.0349 \end{bmatrix}$	$\cdots$
$L_2(k,2)$	$\begin{bmatrix} -0.14590.0575 \\ -0.0112 - 0.1017 \end{bmatrix}$	$\begin{bmatrix} -0.00000.0810 \\ -0.0000 - 0.1190 \end{bmatrix}$	$\begin{bmatrix} 0.00020.5835 \\ 0.0002 - 0.1030 \end{bmatrix}$	$\begin{bmatrix} 0.0113 - 0.2124 \\ -0.0084 - 0.2328 \end{bmatrix}$	$\cdots$
$L_3(k,2)$	$\begin{bmatrix} -0.1125 - 0.0236 \\ 0.0007 - 0.0046 \end{bmatrix}$	$\begin{bmatrix} -0.0428 - 0.0216 \\ -0.0183 - 0.0092 \end{bmatrix}$	$\begin{bmatrix} -0.01230.0117 \\ 0.0724 - 0.0671 \end{bmatrix}$	$\begin{bmatrix} -0.05650.0997 \\ 0.1231 - 0.1598 \end{bmatrix}$	$\cdots$
$Z_1(k,2)$	$\begin{bmatrix} 0.0038 \\ 0.2144 \end{bmatrix}$	$\begin{bmatrix} -0.0623 \\ -0.1415 \end{bmatrix}$	$\begin{bmatrix} -0.1563 \\ -0.1747 \end{bmatrix}$	$\begin{bmatrix} -0.0633 \\ 0.2710 \end{bmatrix}$	$\cdots$
$Z_2(k,2)$	$\begin{bmatrix} -0.2179 \\ -0.0718 \end{bmatrix}$	$\begin{bmatrix} -0.7359 \\ -0.1511 \end{bmatrix}$	$\begin{bmatrix} -0.7810 \\ -0.1447 \end{bmatrix}$	$\begin{bmatrix} -1.1085 \\ -0.2030 \end{bmatrix}$	$\cdots$
$Z_3(k,2)$	$\begin{bmatrix} 0.0390 \\ -0.0617 \end{bmatrix}$	$\begin{bmatrix} -0.0853 \\ 0.1740 \end{bmatrix}$	$\begin{bmatrix} -0.1017 \\ -0.1639 \end{bmatrix}$	$\begin{bmatrix} -0.1040 \\ -0.1397 \end{bmatrix}$	$\cdots$

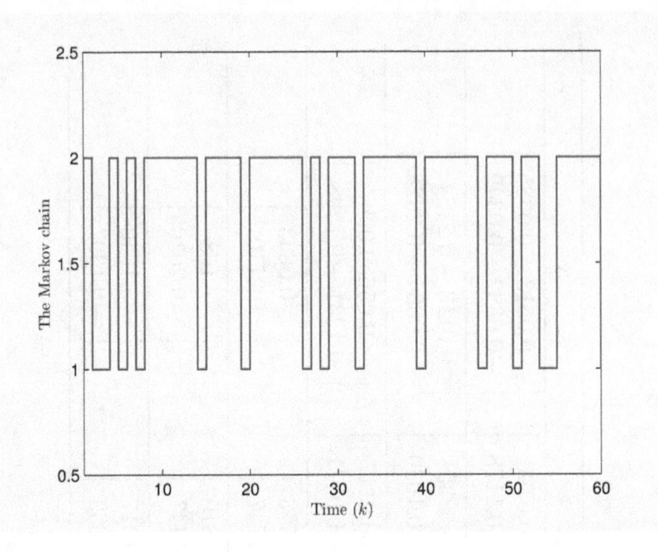

FIGURE 7.2
Modes evolution.

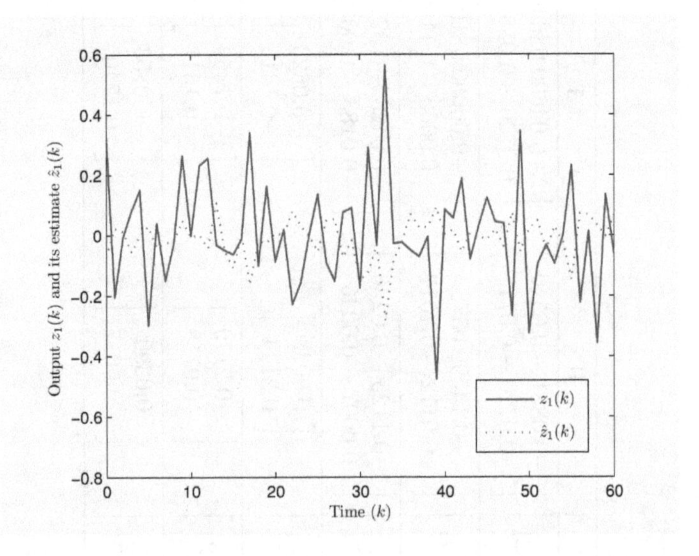

FIGURE 7.3
Output of node 1 and its estimate.

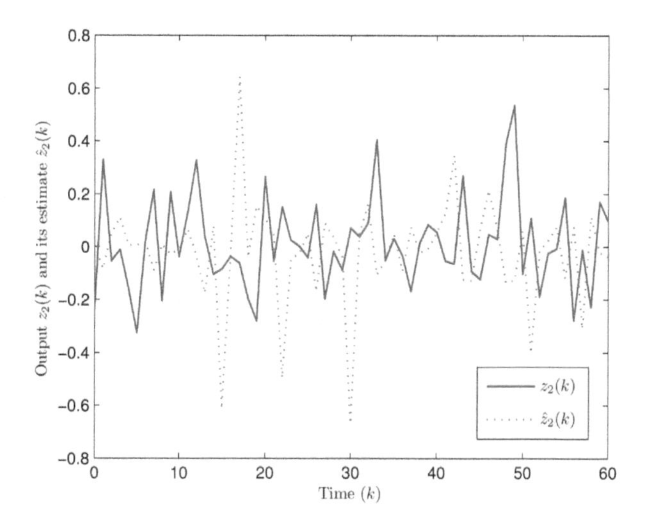

FIGURE 7.4
Output of node 2 and its estimate.

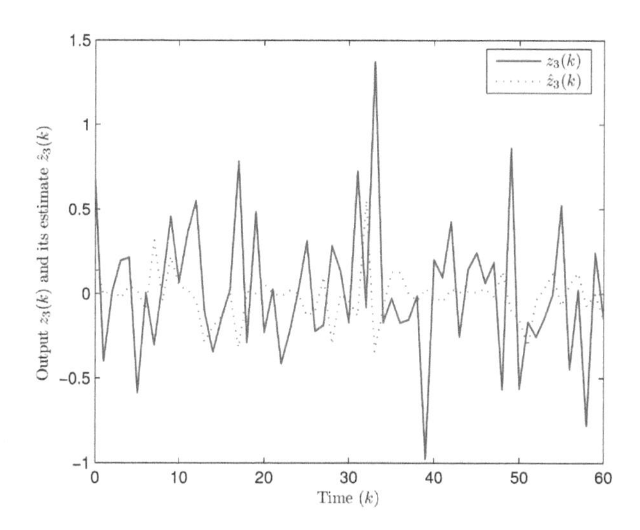

FIGURE 7.5
Output of node 3 and its estimate.

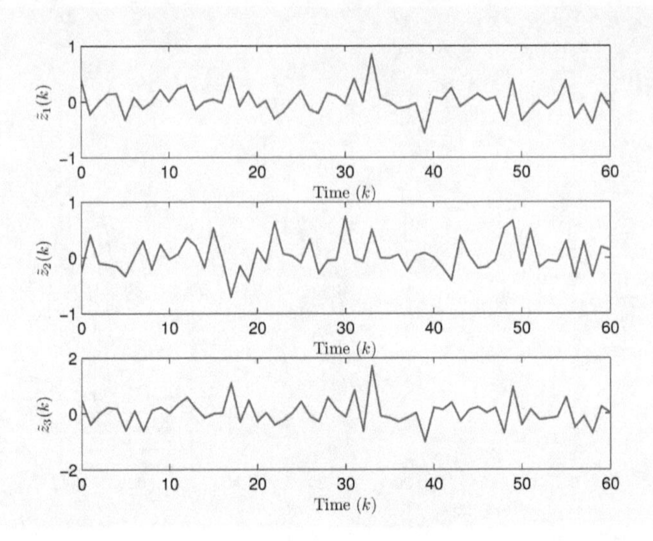

FIGURE 7.6
Output estimation errors of 3 nodes.

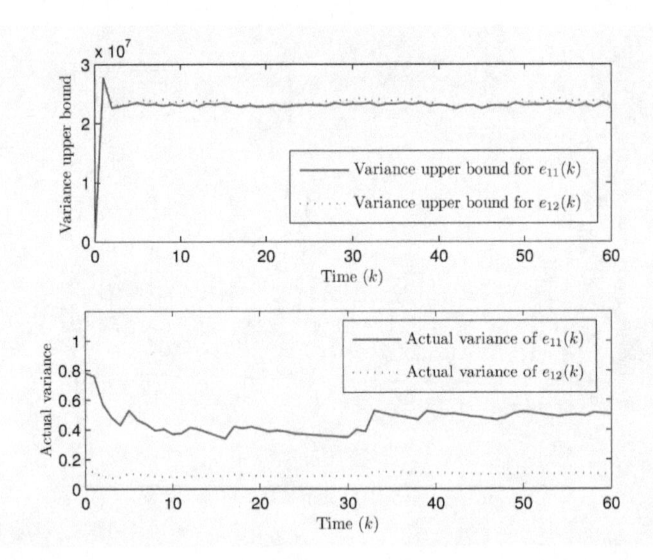

FIGURE 7.7
Error variance upper bound and actual error variance of node 1.

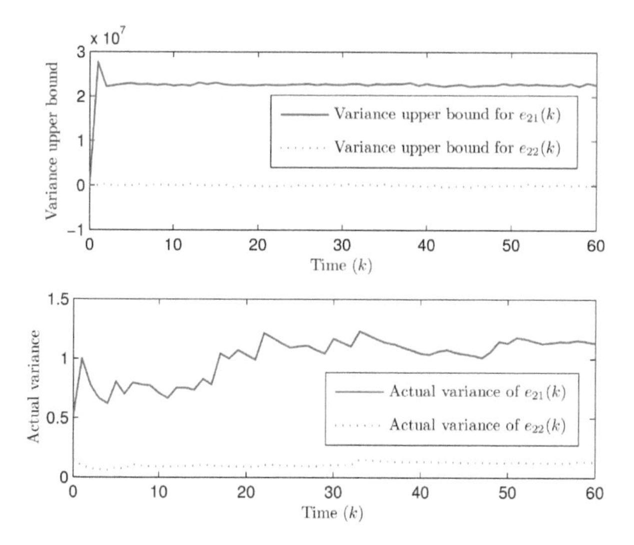

FIGURE 7.8
Error variance upper bound and actual error variance of node 2.

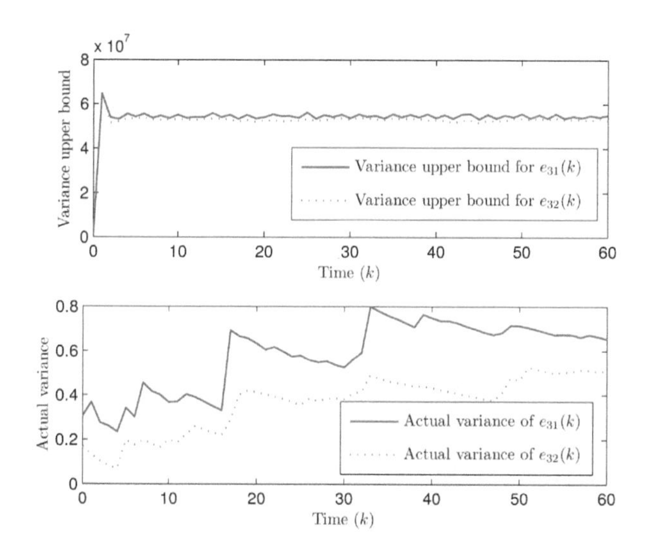

FIGURE 7.9
Error variance upper bound and actual error variance of node 3.

$$
\begin{aligned}
A_5(\hbar) &= \begin{bmatrix} 0.3 & -0.4 \\ 0.2 & -0.1+0.3\sin(2\hbar) \end{bmatrix}, \ \bar{\phi} = 0.4, \\
A_{m1}(\hbar) &= \begin{bmatrix} 0.3\sin(2\hbar) & -0.3 \\ 0.3+0.3\sin(2\hbar) & 0.3 \end{bmatrix}, \ \tilde{\phi}^2 = 0.2, \\
A_{m2}(\hbar) &= \begin{bmatrix} 0.1\sin(4\hbar) & -0.1 \\ -0.2 & 0.2+0.4\sin(2\hbar) \end{bmatrix}, \\
A_{m3}(\hbar) &= \begin{bmatrix} 0.2\sin(5\hbar) & -0.4 \\ 0.2 & -0.4 \end{bmatrix}, \Gamma = \mathrm{diag}\{0.5, 0.5\}, \\
A_{m4}(\hbar) &= \begin{bmatrix} 0.3\sin(2\hbar) & -0.3 \\ -0.2 & 0.2+0.3\sin(2\hbar) \end{bmatrix}, \\
A_{m5}(\hbar) &= \begin{bmatrix} 0.4\sin(2\hbar) & -0.3 \\ 0.3 & -0.4 \end{bmatrix}, D_2(\hbar) = \begin{bmatrix} \sin(4\hbar) \\ -0.1 \end{bmatrix}, \\
E_1(\hbar) &= \begin{bmatrix} -0.2\sin(2\hbar) \\ 0.4 \end{bmatrix}, \ E_2(\hbar) = \begin{bmatrix} 0.2 \\ 0.3\sin(4\hbar) \end{bmatrix}, \\
E_3(\hbar) &= \begin{bmatrix} -0.3\sin(2\hbar) \\ 0.4 \end{bmatrix}, \ C_1(\hbar) = \begin{bmatrix} 0.3 & -\sin(5\hbar) \\ 0.6 & -\sin(3\hbar) \end{bmatrix}, \\
E_4(\hbar) &= \begin{bmatrix} 0.3 \\ 0.4\sin(\hbar) \end{bmatrix}, \ E_5(\hbar) = \begin{bmatrix} -0.5\sin(2\hbar) \\ 0.4 \end{bmatrix}, \\
C_2(\hbar) &= \begin{bmatrix} 0.2 & \sin(4\hbar) \\ 0.5 & \sin(\hbar) \end{bmatrix}, \ D_1(\hbar) = \begin{bmatrix} \sin(5\hbar) \\ -0.3 \end{bmatrix}, \\
H_1(\hbar) &= \begin{bmatrix} \sin(3\hbar) & -0.2 \\ 0.1+0.3\sin(4\hbar) & -0.3 \end{bmatrix}, \ \underline{m} = 1, \\
H_2(\hbar) &= \begin{bmatrix} 3\sin(2\hbar) & -0.1 \\ 0.5+0.4\sin(5\hbar) & -0.2 \end{bmatrix}, \ \bar{m} = 5, \\
H_3(\hbar) &= \begin{bmatrix} 4\sin(3\hbar) & -0.4 \\ 0.3+0.1\sin(2\hbar) & -0.5 \end{bmatrix}, \ L_1 = 0.8, \\
H_4(\hbar) &= \begin{bmatrix} 2\sin(4\hbar) & -0.5 \\ 0.4+0.5\sin(\hbar) & -0.4 \end{bmatrix}, \\
H_5(\hbar) &= \begin{bmatrix} 3\sin(5\hbar) & -0.2 \\ 0.3+0.6\sin(2\hbar) & -0.5 \end{bmatrix}, \\
w_{\imath\imath} &= -0.4, \ w_{\imath\jmath} = 0.1 \ (\imath \neq \jmath, \imath, \jmath = 1, 2, \ldots, \Lambda), \\
m_1(\hbar) &= 3 + 2(-1)^{\hbar}, \ m_2(\hbar) = 3 + (-1)^{\hbar}, \\
a_{\max} &= \begin{bmatrix} 0.06 & 0.06 \end{bmatrix}^T, Q_1(0) = Q_2(0) = I, \\
A_{1\imath}(\hbar) &= F_{\imath}(\hbar) = \mathrm{diag}\{1, 1\}, \ A_{3\imath}(\hbar) = \begin{bmatrix} 0.1 & 0.1 \\ 0.1 & 0.1 \end{bmatrix}, \\
\bar{\theta}_{\imath 1} &= 0.8, \ \bar{\theta}_{\imath 2} = 0.6, \ (\imath = 1, 2, \ldots, \Lambda), \\
Y(u) &= Y(0) = 2I, \ R_1(u) = R_2(u) = 0.1I \\
&\quad (u = -\bar{m}, -\bar{m}+1, \ldots, -1).
\end{aligned}
$$

For $\imath = 1, 2, \ldots, 5$, we choose

$$\zeta(x_\imath(\hbar)) = \begin{bmatrix} -\tanh(0.4x_{\imath 1}(\hbar)) & \tanh(0.6x_{\imath 2}(\hbar)) \end{bmatrix}^T$$

where $x_{\imath t}(\hbar)$ $(t = 1, 2)$ is the t-th element of $x_\imath(\hbar)$. It can be seen that (7.48) holds when $\Omega = \text{diag}\{0.2, 0.3\}$.

By solving the recursive matrix inequalities (7.72), the gain matrices of the desired finite-horizon state estimators can be acquired from (7.73), which are omitted here due to the page limit. Set the initial states as $\rho_1(\jmath) = \begin{bmatrix} 0.1 & 0.2 \end{bmatrix}^T$, $\rho_2(\jmath) = \begin{bmatrix} 0.4 & 0.2 \end{bmatrix}^T$, $\rho_3(\jmath) = \begin{bmatrix} 0.3 & 0.4 \end{bmatrix}^T$, $\rho_4(\jmath) = \begin{bmatrix} 0.2 & 0.15 \end{bmatrix}^T$, $\rho_5(\jmath) = \begin{bmatrix} 0.5 & 0.3 \end{bmatrix}^T$ $(\jmath \in \mathbb{Z})$, $\tilde{y}(-1) = \begin{bmatrix} 0.2 & 0.2 \end{bmatrix}^T$, $\tilde{y}(-2) = \begin{bmatrix} 0.3 & 0.1 \end{bmatrix}^T$, $\hat{x}_{1,1}(0) = \begin{bmatrix} 0.2 & 0.2 \end{bmatrix}^T$, $\hat{x}_{1,2}(0) = \begin{bmatrix} 0.1 & 0.1 \end{bmatrix}^T$, $\hat{x}_{1,3}(0) = \hat{x}_{1,4}(0) = \hat{x}_{1,5}(0) = \hat{x}_{2,1}(0) = \begin{bmatrix} 0.3 & 0.2 \end{bmatrix}^T$, $\hat{x}_{2,2}(0) = \begin{bmatrix} 0.1 & 0.2 \end{bmatrix}^T$. The uncertain matrices are chosen as $A_{21}(\hbar) = \text{diag}\ \{\sin(3\hbar), \sin(3\hbar)\}$, $A_{22}(\hbar) = \text{diag}\{\sin(4\hbar), \sin(4\hbar)\}$, $A_{23}(\hbar) = A_{25}(\hbar) = \text{diag}\{\sin(2\hbar), \sin(2\hbar)\}$, $A_{24}(\hbar) = \text{diag}\{\sin(\hbar), \sin(\hbar)\}$. The disturbance signal is $\varpi(\hbar) = 5\mathrm{e}^{-\hbar}\sin(3\hbar)$. Simulation curves are shown in Figs. 7.10–7.12. Fig. 7.10 plots the mode evolution $r(\hbar)$ of the Markov chain, which illustrates the order of signal transmission of the network nodes. Fig. 7.11 represents the state estimation errors of 5 nodes. Fig. 7.12 describes the actual estimation performance of the estimators (7.58) at each time step

$$\bar{\gamma}(\hbar) = \sqrt{\frac{\sum\limits_{c=0}^{\hbar} \mathbb{E}\{\|z_e(c)\|^2\}}{\sum\limits_{c=0}^{\hbar} \|\varpi(c)\|^2 + \sum\limits_{t=-\bar{m}}^{0} \xi^T(t)Y(t)\xi(t)}}$$

with $\hbar = 0, 1, \ldots, \mathbb{N} - 1$, which is always less than γ over the finite horizon. Thus, it can be seen that the developed state estimation method in Section 7.2 is indeed effective.

7.4 Summary

In this chapter, the variance-constrained H_∞ state estimation problem has been studied for a class of nonlinear time-varying complex networks subject to randomly varying topologies, stochastic inner-coupling and measurement quantization. With the help of a Kronecker delta function, a Markov chain is introduced to characterize the randomly varying features of the network topology. A Gaussian random variable is employed to describe the stochastic disturbances in the inner-coupling matrix. The stochastic nonlinearities are utilized to represent the general nonlinearity occurring in the engineering applications. A new finite-horizon state estimator, which can be derived

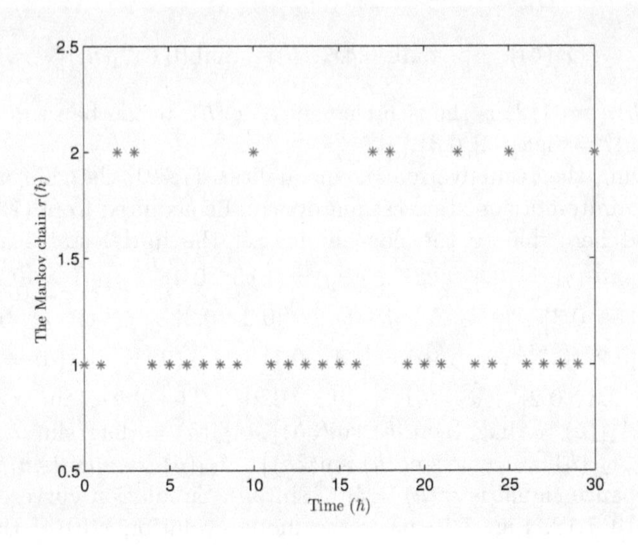

FIGURE 7.10
Order of data transmission of nodes.

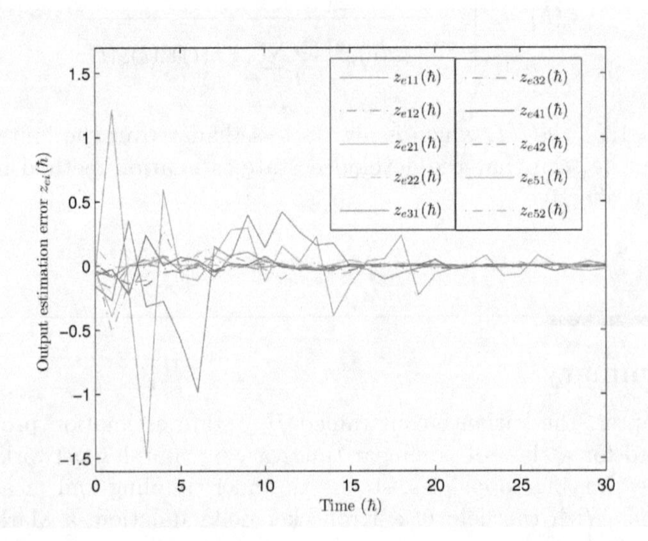

FIGURE 7.11
Output estimation errors of 5 nodes.

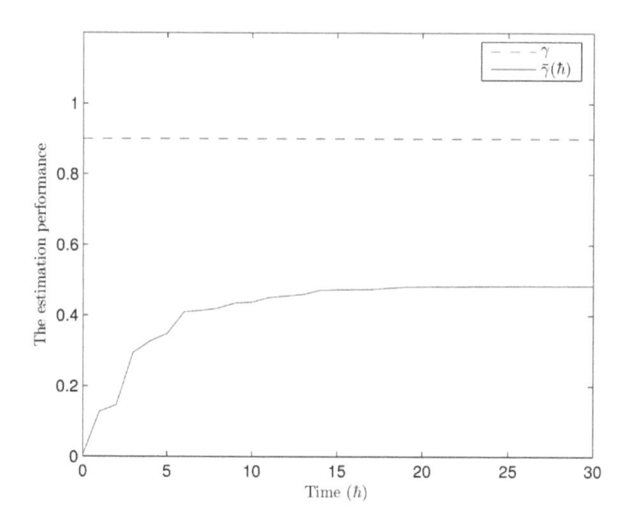

FIGURE 7.12
The estimation performance.

by solving a set of RLMIs, has been proposed such that the prescribed H_∞ performance and the estimation error covariance constraints can be achieved. Moreover, based on the output signals from partial network nodes, a novel set of time-varying robust state estimators has been designed for a class of CNs subject to randomly occurring uncertainties, multiple stochastic time-varying delays, sensor saturations under the scheduling of the random access protocol. The Markov chain has been utilized in the description of the random access protocol to schedule the transmission order of nodes. Two sequences of random variables, which obey the Gaussian distribution and Bernoulli distribution, respectively, have been employed to describe the phenomena of randomly occurring uncertainties and multiple stochastic communication delays within CNs. Through carrying out the stochastic analysis and inequality manipulations, sufficient conditions have been derived for ensuring the existence of the desired finite-horizon partial-nodes-based robust state estimators such that the effects from the noises on the whole estimation error dynamics are restricted below a specified H_∞ disturbance attenuation level. The gain parameters of the estimators have been expressed based on the solutions to the recursive linear matrix inequalities. Finally, the effectiveness of the proposed estimators has been verified by two simulation examples, respectively.

8

Event-Triggered Recursive Filtering for Complex Networks with Random Coupling Strengths

Different from the state estimation methods for complex networks proposed in Chapter 7, in this chapter, the recursive filtering problem is investigated for a class of time-varying complex networks with state saturations and random coupling strengths under an event-triggering transmission mechanism. The coupled strengths among nodes are characterized by a set of random variables obeying the uniform distribution. The event-triggering scheme is employed to mitigate the network data transmission burden. The purpose of the problem addressed is to design a recursive filter such that, in the presence of the state saturations, event-triggering communication mechanism and random coupling strengths, certain locally optimized upper bound is guaranteed on the filtering error covariance. By using the stochastic analysis technique, an upper bound on the filtering error covariance is first derived via the solution to a set of matrix difference equations. Next, the obtained upper bound is minimized by properly parameterizing the filter parameters. Subsequently, the boundedness issue of the filtering error covariance is studied. Finally, simulation examples are provided to illustrate the effectiveness of the proposed algorithm.

8.1 Problem Formulation

Take account of the following kind of discrete-time CN containing N nodes subject to state saturation and random coupling strengths:

$$\begin{cases} x_{r,\ell+1} = \Lambda_r\Big(A_{r,\ell}x_{r,\ell} + h(x_{r,\ell}) + \sum_{s=1}^{N}\alpha_{rs,\ell}\Gamma x_{s,\ell}\Big) + D_{r,\ell}\omega_{r,\ell} \\ y_{r,\ell} = C_{r,\ell}x_{r,\ell} + v_{r,\ell} \end{cases} \tag{8.1}$$

where, for node r, $x_{r,\ell} \in \mathbb{R}^n$ and $y_{r,\ell} \in \mathbb{R}^m$ $(r = 1, 2, \cdots, N)$ represent the state vector and the measurement output, respectively. $h(\cdot)$ describes a given nonlinear function. $\Gamma = \text{diag}\{\gamma_1, \gamma_2, \cdots, \gamma_n\} > 0$ denotes the inner-coupling

DOI: 10.1201/9781003189497-8

matrix with $\gamma_\iota \neq 0$ ($\iota = 1, 2, \cdots, n$) being the coupling strength of the ι-th element of the state vector. $\alpha_{rs,\ell}$ stands for the outer-coupling strength conforming to a uniform stochastic distribution in the domain $[a_{r1}, a_{r2}]$. The mean and variance of the stochastic variable $\alpha_{rs,\ell}$ are $\mu_r \geq 0$ and $\vartheta_r \geq 0$, respectively. $\omega_{r,\ell} \in \mathbb{R}^{\omega_1}$ and $v_{r,\ell} \in \mathbb{R}^{\omega_2}$ represent the process noise and the measurement noise, respectively. $\omega_{r,\ell}$ and $v_{r,\ell}$ are zero-mean Gaussian white noises with covariance matrices $Q_{r\omega,\ell}$ and $Q_{rv,\ell}$, respectively. It is supposed that $\alpha_{rs,\ell}$, $\omega_{r,\ell}$ and $v_{r,\ell}$ are mutually uncorrelated for any r, s, ℓ. $A_{r,\ell}$, $C_{r,\ell}$ and $D_{r,\ell}$ are all given system matrices with suitable dimensions.

The saturation function $\Lambda(\cdot) : \mathbb{R}^n \mapsto \mathbb{R}^n$ is defined as

$$\Lambda_r(q) \triangleq [\Lambda_{r1}(q_{r1}), \Lambda_{r2}(q_{r2}), \cdots, \Lambda_{rn}(q_{rn})]^T \tag{8.2}$$

where

$$\begin{aligned} q &\triangleq [q_{r1}, q_{r2}, \cdots, q_{rn}]^T, \\ \Lambda_{r\iota}(q_{r\iota}) &\triangleq \mathrm{sign}(q_{r\iota})\min\{q_{r\iota,\max}, |q_{r\iota}|\}, \quad \iota = 1, 2, \cdots, n \end{aligned} \tag{8.3}$$

in which $q_{r\iota,\max}$ is the ι-th element of the vector $q_{r,\max}$ with $q_{r,\max}$ being the saturation level vector of the r-th node, and sign$(\cdot)$ denotes a signum function.

Remark 8.1 *It is worth mentioning that the complex network introduced in this chapter is capable of characterizing dynamical behaviours of various dynamical networks that include recurrent neural networks (RNNs) as a typical example. For RNNs, the neurons are highly interconnected dynamical units/nodes, where the connections between neurons form a directed graph along a temporal sequence [13]. It is known from [157] that neuronal activity gives rise to dynamic patterns and processes, and anatomical/structural connections create a structural skeleton for a dynamic regime conducive to flexible and robust neural computation. Note that RNNs have been extensively applied in various domains such as machine translation, time series prediction and speech recognition. Also, techniques/theories for analyzing complex networks have begun to be employed to study deep learning systems (e.g. deep belief networks) in order to gain insights into the structural and functional properties of the computational graph resulting from the learning process [165, 170].*

Generally speaking, under the traditional time-triggering mechanism, a few redundant data could be generated that do not really contribute to the enhancement of the filter performance but increase the burden of communication devices. This is especially true for CNs with a large number of network nodes equipped with limited communication bandwidth and computation capacity. As such, it is of great practical significance to avoid unnecessary communication between the sensors and the remote state filter, thereby decreasing the energy consumption. For this purpose, the following event-triggering mechanism is employed in this chapter:

$$\varphi_r(y_{r,\ell}, y_{r,\ell_t^r}, \theta_r) = (y_{r,\ell} - y_{r,\ell_t^r})^T (y_{r,\ell} - y_{r,\ell_t^r}) - \theta_r, \tag{8.4}$$

where ℓ_t^r is the latest triggering time of the r-th node corresponding to the current sampling instant ℓ, and $\theta_r > 0$ is a given scalar serving as an adjustable threshold that determines the triggering frequency. Based on the event-triggering function (8.4), the measurement output of the r-th node will be transmitted to the corresponding filter (or estimator) if and only if the triggering function $\varphi_r(y_{r,\ell}, y_{r,\ell_t^r}, \theta_r) > 0$.

Denote the triggering instant series of the r-th node as $T_0^r < T_1^r < T_2^r < \cdots < T_m^r < T_{m+1}^r < \cdots$, and let $\ell_t^r = T_m^r$ when $\ell \in [T_m^r, T_{m+1}^r)$. Then, the new triggering instant of the r-th node can be determined iteratively as

$$T_{m+1}^r = \min\{\ell | \ell > T_m^r, \varphi_r(y_{r,\ell}, y_{r,\ell_t^r}, \theta_r) > 0\}. \tag{8.5}$$

Let $\hat{x}_{r,\ell+1|\ell}$ and $\hat{x}_{r,\ell+1|\ell+1}$ denote the one-step prediction and estimation of the state $x_{r,\ell}$ for the r-th node at time instant $\ell + 1$, respectively. The following recursive filter is designed for the r-th node:

$$\begin{cases} \hat{x}_{r,\ell+1|\ell} = \Lambda_r\Big(A_{r,\ell}\hat{x}_{r,\ell|\ell} + h(\hat{x}_{r,\ell|\ell}) + \sum_{s=1}^{N} \alpha_{rs,\ell}\Gamma\hat{x}_{s,\ell|\ell}\Big) \\ \hat{x}_{r,\ell+1|\ell+1} = \hat{x}_{r,\ell+1|\ell} + K_{r,\ell+1}\Big(y_{r,(\ell+1)_t^r} - C_{r,\ell+1}\hat{x}_{r,\ell+1|\ell}\Big) \end{cases} \tag{8.6}$$

where $K_{r,\ell+1}$ is the gain matrix to be determined for the r-th node at time instant $\ell + 1$.

Define $\xi_{r,\ell+1} \triangleq y_{r,(\ell+1)_t^r} - y_{r,\ell+1}$ as the difference of the r-th sensor's measurements between the latest triggering instant and current sampling instant. Then, the filter (8.6) can be rewritten as follows:

$$\begin{cases} \hat{x}_{r,\ell+1|\ell} = \Lambda_r\Big(A_{r,\ell}\hat{x}_{r,\ell|\ell} + h(\hat{x}_{r,\ell|\ell}) + \sum_{s=1}^{N} \mu_r\Gamma\hat{x}_{s,\ell|\ell}\Big) \\ \hat{x}_{r,\ell+1|\ell+1} = \hat{x}_{r,\ell+1|\ell} + K_{r,\ell+1}\Big(C_{r,\ell+1}x_{r,\ell+1} - C_{r,\ell+1}\hat{x}_{r,\ell+1|\ell} \\ \qquad + \xi_{r,\ell+1} + v_{r,\ell+1}\Big). \end{cases} \tag{8.7}$$

Furthermore, for analysis convenience, we denote the one-step prediction error, the filtering error and corresponding covariance matrices of the r-th node as $e_{r,\ell+1|\ell} \triangleq x_{r,\ell+1} - \hat{x}_{r,\ell+1|\ell}$, $e_{r,\ell+1|\ell+1} \triangleq x_{r,\ell+1} - \hat{x}_{r,\ell+1|\ell+1}$, $P_{r,\ell+1|\ell} \triangleq \mathbb{E}\{e_{r,\ell+1|\ell}e_{r,\ell+1|\ell}^T\}$ and $P_{r,\ell+1|\ell+1} \triangleq \mathbb{E}\{e_{r,\ell+1|\ell+1}e_{r,\ell+1|\ell+1}^T\}$, respectively.

The main objectives of this chapter are to (1) design a filter of the form (8.7) for system (8.1) such that an upper bound of $P_{r,\ell+1|\ell+1}$ can be obtained, (2) determine the filter gain to minimize the trace of the acquired upper bound, and (3) analyze the stability of the proposed recursive filter by looking into the boundedness of the estimation errors.

8.2 Main Results

In this section, a novel filtering algorithm is presented to address the filtering problem for state-saturated CNs with random coupling strengths. Then, an upper bound of the filtering error covariance matrix $P_{r,\ell+1|\ell+1}$ is proposed. Subsequently, the filter gain $K_{\ell+1}$ is designed to minimize the obtained upper bound at each time step. At last, the stability of the proposed filter is analyzed.

The following lemmas will be used in establishing the main results of this chapter.

Lemma 8.1 *[191] For $\forall x, y \in \mathbb{R}$, there exists a real number $\varepsilon_\iota \in [0,1]$ such that*

$$\Lambda_\iota(x) - \Lambda_\iota(y) = \varepsilon_\iota(x-y), \quad \iota = 1, 2, \cdots, n, \tag{8.8}$$

where the saturation function $\Lambda_\iota(\cdot)$ is defined in (8.2)–(8.3).

Lemma 8.2 *[130] Given matrices A, B, C and D with appropriate dimensions such that $CC^T < I$. Let M be a symmetric positive definite matrix and $\eta > 0$ be an arbitrary positive constant such that $\eta^{-1}I - DMD^T > 0$. Then, the following matrix inequality holds:*

$$(A+BCD)M(A+BCD)^T \leq A(M^{-1} - \eta D^T D)^{-1}A^T + \eta^{-1}BB^T. \tag{8.9}$$

Lemma 8.3 *[178] For $0 \leq \ell \leq n$, suppose that $X = X^T > 0$, $\mathcal{A}_\ell(X) = \mathcal{A}_\ell^T(X) \in \mathbb{R}^{L\times L}$, and $\mathcal{B}_\ell(X) = \mathcal{B}_\ell^T(X) \in \mathbb{R}^{L\times L}$. If there exits $\mathcal{C} = \mathcal{C}^T > X$ such that*

$$\mathcal{A}_\ell(X) \geq \mathcal{A}_\ell(\mathcal{C}), \tag{8.10}$$

$$\mathcal{B}_\ell(X) \geq \mathcal{A}_\ell(X), \tag{8.11}$$

then the solutions $\mathcal{R}_\ell$ and $\mathcal{S}_\ell$ to the following difference equations

$$\mathcal{R}_\ell = \mathcal{A}_\ell(\mathcal{R}_{\ell-1}), \mathcal{S}_\ell = \mathcal{B}_\ell(\mathcal{S}_{\ell-1}), \mathcal{R}_0 = \mathcal{S}_0 > 0 \tag{8.12}$$

satisfy $\mathcal{R}_\ell \leq \mathcal{S}_\ell$.

Lemma 8.4 *[138] Assume that there is a stochastic process $V_\ell(\varsigma_\ell)$ as well as real numbers $v_{\min}, v_{\max}, c > 0$ and $0 < \beta \leq 1$ such that*

$$v_{\min}\|\varsigma_\ell\|^2 \leq V_\ell(\varsigma_\ell) \leq v_{\max}\|\varsigma_\ell\|^2, \tag{8.13}$$

and

$$\mathbb{E}\{V_\ell(\varsigma_\ell)|\varsigma_{\ell-1}\} \leq (1-\beta)V_{\ell-1}(\varsigma_{\ell-1}) + c. \tag{8.14}$$

Then, ς_ℓ is exponentially bounded in the mean square, i.e.,

$$\mathbb{E}\{\|\varsigma_\ell\|^2\} \leq \frac{\upsilon_{\max}}{\upsilon_{\min}}\mathbb{E}\{\|\varsigma_0\|^2\}(1-\beta)^\ell + \frac{c}{\upsilon_{\min}}\sum_{r=1}^{\ell}(1-\beta)^r. \tag{8.15}$$

In the following theorem, the explicit expressions on the one-step prediction error covariance matrix $P_{r,\ell+1|\ell}$ and the filtering error covariance matrix $P_{r,\ell+1|\ell+1}$ are given.

Theorem 8.1 *The one-step prediction error covariance matrix $P_{r,\ell+1|\ell}$ and the filtering error covariance matrix $P_{r,\ell+1|\ell+1}$ satisfy the following two recursive equations:*

$$\begin{aligned} P_{r,\ell+1|\ell} =& D_{r,\ell}Q_{r\omega,\ell}D_{r,\ell}^T + \mathbb{E}\Big\{\Xi_{r,\ell}\tilde{A}_{r,\ell}P_{r,\ell|\ell}\tilde{A}_{r,\ell}^T\Xi_{r,\ell}^T\Big\} \\ &+ \sum_{s=1}^{N}\mu_r\mathbb{E}\Big\{\Xi_{r,\ell}(\tilde{A}_{r,\ell}e_{r,\ell|\ell}e_{s,\ell|\ell}^T\Gamma^T + \Gamma e_{s,\ell|\ell}e_{r,\ell|\ell}^T\tilde{A}_{r,\ell}^T)\Xi_{r,\ell}^T\Big\} \\ &+ \sum_{s=1}^{N}\vartheta_r\mathbb{E}\Big\{\Xi_{r,\ell}\Gamma x_{s,\ell|\ell}x_{s,\ell|\ell}^T\Xi_{r,\ell}^T\Big\} \\ &+ \sum_{s=1}^{N}\sum_{d=1}^{N}\mu_r^2\mathbb{E}\Big\{\Xi_{r,\ell}\Gamma e_{s,\ell|\ell}e_{d,\ell|\ell}^T\Gamma^T\Xi_{r,\ell}^T\Big\}, \end{aligned} \tag{8.16}$$

and

$$\begin{aligned} P_{r,\ell+1|\ell+1} =& (I - K_{r,\ell+1}C_{r,\ell})P_{r,\ell+1|\ell}(I - K_{r,\ell+1}C_{r,\ell})^T \\ &+ K_{r,\ell+1}\mathbb{E}\{\xi_{r,\ell+1}\xi_{r,\ell+1}^T\}K_{r,\ell+1}^T \\ &- (I - K_{r,\ell+1}C_{r,\ell})\mathbb{E}\{e_{r,\ell+1|\ell}\xi_{r,\ell+1}^T\}K_{r,\ell+1}^T \\ &- K_{r,\ell+1}\mathbb{E}\{\xi_{r,\ell+1}e_{r,\ell+1|\ell}^T\}(I - K_{r,\ell+1}C_{r,\ell})^T \\ &+ K_{r,\ell+1}\mathbb{E}\{\xi_{r,\ell+1}v_{r,\ell+1}^T\}K_{r,\ell+1}^T \\ &+ K_{r,\ell+1}\mathbb{E}\{v_{r,\ell+1}\xi_{r,\ell+1}^T\}K_{r,\ell+1}^T + K_{r,\ell+1}Q_{rv,\ell+1}K_{r,\ell+1}^T \end{aligned} \tag{8.17}$$

where $\Xi_{r,\ell} \triangleq diag\{\varepsilon_{r,\ell}^{(1)}, \varepsilon_{r,\ell}^{(2)}, \cdots, \varepsilon_{r,\ell}^{(n)}\}$, $\varepsilon_{r,\ell}^{(\nu)} \in [0,1]$ $(\nu = 1,2,\cdots,n)$, $\tilde{A}_{r,\ell} \triangleq A_{r,\ell} + (H_{r,\ell} + U_{r,\ell}\Delta_{r,\ell})$, $H_{r,\ell} = \frac{\partial h(x_{r,\ell})}{\partial x_{r,\ell}}|_{x_{r,\ell}=\hat{x}_{r,\ell|\ell}}$, and $U_{r,\ell}$ is a problem-dependent scaling matrix and $\Delta_{r,\ell}$ is an unknown matrix accounting for the linearization error satisfying $\Delta_{r,\ell}\Delta_{r,\ell}^T \leq I$.

Proof *It follows from (8.1) and Lemma 8.1 that*

$$\begin{aligned} e_{r,\ell+1|\ell} =& x_{r,\ell+1} - \hat{x}_{r,\ell+1|\ell} \\ =& D_{r,\ell}\omega_{r,\ell} + \Lambda_r\left(A_{r,\ell}x_{r,\ell} + h(x_{r,\ell}) + \sum_{s=1}^{N}\alpha_{rs,\ell}\Gamma x_{s,\ell}\right) \end{aligned}$$

$$
\begin{aligned}
&- \Lambda_r \left(A_{r,\ell}\hat{x}_{r,\ell|\ell} + h(\hat{x}_{r,\ell|\ell}) + \sum_{s=1}^{N} \mu_r \Gamma \hat{x}_{s,\ell|\ell} \right) \\
=& D_{r,\ell}\omega_{r,\ell} + \Xi_{r,\ell} \left(\left(A_{r,\ell}x_{r,\ell} + h(x_{r,\ell}) + \sum_{s=1}^{N} \alpha_{rs,\ell}\Gamma x_{s,\ell} \right) \right. \\
&\left. - \left(A_{r,\ell}\hat{x}_{r,\ell|\ell} + h(\hat{x}_{r,\ell|\ell}) + \sum_{s=1}^{N} \mu_r \Gamma \hat{x}_{s,\ell|\ell} \right) \right) \\
=& D_{r,\ell}\omega_{r,\ell} + \Xi_{r,\ell} \Big(A_{r,\ell}e_{r,\ell|\ell} + h(x_{r,\ell}) - h(\hat{x}_{r,\ell|\ell}) \\
&+ \sum_{s=1}^{N} (\alpha_{rs,\ell} - \mu_r)\Gamma x_{s,\ell|\ell} + \sum_{s=1}^{N} \mu_r \Gamma e_{s,\ell|\ell} \Big).
\end{aligned} \tag{8.18}
$$

Expanding the nonlinear function in a Taylor series about $x_{r,\ell}$, we have

$$
\begin{aligned}
h(x_{r,\ell}) =& h(\hat{x}_{r,\ell|\ell}) + H_{r,\ell}(x_{r,\ell} - \hat{x}_{r,\ell|\ell}) + o(|e_{r,\ell|\ell}|) \\
=& h(\hat{x}_{r,\ell|\ell}) + H_{r,\ell}e_{r,\ell|\ell} + o(|e_{r,\ell|\ell}|)
\end{aligned} \tag{8.19}
$$

where $H_{r,\ell} = \frac{\partial h(x_{r,\ell})}{\partial x_{r,\ell}}|_{x_{r,\ell}=\hat{x}_{r,\ell|\ell}}$ denotes the Jacobian matrix and $o(|e_{r,\ell|\ell}|)$ represents the high-order terms of Taylor series expression. Referring to [80, 89], the high-order terms can be represented by

$$
o(|e_{r,\ell|\ell}|) = U_{r,\ell}\Delta_{r,\ell}e_{r,\ell|\ell}. \tag{8.20}
$$

Substituting (8.19)-(8.20) into (8.18) yields

$$
\begin{aligned}
e_{r,\ell+1|\ell} =& x_{r,\ell+1} - \hat{x}_{r,\ell+1|\ell} \\
=& \Xi_{r,\ell}\tilde{A}_{r,\ell}e_{r,\ell|\ell} + \Xi_{r,\ell}\sum_{s=1}^{N}(\alpha_{rs,\ell} - \mu_r)\Gamma x_{s,\ell|\ell} \\
&+ \Xi_{r,\ell}\sum_{s=1}^{N}\mu_r\Gamma e_{s,\ell|\ell} + D_{r,\ell}\omega_{r,\ell}.
\end{aligned} \tag{8.21}
$$

In view of $P_{r,\ell+1|\ell} = \mathbb{E}\{e_{r,\ell+1|\ell}e^T_{r,\ell+1|\ell}\}$, we have

$$
\begin{aligned}
P_{r,\ell+1|\ell} =& \mathbb{E}\Big\{ \Big[\Xi_{r,\ell}\tilde{A}_{r,\ell}e_{r,\ell|\ell} + \Xi_{r,\ell}\sum_{s=1}^{N}(\alpha_{rs,\ell} - \mu_r)\Gamma x_{s,\ell|\ell} \\
&+ \Xi_{r,\ell}\sum_{s=1}^{N}\mu_r\Gamma e_{s,\ell|\ell} + D_{r,\ell}\omega_{r,\ell} \Big] \\
&\times \Big[\Xi_{r,\ell}\tilde{A}_{r,\ell}e_{r,\ell|\ell} + \Xi_{r,\ell}\sum_{s=1}^{N}(\alpha_{rs,\ell} - \mu_r)\Gamma x_{s,\ell|\ell}
\end{aligned} \tag{8.22}
$$

$$+ \Xi_{r,\ell} \sum_{s=1}^{N} \mu_r \Gamma e_{s,\ell|\ell} + D_{r,\ell}\omega_{r,\ell}\Big]^T \Big\}.$$

Since $e_{r,\ell|\ell}$ is independent of the process noise $\omega_{r,\ell}$, (8.16) can be derived directly from (8.22). Next, let us focus on the filtering error covariance matrix $P_{r,\ell+1|\ell+1}$. From (8.1) and (8.7), the filtering error $e_{r,\ell+1|\ell+1}$ can be written as

$$\begin{aligned} e_{r,\ell+1|\ell+1} =& x_{r,\ell+1} - \hat{x}_{r,\ell+1|\ell+1} \\ =& x_{r,\ell+1} - \hat{x}_{r,\ell+1|\ell} - K_{r,\ell+1}(C_{r,\ell}x_{r,\ell+1} \\ & - C_{r,\ell}\hat{x}_{r,\ell+1|\ell} + \xi_{r,\ell+1} + v_{r,\ell}) \\ =& (I - K_{r,\ell+1}C_{r,\ell})e_{r,\ell+1|\ell} - K_{r,\ell+1}\xi_{r,\ell+1} - K_{r,\ell+1}v_{r,\ell+1}, \end{aligned} \tag{8.23}$$

based on which the filtering error covariance matrix can be derived by

$$\begin{aligned} P_{r,\ell+1|\ell+1} =& \mathbb{E}\{e_{r,\ell+1|\ell+1}e^T_{r,\ell+1|\ell+1}\} \\ =& \mathbb{E}\{[(I - K_{r,\ell+1}C_{r,\ell})e_{r,\ell+1|\ell} - K_{r,\ell+1}\xi_{r,\ell+1} \\ & - K_{r,\ell+1}v_{r,\ell+1}][(I - K_{r,\ell+1}C_{r,\ell})e_{r,\ell+1|\ell} \\ & - K_{r,\ell+1}\xi_{r,\ell+1} - K_{r,\ell+1}v_{r,\ell+1}]^T\} \\ =& (I - K_{r,\ell+1}C_{r,\ell})P_{r,\ell+1|\ell}(I - K_{r,\ell+1}C_{r,\ell})^T \\ & + \mathbb{E}\{K_{r,\ell+1}\xi_{r,\ell+1}\xi^T_{r,\ell+1}K^T_{r,\ell+1}\} \\ & - \mathbb{E}\{(I - K_{r,\ell+1}C_{r,\ell})e_{r,\ell+1|\ell}\xi^T_{r,\ell+1}K^T_{r,\ell+1}\} \\ & - \mathbb{E}\{K_{r,\ell+1}\xi_{r,\ell+1}e^T_{r,\ell+1|\ell}(I - K_{r,\ell+1}C_{r,\ell})^T\} \\ & + \mathbb{E}\{K_{r,\ell+1}\xi_{r,\ell+1}v^T_{r,\ell+1}K^T_{r,\ell+1}\} \\ & + \mathbb{E}\{K_{r,\ell+1}v_{r,\ell+1}\xi^T_{r,\ell+1}K^T_{r,\ell+1}\} + \mathbb{E}\{K_{r,\ell+1}Q_{rv,\ell+1}K^T_{r,\ell+1}\}, \end{aligned} \tag{8.24}$$

which concludes the proof.

It should be mentioned that, some uncertain terms have appeared in the filtering error covariance matrix (8.17), which is mainly due to the introduction of the event-triggering mechanism and the noises. In this case, it is almost impossible to obtain the *exact* value of the filtering error covariance which, in turn, renders the impossibility to calculate the filter gain. In order to resolve this issue, a practical way is to find an upper bound of the filtering error covariance matrix and then determine the filter gain by minimizing the trace of such an upper bound matrix.

Theorem 8.2 *Consider one-step prediction error covariance matrix $P_{r,\ell+1|\ell}$ and filtering error covariance matrix $P_{r,\ell+1|\ell+1}$ in (8.16) and (8.17), respectively. Let λ_1, λ_2 and ϵ_1 be positive scalars. Assume that the following two recursive matrix equations*

$$\Psi_{r,\ell+1|\ell} = 2(1 + N\mu_r\lambda_1)[A_{r,\ell}\Psi_{r,\ell|\ell}A^T_{r,\ell} + H_{r,\ell}(\Psi^{-1}_{r,\ell|\ell}$$

$$
\begin{aligned}
&- (1+\epsilon_1 I))^{-1}H_{r,\ell}^T + \epsilon_1^{-1}U_{r,\ell}U_{r,\ell}^T] \\
&+\sum_{s=1}^{N}(\mu_r\lambda_1^{-1}+\vartheta_r+\mu_r^2 N)\Gamma\Psi_{s,\ell|\ell}\Gamma^T \\
&+\sum_{s=1}^{N}\vartheta_r\Gamma\hat{x}_{s,\ell|\ell}\hat{x}_{s,\ell|\ell}^T\Gamma^T + D_{r,\ell}Q_{r\omega,\ell}D_{r,\ell}^T
\end{aligned} \tag{8.25}
$$

and

$$
\begin{aligned}
\Psi_{r,\ell+1|\ell+1} =&(1+\lambda_2)(I-K_{r,\ell+1}C_{r,\ell})\Psi_{r,\ell+1|\ell}(I-K_{r,\ell+1}C_{r,\ell})^T \\
&+ K_{r,\ell+1}[\theta_r(1+\lambda_2^{-1})I+\delta_{\ell+1,T_{m+1}^r}Q_{rv,\ell+1}]K_{r,\ell+1}^T
\end{aligned} \tag{8.26}
$$

have positive define solutions $\Psi_{r,\ell+1|\ell}$ *and* $\Psi_{r,\ell+1|\ell+1}$ *with* $\epsilon_1^{-1}I - \Psi_{r,\ell|\ell} > 0$ *and the initial condition* $P_{0|0} \leq \Psi_{0|0}$. *Then, the matrix* $\Psi_{r,\ell+1|\ell}$ *is an upper bound of* $P_{r,\ell+1|\ell}$ *and the matrix* $\Psi_{r,\ell+1|\ell+1}$ *is an upper bound of* $P_{r,\ell+1|\ell+1}$, *i.e.*

$$
P_{r,\ell+1|\ell} \leq \Psi_{r,\ell+1|\ell}, \tag{8.27}
$$

$$
P_{r,\ell+1|\ell+1} \leq \Psi_{r,\ell+1|\ell+1}. \tag{8.28}
$$

Furthermore, the filter gain matrix is obtained by minimizing the trace of the upper bound matrix $\Psi_{r,\ell+1|\ell+1}$, *that is*

$$
\begin{aligned}
K_{r,\ell+1} =&(1+\lambda_2)\Psi_{r,\ell+1|\ell}C_{r,\ell}^T[(1+\lambda_2)C_{r,\ell}\Psi_{r,\ell+1|\ell}C_{r,\ell}^T \\
&+\theta_r(1+\lambda_2^{-1})I+\delta_{\ell+1,T_{m+1}^r}Q_{rv,\ell+1}]^{-1}
\end{aligned} \tag{8.29}
$$

where $\delta_{\ell+1,T_{m+1}^r}$ *is the Kronecker delta function.*

Proof *This theorem can be proved by mathematical induction. Consider the initial condition* $P_{0|0} \leq \Psi_{0|0}$ *at time step* ℓ *and assume that* $P_{r,\ell|\ell} \leq \Psi_{r,\ell|\ell}$.

Applying Lemma 3.2 to the third term of the right-hand side of (8.16), we have

$$
\begin{aligned}
&\sum_{s=1}^{N}\mu_r\mathbb{E}\Big\{\Xi_{r,\ell}(\tilde{A}_{r,\ell}e_{r,\ell|\ell}e_{s,\ell|\ell}^T\Gamma^T+\Gamma e_{s,\ell|\ell}e_{r,\ell|\ell}^T\tilde{A}_{r,\ell}^T)\Xi_{r,\ell}^T\Big\} \\
\leq&\sum_{s=1}^{N}\mu_r\mathbb{E}\Big\{\Xi_{r,\ell}[\lambda_1\tilde{A}_{r,\ell}P_{r,\ell|\ell}\tilde{A}_{r,\ell}^T+\lambda_1^{-1}\Gamma P_{s,\ell|\ell}\Gamma^T]\Xi_{r,\ell}^T\Big\}.
\end{aligned} \tag{8.30}
$$

The second and third terms of the right-hand side of (8.16) satisfy

$$
\mathbb{E}\Big\{\Xi_{r,\ell}\tilde{A}_{r,\ell}P_{r,\ell|\ell}\tilde{A}_{r,\ell}^T\Xi_{r,\ell}^T\Big\}+\sum_{s=1}^{N}\mu_r\mathbb{E}\Big\{\Xi_{r,\ell}(\tilde{A}_{r,\ell}e_{r,\ell|\ell}e_{s,\ell|\ell}^T\Gamma^T
$$

$$
\begin{aligned}
&+\Gamma e_{s,\ell|\ell}e_{r,\ell|\ell}^{T}\tilde{A}_{r,\ell}^{T})\Xi_{r,\ell}^{T}\Big\}\\
\leq&\mathbb{E}\Big\{\Xi_{r,\ell}\tilde{A}_{r,\ell}P_{r,\ell|\ell}\tilde{A}_{r,\ell}^{T}\Xi_{r,\ell}^{T}\Big\}\\
&+\sum_{s=1}^{N}\mu_r\mathbb{E}\Big\{\Xi_{r,\ell}[\lambda_1\tilde{A}_{r,\ell}P_{r,\ell|\ell}\tilde{A}_{r,\ell}^{T}+\lambda_1^{-1}\Gamma P_{s,\ell|\ell}\Gamma^{T}]\Xi_{r,\ell}^{T}\Big\}\\
=&(1+N\mu_r\lambda_1)\mathbb{E}\Big\{\Xi_{r,\ell}\tilde{A}_{r,\ell}P_{r,\ell|\ell}\tilde{A}_{r,\ell}^{T}\Xi_{r,\ell}^{T}\Big\}\\
&+\sum_{s=1}^{N}\mu_r\lambda_1^{-1}\mathbb{E}\Big\{\Xi_{r,\ell}\Gamma P_{s,\ell|\ell}\Gamma^{T}\Xi_{r,\ell}^{T}\Big\}.
\end{aligned}
\tag{8.31}
$$

The fourth term of the right-hand side of (8.16) can be derived by

$$
\begin{aligned}
&\sum_{s=1}^{N}\vartheta_r\mathbb{E}\Big\{\Xi_{r,\ell}\Gamma x_{s,\ell|\ell}x_{s,\ell|\ell}^{T}\Xi_{r,\ell}^{T}\Big\}\\
=&\sum_{s=1}^{N}\vartheta_r\mathbb{E}\Big\{\Xi_{r,\ell}\Gamma(\hat{x}_{s,\ell|\ell}+e_{s,\ell|\ell})(\hat{x}_{s,\ell|\ell}+e_{s,\ell|\ell})^{T}\Xi_{r,\ell}^{T}\Big\}\\
=&\sum_{s=1}^{N}\vartheta_r\mathbb{E}\Big\{\Xi_{r,\ell}\Gamma\hat{x}_{s,\ell|\ell}\hat{x}_{s,\ell|\ell}^{T}\Xi_{r,\ell}^{T}\Big\}+\sum_{s=1}^{N}\vartheta_r\mathbb{E}\Big\{\Xi_{r,\ell}\Gamma P_{s,\ell|\ell}\Xi_{r,\ell}^{T}\Big\}.
\end{aligned}
\tag{8.32}
$$

The fifth term of the right-hand side of (8.16) can be further shown as

$$
\begin{aligned}
&\sum_{s=1}^{N}\sum_{d=1}^{N}\mu_r^2\mathbb{E}\Big\{\Xi_{r,\ell}\Gamma e_{s,\ell|\ell}e_{d,\ell|\ell}^{T}\Gamma^{T}\Xi_{r,\ell}^{T}\Big\}\\
=&\frac{1}{2}\mu_r^2\sum_{s=1}^{N}\sum_{d=1}^{N}\mathbb{E}\Big\{\Xi_{r,\ell}(\Gamma e_{s,\ell|\ell}e_{d,\ell|\ell}^{T}\Gamma^{T}+\Gamma e_{d,\ell|\ell}e_{s,\ell|\ell}^{T}\Gamma^{T})\Xi_{r,\ell}^{T}\Big\}\\
\leq&\frac{1}{2}\mu_r^2\sum_{s=1}^{N}\sum_{d=1}^{N}\mathbb{E}\Big\{\Xi_{r,\ell}(\Gamma e_{s,\ell|\ell}e_{s,\ell|\ell}^{T}\Gamma^{T}+\Gamma e_{d,\ell|\ell}e_{d,\ell|\ell}^{T}\Gamma^{T})\Xi_{r,\ell}^{T}\Big\}\\
\leq&\mu_r^2N\sum_{s=1}^{N}\mathbb{E}\Big\{\Xi_{r,\ell}\Gamma P_{s,\ell|\ell}\Gamma^{T}\Xi_{r,\ell}^{T}\Big\}.
\end{aligned}
\tag{8.33}
$$

Substituting (8.30)–(8.33) into (8.16) leads to

$$
\begin{aligned}
P_{r,\ell+1|\ell}\leq&(1+N\mu_r\lambda_1)\mathbb{E}\Big\{\Xi_{r,\ell}\tilde{A}_{r,\ell}P_{r,\ell|\ell}\tilde{A}_{r,\ell}^{T}\Xi_{r,\ell}^{T}\Big\}\\
&+\sum_{s=1}^{N}(\mu_r\lambda_1^{-1}+\vartheta_r+\mu_r^2N)\mathbb{E}\Big\{\Xi_{r,\ell}\Gamma P_{s,\ell|\ell}\Gamma^{T}\Xi_{r,\ell}^{T}\Big\}\\
&+\sum_{s=1}^{N}\vartheta_r\mathbb{E}\Big\{\Xi_{r,\ell}\Gamma\hat{x}_{s,\ell|\ell}\hat{x}_{s,\ell|\ell}^{T}\Xi_{r,\ell}^{T}\Big\}+D_{r,\ell}Q_{r\omega,\ell}D_{r,\ell}^{T}
\end{aligned}
\tag{8.34}
$$

where

$$\mathbb{E}\{\Xi_{r,\ell}\tilde{A}_{r,\ell}P_{r,\ell|\ell}\tilde{A}_{r,\ell}^T\Xi_{r,\ell}^T\} \leq \mathbb{E}\{\tilde{A}_{r,\ell}P_{r,\ell|\ell}\tilde{A}_{r,\ell}^T\} = \tilde{A}_{r,\ell}P_{r,\ell|\ell}\tilde{A}_{r,\ell}^T. \tag{8.35}$$

Similarly, it is easily obtained that

$$\sum_{s=1}^{N}\mathbb{E}\{\Xi_{r,\ell}\Gamma P_{s,\ell|\ell}\Gamma^T\Xi_{r,\ell}^T\} \leq \sum_{s=1}^{N}\Gamma P_{s,\ell|\ell}\Gamma^T \tag{8.36}$$

and

$$\sum_{s=1}^{N}\mathbb{E}\{\Xi_{r,\ell}\Gamma\hat{x}_{s,\ell|\ell}\hat{x}_{s,\ell|\ell}^T\Gamma^T\Xi_{r,\ell}^T\} \leq \sum_{s=1}^{N}\Gamma\hat{x}_{s,\ell|\ell}\hat{x}_{s,\ell|\ell}^T\Gamma^T. \tag{8.37}$$

Substituting (8.35)–(8.37) into (8.34) and using the elementary inequality $xy^T + yx^T \leq xx^T + yy^T$*, it follows from Lemma 8.2 that*

$$\begin{aligned}
P_{r,\ell+1|\ell} \leq & (1+N\mu_r\lambda_1)[A_{r,\ell} + (H_{r,\ell} + U_{r,\ell}\Delta_{r,\ell})]P_{r,\ell|\ell}[A_{r,\ell} + (H_{r,\ell} \\
& + U_{r,\ell}\Delta_{r,\ell})]^T + \sum_{s=1}^{N}(\mu_r\lambda_1^{-1} + \vartheta_r + \mu_r^2 N)\Gamma P_{s,\ell|\ell}\Gamma^T \\
& + \sum_{s=1}^{N}\vartheta_r\Gamma\hat{x}_{s,\ell|\ell}\hat{x}_{s,\ell|\ell}^T\Gamma^T + D_{r,\ell}Q_{r\omega,\ell}D_{r,\ell}^T \\
\leq & 2(1+N\mu_r\lambda_1)[A_{r,\ell}P_{r,\ell|\ell}A_{r,\ell}^T + (H_{r,\ell} + U_{r,\ell}\Delta_{r,\ell}) \\
& \times P_{r,\ell|\ell}(H_{r,\ell} + U_{r,\ell}\Delta_{r,\ell})^T] + \sum_{s=1}^{N}(\mu_r\lambda_1^{-1} + \vartheta_r + \mu_r^2 N) \\
& \times \Gamma P_{s,\ell|\ell}\Gamma^T + \sum_{s=1}^{N}\vartheta_r\Gamma\hat{x}_{s,\ell|\ell}\hat{x}_{s,\ell|\ell}^T\Gamma^T + D_{r,\ell}Q_{r\omega,\ell}D_{r,\ell}^T \\
\leq & 2(1+N\mu_r\lambda_1)[A_{r,\ell}P_{r,\ell|\ell}A_{r,\ell}^T + H_{r,\ell}(P_{r,\ell|\ell}^{-1} - \epsilon_1 I)^{-1}H_{r,\ell}^T \\
& + \epsilon_1^{-1}U_{r,\ell}U_{r,\ell}^T] + \sum_{s=1}^{N}(\mu_r\lambda_1^{-1} + \vartheta_r + \mu_r^2 N)\Gamma P_{s,\ell|\ell}\Gamma^T \\
& + \sum_{s=1}^{N}\vartheta_r\Gamma\hat{x}_{s,\ell|\ell}\hat{x}_{s,\ell|\ell}^T\Gamma^T + D_{r,\ell}Q_{r\omega,\ell}D_{r,\ell}^T.
\end{aligned} \tag{8.38}$$

To this end, the filtering error covariance matrix (8.17) can be derived by applying Lemma 3.2 as follows:

$$\begin{aligned}
& P_{r,\ell+1|\ell+1} \\
\leq & (I - K_{r,\ell+1}C_{r,\ell})P_{r,\ell+1|\ell}(I - K_{r,\ell+1}C_{r,\ell})^T \\
& + \mathbb{E}\{K_{r,\ell+1}\xi_{r,\ell+1}\xi_{r,\ell+1}^T K_{r,\ell+1}^T\}
\end{aligned}$$

$$+\mathbb{E}\{\lambda_2(I-K_{r,\ell+1}C_{r,\ell})P_{r,\ell+1|\ell}(I-K_{r,\ell+1}C_{r,\ell})^T$$
$$+\lambda_2^{-1}K_{r,\ell+1}\xi_{r,\ell+1}\xi_{r,\ell+1}^TK_{r,\ell+1}^T\}+\mathbb{E}\{K_{r,\ell+1}\xi_{r,\ell+1}v_{r,\ell+1}^TK_{r,\ell+1}^T\}$$
$$+\mathbb{E}\{K_{r,\ell+1}v_{r,\ell+1}\xi_{r,\ell+1}^TK_{r,\ell+1}^T\}+\mathbb{E}\{K_{r,\ell+1}Q_{rv,\ell+1}K_{r,\ell+1}^T\}. \tag{8.39}$$

Recall the definition of the difference $\xi_{r,\ell+1}=y_{r,(\ell+1)_t^r}-y_{r,\ell+1}$. *If* $\ell=T_m^r$ *(i.e. the current sampling instant is precisely the event-triggering instant of the r-th node), at this instant,* $\xi_{r,\ell+1}=0$ *and thus* $\mathbb{E}\{\xi_{r,\ell+1}v_{r,\ell+1}^T\}=0$, *otherwise we have* $\xi_{r,\ell+1}=y_{r,(\ell+1)_t^r}-y_{r,\ell+1}\neq 0$ *and then* $\mathbb{E}\{\xi_{r,\ell+1}v_{r,\ell+1}^T\}=\mathbb{E}\{[y_{r,(\ell+1)_t^r}-(C_{r,\ell}x_{r,\ell}+v_{r,\ell})]v_{r,\ell+1}^T\}=-Q_{rv,\ell+1}$. *Therefore, it is obtained that*

$$\mathbb{E}\{\xi_{r,\ell+1}v_{s,\ell+1}^T\}=-Q_{rv,\ell+1}\delta_{r,s}(1-\delta_{\ell+1,T_{m+1}^r}). \tag{8.40}$$

Furthermore, it is known that

$$\mathbb{E}\{\xi_{r,\ell+1}v_{r,\ell+1}^T\}=-(1-\delta_{\ell+1,T_{m+1}^r})Q_{rv,\ell+1}. \tag{8.41}$$

Substituting (8.41) into (8.39) results in

$$\begin{aligned}P_{r,\ell+1|\ell+1}\leq&(1+\lambda_2)(I-K_{r,\ell+1}C_{r,\ell})P_{r,\ell+1|\ell}(I-K_{r,\ell+1}C_{r,\ell})^T\\&+(1+\lambda_2^{-1})K_{r,\ell+1}\xi_{r,\ell+1}\xi_{r,\ell+1}^TK_{r,\ell+1}^T\\&+\delta_{\ell+1,T_{m+1}^r}K_{r,\ell+1}Q_{rv,\ell+1}K_{r,\ell+1}^T.\end{aligned} \tag{8.42}$$

Based on the event-triggering condition (8.4) and the definition of $\xi_{r,\ell}$, *for any* $\ell\in\mathbb{N}$, *one has*

$$\xi_{r,\ell+1}\xi_{r,\ell+1}^T\leq\theta_rI. \tag{8.43}$$

From (8.42), it follows that

$$\begin{aligned}P_{r,\ell+1|\ell+1}\leq&(1+\lambda_2)(I-K_{r,\ell+1}C_{r,\ell})P_{r,\ell+1|\ell}(I-K_{r,\ell+1}C_{r,\ell})^T\\&+K_{r,\ell+1}[\theta_r(1+\lambda_2^{-1})I+\delta_{\ell+1,T_{m+1}^r}Q_{rv,\ell+1}]K_{r,\ell+1}^T.\end{aligned} \tag{8.44}$$

Applying Lemma 8.3 to (8.25), (8.26), (8.38) and (8.44) and recalling the assumption $P_{r,\ell|\ell}\leq\Psi_{r,\ell|\ell}$, *we conclude that*

$$P_{r,\ell+1|\ell}\leq\Psi_{r,\ell+1|\ell}, \tag{8.45}$$
$$P_{r,\ell+1|\ell+1}\leq\Psi_{r,\ell+1|\ell+1}. \tag{8.46}$$

Based on the above procedures, an upper bound of filtering error covariance matrix $P_{r,\ell+1|\ell+1}$ *has been established. Now, taking the partial derivative of the trace of* $\Psi_{r,\ell+1|\ell+1}$ *with respect to* $K_{r,\ell+1}$ *and setting it to be zero, we have*

$$\frac{\partial tr(\Psi_{r,\ell+1|\ell+1})}{\partial K_{r,\ell+1}}=(1+\lambda_2)[-2\Psi_{r,\ell+1|\ell}C_{r,\ell}^T+2K_{r,\ell+1}C_{r,\ell}$$

$$\times \Psi_{r,\ell+1|\ell}C_{r,\ell}^T] + 2\theta_r(1+\lambda_2^{-1})K_{r,\ell+1} \quad (8.47)$$
$$+ 2\delta_{\ell+1,T_{m+1}^r}K_{r,\ell+1}Q_{rv,\ell+1}$$
$$=0$$

from which we can calculate the filter gain matrix as in (8.29) by noting that the term $(1+\lambda_2)C_{r,\ell}\Psi_{r,\ell+1|\ell}C_{r,\ell}^T + \theta_r(1+\lambda_2^{-1})I + \delta_{\ell+1,T_{m+1}^r}Q_{rv,\ell+1}$ *in (8.29) is invertible. The proof is now complete.*

In Theorem 8.2, the upper bound of the filtering error covariance has been computed and the filter gain has been obtained by minimizing the trace of such an upper bound. In the next step, we shall analyze the performance of the designed filter by discussing the exponential boundedness of filtering errors, for which the following assumption is needed.

Assumption 8.1 *There exist positive scalars* $\underline{q}_{rw}$, $\bar{q}_{rw}$, $\underline{q}_{rv}$, $\bar{q}_{rv}$, $\underline{\chi}$, $\bar{\chi}$, $\underline{c}_r$, $\bar{c}_r$, $\underline{d}_r$, *and* $\bar{d}_r$ *such that*

$$\begin{aligned}
&\underline{q}_{rw} \leq \|Q_{iw,\ell+1}\| \leq \bar{q}_{rw}, \quad \underline{q}_{rv} \leq \|Q_{rv,\ell+1}\| \leq \bar{q}_{rv} \\
&\underline{\chi}^2 I \leq \hat{x}_{r,\ell|\ell}\hat{x}_{r,\ell|\ell}^T \leq \bar{\chi}^2 I, \quad \underline{c}_r \leq \|C_{r,\ell}\| \leq \bar{c}_r \\
&\underline{d}_r \leq \|D_{r,\ell}\| \leq \bar{d}_r
\end{aligned}$$

for any r *and* ℓ.

Remark 8.2 *In the prediction step of the filter (8.6), the state estimate is made saturated according to the same saturation function in the state equation (8.1), which would render the unbiasedness of the estimation errors possible. On the other hand, the saturated state estimates in (8.6) justify the assumption of* $\underline{\chi}^2 I \leq \hat{x}_{r,\ell|\ell}\hat{x}_{r,\ell|\ell}^T \leq \bar{\chi}^2 I$, *thereby facilitating the proof of the exponential boundedness of the filtering errors in the sense of mean square.*

Theorem 8.3 *Consider the time-varying CN with* N *nodes described by (8.1). Under the conditions of Theorem 8.2 with* $\lambda_1 = 1$, *if there exist positive scalars* λ_2 *and* λ_3 *such that*

$$N < \lambda_2 - 2\lambda_3 - 1 \quad (8.48)$$

for any r *and* ℓ, *then the filtering error* $e_{r,\ell|\ell}$ *is exponentially bounded in the mean square.*

Proof *For notational simplicity, we denote*

$$e_{\ell+1|\ell} \triangleq \begin{bmatrix} e_{1,\ell+1|\ell}^T & e_{2,\ell+1|\ell}^T & \cdots & e_{N,\ell+1|\ell}^T \end{bmatrix}^T,$$
$$e_{\ell+1|\ell+1} \triangleq \begin{bmatrix} e_{1,\ell+1|\ell+1}^T & e_{2,\ell+1|\ell+1}^T & \cdots & e_{N,\ell+1|\ell+1}^T \end{bmatrix}^T.$$

Then, we consider the following quadratic function:

$$V_\ell(e_{\ell|\ell}) = \sum_{r=1}^{N} e_{r,\ell|\ell}^T \Psi_{r,\ell|\ell}^{-1} e_{r,\ell|\ell}. \tag{8.49}$$

Substituting the one-step prediction error (8.21) and filtering error (8.23) into (8.49), it follows that

$$\mathbb{E}\{V_{\ell+1}(e_{\ell+1|\ell+1})|e_{\ell|\ell}\} = \sum_{r=1}^{N} e_{r,\ell+1|\ell+1}^T \Psi_{r,\ell+1|\ell+1}^{-1} e_{r,\ell+1|\ell+1}. \tag{8.50}$$

In view of

$$\begin{aligned} & e_{r,\ell+1|\ell+1} \\ =& \mathscr{L}_{r,\ell+1}[\Xi_{r,\ell}\tilde{A}_{r,\ell}e_{r,\ell|\ell} + \Xi_{r,\ell}\sum_{s=1}^{N}(\alpha_{rs,\ell} - \mu_r)\Gamma x_{s,\ell|\ell} \\ & + \Xi_{r,\ell}\sum_{s=1}^{N}\mu_r\Gamma e_{s,\ell|\ell} + D_{r,\ell}\omega_{r,\ell}] - K_{r,\ell+1}\xi_{r,\ell+1} - K_{r,\ell+1}v_{r,\ell+1}, \end{aligned} \tag{8.51}$$

and

$$x_{r,\ell|\ell} = e_{r,\ell|\ell} + \hat{x}_{r,\ell|\ell} \tag{8.52}$$

where $\mathscr{L}_{r,\ell+1} \triangleq I - K_{r,\ell+1}C_{r,\ell}$, *it can be obtained that*

$$\begin{aligned} & \mathbb{E}\{V_{\ell+1}(e_{\ell+1|\ell+1})|e_{\ell|\ell}\} \\ =& \sum_{r=1}^{N} e_{r,\ell|\ell}^T \tilde{A}_{r,\ell}^T \Xi_{r,\ell}^T \mathscr{R}_{r,\ell+1} \Xi_{r,\ell} \tilde{A}_{r,\ell} e_{r,\ell|\ell} \\ & + 2\sum_{r=1}^{N}\sum_{s=1}^{N} \mu_r e_{r,\ell|\ell}^T \tilde{A}_{r,\ell}^T \Xi_{r,\ell}^T \mathscr{R}_{r,\ell+1} \Xi_{r,\ell} \Gamma e_{s,\ell|\ell} \\ & + \sum_{r=1}^{N}\sum_{s=1}^{N}\sum_{d=1}^{N} \mu_r^2 e_{s,\ell|\ell}^T \Gamma^T \Xi_{r,\ell}^T \mathscr{R}_{r,\ell+1} \Xi_{r,\ell} \Gamma e_{d,\ell|\ell} \\ & + \sum_{r=1}^{N}\sum_{s=1}^{N} \vartheta_r e_{s,\ell|\ell}^T \Gamma^T \Xi_{r,\ell}^T \mathscr{R}_{r,\ell+1} \Xi_{r,\ell} \Gamma e_{s,\ell|\ell} \\ & + \sum_{r=1}^{N}\sum_{s=1}^{N} \vartheta_r \hat{x}_{s,\ell|\ell}^T \Gamma^T \Xi_{r,\ell}^T \mathscr{R}_{r,\ell+1} \Xi_{r,\ell} \Gamma \hat{x}_{s,\ell|\ell} \\ & + \sum_{r=1}^{N} \mathbb{E}\{\omega_{r,\ell}^T D_{r,\ell}^T \mathscr{R}_{r,\ell+1} D_{r,\ell} \omega_{r,\ell} \end{aligned} \tag{8.53}$$

$$+ v_{r,\ell+1}^T K_{r,\ell+1}^T \Psi_{r,\ell+1|\ell+1}^{-1} K_{r,\ell+1} v_{r,\ell+1}\}$$
$$+ \sum_{r=1}^{N} \mathbb{E}\{\xi_{r,\ell+1}^T K_{r,\ell+1}^T \Psi_{r,\ell+1|\ell+1}^{-1} K_{r,\ell+1}\xi_{r,\ell+1}$$
$$- e_{r,\ell|\ell}^T \tilde{A}_{r,\ell}^T \Xi_{r,\ell}^T \mathscr{L}_{r,\ell+1}^T \Psi_{r,\ell+1|\ell+1}^{-1} K_{r,\ell+1}\xi_{r,\ell+1}$$
$$- \xi_{r,\ell+1}^T K_{r,\ell+1}^T \Psi_{r,\ell+1|\ell+1}^{-1} \mathscr{L}_{r,\ell+1} \Xi_\ell \tilde{A}_\ell e_{\ell|\ell}\}$$

where $\mathscr{R}_{r,\ell+1} \triangleq \mathscr{L}_{r,\ell+1}^T \Psi_{r,\ell+1|\ell+1}^{-1} \mathscr{L}_{r,\ell+1}$.

Now, we shall deal with the cross terms in (8.53) by employing the elementary inequality $xy^T + yx^T \leq xx^T + yy^T$ *where* x *and* y *are vectors of compatible dimensions. First, the second term on the right-hand side of (8.53) satisfies*

$$2\sum_{r=1}^{N}\sum_{s=1}^{N} \mu_r e_{r,\ell|\ell}^T \tilde{A}_{r,\ell}^T \Xi_{r,\ell}^T \mathscr{R}_{r,\ell+1} \Xi_{r,\ell} \Gamma e_{s,\ell|\ell}$$
$$\leq \sum_{r=1}^{N} \mu_r N e_{r,\ell|\ell}^T \tilde{A}_{r,\ell}^T \Xi_{r,\ell}^T \mathscr{R}_{r,\ell+1} \Xi_{r,\ell} \tilde{A}_{r,\ell} e_{r,\ell|\ell}$$
$$+ \sum_{r=1}^{N}\sum_{s=1}^{N} \mu_r e_{s,\ell|\ell}^T \Gamma^T \Xi_{r,\ell}^T \mathscr{R}_{r,\ell+1} \Xi_{r,\ell} \Gamma e_{s,\ell|\ell} \tag{8.54}$$

and the third term on the right-hand side of (8.53) is manipulated as

$$\sum_{r=1}^{N}\sum_{s=1}^{N}\sum_{d=1}^{N} \mu_r^2 e_{s,\ell|\ell}^T \Gamma^T \Xi_{r,\ell}^T \mathscr{R}_{r,\ell+1} \Xi_{r,\ell} \Gamma e_{d,\ell|\ell}$$
$$\leq \sum_{r=1}^{N}\sum_{s=1}^{N} \mu_r^2 N e_{s,\ell|\ell}^T \Gamma^T \Xi_{r,\ell}^T \mathscr{R}_{r,\ell+1} \Xi_{r,\ell} \Gamma e_{s,\ell|\ell}. \tag{8.55}$$

Furthermore, applying Lemma 3.2 to the last term of the right-hand side of (8.53) gives

$$\sum_{r=1}^{N} \mathbb{E}\{\xi_{r,\ell+1}^T K_{r,\ell+1}^T \Psi_{r,\ell+1|\ell+1}^{-1} K_{r,\ell+1}\xi_{r,\ell+1}$$
$$- e_{r,\ell|\ell}^T \tilde{A}_{r,\ell}^T \Xi_{r,\ell}^T \mathscr{L}_{r,\ell+1}^T \Psi_{r,\ell+1|\ell+1}^{-1} K_{r,\ell+1}\xi_{r,\ell+1}$$
$$- \xi_{r,\ell+1}^T K_{r,\ell+1}^T \Psi_{r,\ell+1|\ell+1}^{-1} \mathscr{L}_{r,\ell+1} \Xi_\ell \tilde{A}_\ell e_{r,\ell|\ell}\}$$
$$\leq \sum_{r=1}^{N} \mathbb{E}\{(1+\lambda_3^{-1})\xi_{r,\ell+1}^T K_{r,\ell+1}^T \Psi_{r,\ell+1|\ell+1}^{-1} K_{r,\ell+1}\xi_{r,\ell+1}\}$$
$$+ \sum_{r=1}^{N} \lambda_3 e_{r,\ell|\ell}^T \tilde{A}_{r,\ell}^T \Xi_{r,\ell}^T \mathscr{R}_{r,\ell+1} \Xi_{r,\ell} \tilde{A}_{r,\ell} e_{r,\ell|\ell}. \tag{8.56}$$

As a result, substituting (8.54)-(8.56) into (8.53) yields

$$\begin{aligned}
&\mathbb{E}\{V_{\ell+1}(e_{\ell+1|\ell+1})|e_{\ell|\ell}\} \\
\leq &\sum_{r=1}^{N}(1+\mu_r N+\lambda_3)e_{r,\ell|\ell}^T\tilde{A}_{r,\ell}^T\Xi_{r,\ell}^T\mathscr{R}_{r,\ell+1}\Xi_{r,\ell}\tilde{A}_{r,\ell}e_{r,\ell|\ell} \\
&+\sum_{r=1}^{N}\sum_{s=1}^{N}(\mu_r+\vartheta_r+\mu_r^2 N)e_{s,\ell|\ell}^T\Gamma^T\Xi_{r,\ell}^T\mathscr{R}_{r,\ell+1}\Xi_{r,\ell}\Gamma e_{s,\ell|\ell} \\
&+\sum_{r=1}^{N}\sum_{s=1}^{N}\vartheta_r\hat{x}_{s,\ell|\ell}^T\Gamma^T\Xi_{r,\ell}^T\mathscr{R}_{r,\ell+1}\Xi_{r,\ell}\Gamma\hat{x}_{s,\ell|\ell} \\
&+\sum_{r=1}^{N}\mathbb{E}\{(1+\lambda_3^{-1})\xi_{r,\ell+1}^T K_{r,\ell+1}^T\Psi_{r,\ell+1|\ell+1}^{-1}K_{r,\ell+1}\xi_{r,\ell+1} \\
&+\omega_{r,\ell}^T D_{r,\ell}^T\mathscr{R}_{r,\ell+1}D_{r,\ell}\omega_{r,\ell}+v_{r,\ell+1}^T K_{r,\ell+1}^T\Psi_{r,\ell+1|\ell+1}^{-1}K_{r,\ell+1}v_{r,\ell+1}\}.
\end{aligned} \tag{8.57}$$

According to (8.26) and the denotations of $\mathscr{L}_{r,\ell+1}$ and $\mathscr{R}_{r,\ell+1}$, one has

$$\Psi_{r,\ell+1|\ell+1}\geq(1+\lambda_2)\mathscr{L}_{r,\ell+1}\Psi_{r,\ell+1|\ell}\mathscr{L}_{r,\ell+1}^T. \tag{8.58}$$

Next, via Lemma 2.1, the inequality (8.58) becomes

$$\mathscr{R}_{r,\ell+1}\leq(1+\lambda_2)^{-1}\Psi_{r,\ell+1|\ell}^{-1}. \tag{8.59}$$

Then, substituting (8.59) into (8.57), we obtain

$$\begin{aligned}
&\mathbb{E}\{V_{\ell+1}(e_{\ell+1|\ell+1})|e_{\ell|\ell}\} \\
\leq &\sum_{r=1}^{N}(1+\mu_r N+\lambda_3)(1+\lambda_2)^{-1}e_{r,\ell|\ell}^T\tilde{A}_{r,\ell}^T\Psi_{r,\ell+1|\ell}^{-1}\tilde{A}_{r,\ell}e_{r,\ell|\ell} \\
&+\sum_{r=1}^{N}\sum_{s=1}^{N}(\mu_r+\vartheta_r+\mu_r^2 N)(1+\lambda_2)^{-1}e_{s,\ell|\ell}^T\mathscr{W}_{r,\ell+1}e_{s,\ell|\ell} \\
&+\sum_{r=1}^{N}\sum_{s=1}^{N}\vartheta_r(1+\lambda_2)^{-1}\hat{x}_{s,\ell|\ell}^T\mathscr{W}_{r,\ell+1}\hat{x}_{s,\ell|\ell} \\
&+\sum_{r=1}^{N}\mathbb{E}\{(1+\lambda_3^{-1})\xi_{r,\ell+1}^T K_{r,\ell+1}^T\Psi_{r,\ell+1|\ell+1}^{-1}K_{r,\ell+1}\xi_{r,\ell+1} \\
&+\omega_{r,\ell}^T D_{r,\ell}^T\mathscr{R}_{r,\ell+1}D_{r,\ell}\omega_{r,\ell}+v_{r,\ell+1}^T K_{r,\ell+1}^T\Psi_{r,\ell+1|\ell+1}^{-1}K_{r,\ell+1}v_{r,\ell+1}\} \\
\leq &\sum_{r=1}^{N}2(1+\mu_r N+\lambda_3)(1+\lambda_2)^{-1}e_{r,\ell|\ell}^T[A_{r,\ell}^T\Psi_{r,\ell+1|\ell}^{-1}A_{r,\ell} \\
&+(H_{r,\ell}+U_{r,\ell}\Delta_{r,\ell})^T\Psi_{r,\ell+1|\ell}^{-1}(H_{r,\ell}+U_{r,\ell}\Delta_{r,\ell})]e_{r,\ell|\ell}
\end{aligned}$$

$$
\begin{aligned}
&+\sum_{r=1}^{N}\sum_{s=1}^{N}(\mu_r+\vartheta_r+\mu_r^2N)(1+\lambda_2)^{-1}e_{s,\ell|\ell}^T\mathscr{W}_{r,\ell+1}e_{s,\ell|\ell}\\
&+\sum_{r=1}^{N}\sum_{s=1}^{N}\vartheta_r(1+\lambda_2)^{-1}\hat{x}_{s,\ell|\ell}^T\mathscr{W}_{r,\ell+1}\hat{x}_{s,\ell|\ell}\\
&+\sum_{r=1}^{N}\mathbb{E}\{(1+\lambda_3^{-1})\xi_{r,\ell+1}^TK_{r,\ell+1}^T\Psi_{r,\ell+1|\ell+1}^{-1}K_{r,\ell+1}\xi_{r,\ell+1}\\
&+\omega_{r,\ell}^TD_{r,\ell}^T\mathscr{R}_{r,\ell+1}D_{r,\ell}\omega_{r,\ell}v_{r,\ell+1}^TK_{r,\ell+1}^T\Psi_{r,\ell+1|\ell+1}^{-1}K_{r,\ell+1}v_{r,\ell+1}\}
\end{aligned}
\tag{8.60}
$$

where $\mathscr{W}_{r,\ell+1}\triangleq\Gamma^T\Psi_{r,\ell+1|\ell}^{-1}\Gamma$.

Based on Lemma 8.2 and (8.25), we know that

$$
\Psi_{r,\ell+1|\ell}\geq 2(1+N\mu_r\lambda_1)A_{r,\ell}\Psi_{r,\ell|\ell}A_{r,\ell}^T, \tag{8.61}
$$

$$
\Psi_{r,\ell+1|\ell}\geq 2(1+N\mu_r\lambda_1)(H_{r,\ell}+U_{r,\ell}\Delta_{r,\ell})\Psi_{r,\ell|\ell}(H_{r,\ell}+U_{r,\ell}\Delta_{r,\ell})^T, \tag{8.62}
$$

$$
\Psi_{r,\ell+1|\ell}\geq\sum_{s=1}^{N}(\mu_r\lambda_1^{-1}+\vartheta_r+\mu_r^2N)\Gamma\Psi_{s,\ell|\ell}\Gamma^T, \tag{8.63}
$$

$$
\Psi_{r,\ell+1|\ell}\geq\sum_{s=1}^{N}\vartheta_r\Gamma\hat{x}_{s,\ell|\ell}\hat{x}_{s,\ell|\ell}^T\Gamma^T. \tag{8.64}
$$

It follows from (8.61)–(8.64) and Lemma 2.1 that

$$
A_{r,\ell}^T\Psi_{r,\ell+1|\ell}^{-1}A_{r,\ell}\leq\frac{1}{2(1+N\mu_r\lambda_1)}\Psi_{r,\ell|\ell}^{-1}, \tag{8.65}
$$

$$
(H_{r,\ell}+U_{r,\ell}\Delta_{r,\ell})^T\Psi_{r,\ell+1|\ell}^{-1}(H_{r,\ell}+U_{r,\ell}\Delta_{r,\ell})\leq\frac{1}{2(1+N\mu_r\lambda_1)}\Psi_{r,\ell|\ell}^{-1}, \tag{8.66}
$$

$$
\Gamma^T\Psi_{r,\ell+1|\ell}^{-1}\Gamma\leq\frac{1}{\mu_r\lambda_1^{-1}+\vartheta_r+\mu_r^2N}(\sum_{s=1}^{N}\Psi_{s,\ell|\ell})^{-1}, \tag{8.67}
$$

$$
\Gamma^T\Psi_{r,\ell+1|\ell}^{-1}\Gamma\leq\frac{1}{\vartheta_r}(\sum_{s=1}^{N}\hat{x}_{s,\ell|\ell}\hat{x}_{s,\ell|\ell}^T)^{-1}\leq\frac{1}{N\vartheta_r\underline{\chi}^2}I. \tag{8.68}
$$

Substituting (8.65)–(8.68) into (8.60), we conclude that

$$
\begin{aligned}
&\mathbb{E}\{V_{\ell+1}(e_{\ell+1|\ell+1})|e_{\ell|\ell}\}\\
\leq&\frac{2}{1+\lambda_2}\sum_{r=1}^{N}\frac{1+\mu_rN+\lambda_3}{1+N\mu_r\lambda_1}e_{r,\ell|\ell}^T\Psi_{r,\ell|\ell}^{-1}e_{r,\ell|\ell}\\
&+\frac{1}{1+\lambda_2}\sum_{r=1}^{N}\sum_{s=1}^{N}\frac{\mu_r+\vartheta_r+\mu_r^2N}{\mu_r\lambda_1^{-1}+\vartheta_r+\mu_r^2N}e_{s,\ell|\ell}^T(\sum_{d=1}^{N}\Psi_{d,\ell|\ell})^{-1}e_{s,\ell|\ell}\\
&+\frac{1}{1+\lambda_2}\sum_{r=1}^{N}\sum_{s=1}^{N}\vartheta_r\frac{1}{N\vartheta_r\underline{\chi}^2}\hat{x}_{s,\ell|\ell}^T\hat{x}_{s,\ell|\ell}
\end{aligned}
$$

$$
\begin{aligned}
&+\sum_{r=1}^{N}\mathbb{E}\{(1+\lambda_3^{-1})\xi_{r,\ell+1}^T K_{r,\ell+1}^T \Psi_{r,\ell+1|\ell+1}^{-1} K_{r,\ell+1}\xi_{r,\ell+1}\\
&+\omega_{r,\ell}^T D_{r,\ell}^T \mathscr{R}_{r,\ell+1} D_{r,\ell}\omega_{r,\ell} + v_{r,\ell+1}^T K_{r,\ell+1}^T \Psi_{r,\ell+1|\ell+1}^{-1} K_{r,\ell+1} v_{r,\ell+1}\} \qquad (8.69)\\
\le&\frac{2}{1+\lambda_2}\sum_{r=1}^{N}\frac{1+\mu_r N+\lambda_3}{1+N\mu_r\lambda_1} e_{r,\ell|\ell}^T \Psi_{r,\ell|\ell}^{-1} e_{r,\ell|\ell} + \frac{N\bar{\chi}^2}{\underline{\chi}^2(1+\lambda_2)}\\
&+\frac{1}{1+\lambda_2}\sum_{r=1}^{N}\sum_{s=1}^{N}\frac{\mu_r+\vartheta_r+\mu_r^2 N}{\mu_r\lambda_1^{-1}+\vartheta_r+\mu_r^2 N} e_{s,\ell|\ell}^T \Psi_{s,\ell|\ell}^{-1} e_{s,\ell|\ell}\\
&+\sum_{r=1}^{N}\mathbb{E}\{(1+\lambda_3^{-1})\xi_{r,\ell+1}^T K_{r,\ell+1}^T \Psi_{r,\ell+1|\ell+1}^{-1} K_{r,\ell+1}\xi_{r,\ell+1}\\
&+\omega_{r,\ell}^T D_{r,\ell}^T \mathscr{R}_{r,\ell+1} D_{r,\ell}\omega_{r,\ell} + v_{r,\ell+1}^T K_{r,\ell+1}^T \Psi_{r,\ell+1|\ell+1}^{-1} K_{r,\ell+1} v_{r,\ell+1}\}.
\end{aligned}
$$

Furthermore, letting $\lambda_1 = 1$*, one has*

$$
\begin{aligned}
&\mathbb{E}\{V_{\ell+1}(e_{\ell+1|\ell+1})|e_{\ell|\ell}\}\\
\le&\frac{2}{1+\lambda_2}\sum_{r=1}^{N}\frac{1+\mu_r N+\lambda_3}{1+N\mu_r} e_{r,\ell|\ell}^T \Psi_{r,\ell|\ell}^{-1} e_{r,\ell|\ell} + \frac{N}{1+\lambda_2}\sum_{r=1}^{N} e_{r,\ell|\ell}^T \Psi_{r,\ell|\ell}^{-1} e_{r,\ell|\ell}\\
&+\frac{N\bar{\chi}^2}{\underline{\chi}^2(1+\lambda_2)}+\sum_{r=1}^{N}\mathbb{E}\{(1+\lambda_3^{-1})\xi_{r,\ell+1}^T K_{r,\ell+1}^T \Psi_{r,\ell+1|\ell+1}^{-1} K_{r,\ell+1}\xi_{r,\ell+1}\\
&+\omega_{r,\ell}^T D_{r,\ell}^T \mathscr{R}_{r,\ell+1} D_{r,\ell}\omega_{r,\ell} + v_{r,\ell+1}^T K_{r,\ell+1}^T \Psi_{r,\ell+1|\ell+1}^{-1} K_{r,\ell+1} v_{r,\ell+1}\} \qquad (8.70)\\
\le&\sum_{r=1}^{N}\Pi_r e_{r,\ell|\ell}^T \Psi_{r,\ell|\ell}^{-1} e_{r,\ell|\ell} + \frac{N\bar{\chi}^2}{\underline{\chi}^2(1+\lambda_2)}\\
&+\sum_{r=1}^{N}\mathbb{E}\{(1+\lambda_3^{-1})\xi_{r,\ell+1}^T K_{r,\ell+1}^T \Psi_{r,\ell+1|\ell+1}^{-1} K_{r,\ell+1}\xi_{r,\ell+1}\\
&+\omega_{r,\ell}^T D_{r,\ell}^T \mathscr{R}_{r,\ell+1} D_{r,\ell}\omega_{r,\ell} + v_{r,\ell+1}^T K_{r,\ell+1}^T \Psi_{r,\ell+1|\ell+1}^{-1} K_{r,\ell+1} v_{r,\ell+1}\}
\end{aligned}
$$

where $\Pi_r \triangleq \frac{1}{1+\lambda_2}(2+N+\frac{2\lambda_3}{1+\mu_r N})$.

For the last term of the right-hand side of (8.70), it follows from (8.25) and Assumption 8.1 that

$$
\Psi_{r,\ell+1|\ell} \ge D_{r,\ell} Q_{r\omega,\ell} D_{r,\ell}^T \ge \underline{q}_{rw}\underline{d}_r^2 I. \qquad (8.71)
$$

Also, from (8.29), it follows that

$$
\|K_{r,\ell+1}\| \le \|(1+\lambda_2)\Psi_{r,\ell+1|\ell} C_{r,\ell}^T[(1+\lambda_2)C_{r,\ell}\Psi_{r,\ell+1|\ell}C_{r,\ell}^T]^{-1}\| \le \Upsilon_r \qquad (8.72)
$$

and

$$
K_{r,\ell+1} \ge (1+\lambda_2)\underline{c}_r \Psi_{r,\ell+1|\ell}[(1+\lambda_2)\bar{c}_r^2\Psi_{r,\ell+1|\ell}+\theta_r(1+\lambda_2^{-1}+\bar{q}_{rv})I]^{-1}
$$

$$\geq \underline{\kappa}_r I \tag{8.73}$$

where

$$\underline{\kappa}_r \triangleq \frac{(1+\lambda_2)\underline{c}_r \underline{q}_{rw} \underline{d}_r^2}{(1+\lambda_2)^2 \bar{c}_r^2 (h_{r,\ell} + \bar{q}_{rw} \bar{d}_r^2) + \theta_r (1+\lambda_2^{-1} + \bar{q}_{rv})}, \quad \Upsilon_r \triangleq \frac{\bar{c}_r}{\underline{c}_r^2}.$$

Moreover, based on (8.26), it is seen that

$$\Psi_{r,\ell+1|\ell+1} \geq \theta_r (1+\lambda_2^{-1}) K_{r,\ell+1} K_{r,\ell+1}^T \geq \underline{\psi}_r I \tag{8.74}$$

where $\underline{\psi}_r I$ *is the lower bound of* $\Psi_{r,\ell+1|\ell+1}$.

According to (8.72) and (8.73), we know that the filter gain $K_{r,\ell+1}$ *is bounded, and hence there exists* $\underline{\psi}_r$ *such that the inequality (8.74) holds. Furthermore, the following inequality can be obtained:*

$$\Psi_{r,\ell+1|\ell+1}^{-1} \leq \frac{1}{\underline{\psi}_r} I. \tag{8.75}$$

According to the event-triggering condition (8.4), inequalities (8.71)-(8.75) and Assumption 8.1, the last item of the right-hand side of (8.70) is bounded, i.e., there exists a positive scalar η_0 *such that*

$$\sum_{r=1}^{N} \mathbb{E}\{(1+\lambda_3^{-1})\xi_{r,\ell+1}^T K_{r,\ell+1}^T \Psi_{r,\ell+1|\ell+1}^{-1} K_{r,\ell+1}\xi_{r,\ell+1} + \omega_{r,\ell}^T D_{r,\ell}^T \mathscr{R}_{r,\ell+1}$$
$$\times D_{r,\ell}\omega_{r,\ell} + v_{r,\ell+1}^T K_{r,\ell+1}^T \Psi_{r,\ell+1|\ell+1}^{-1} K_{r,\ell+1} v_{r,\ell+1}\} \leq \eta_0. \tag{8.76}$$

Thus, we have

$$\mathbb{E}\{V_{\ell+1}(e_{\ell+1|\ell+1})|e_{\ell|\ell}\} \leq \sum_{r=1}^{N} \Pi_r e_{r,\ell|\ell}^T \Psi_{r,\ell|\ell}^{-1} e_{r,\ell|\ell} + \frac{N\bar{\chi}^2}{\underline{\chi}^2(1+\lambda_2)} + \eta_0. \tag{8.77}$$

In addition, we can also derive

$$\Pi_r = \frac{1}{1+\lambda_2}(2+N+\frac{2\lambda_3}{1+\mu_r N}) \leq \frac{1}{1+\lambda_2}(2+N+2\lambda_3). \tag{8.78}$$

From (8.76)-(8.78), it is obtained that

$$\mathbb{E}\{V_{\ell+1}(e_{\ell+1|\ell+1})|e_{\ell|\ell}\} \leq \varrho V_\ell(e_{\ell|\ell}) + \epsilon + \eta_0 \tag{8.79}$$

where

$$\varrho \triangleq \frac{2+N+2\lambda_3}{1+\lambda_2}, \quad \epsilon \triangleq \frac{N\bar{\chi}^2}{\underline{\chi}^2(1+\lambda_2)}. \tag{8.80}$$

It follows from (8.48) that $0 \leq \varrho < 1$, and then we can conclude from Lemma 8.4 that the filtering error is bounded in mean square. The proof is complete.

In this chapter, the main results are stated in Theorems 8.2 and 8.3, where Theorem 8.2 presents a novel recursive filtering method and an upper bound of filtering error covariance matrix, and Theorem 8.3 analyzes the bounded stability of the designed filter by establishing a parameterized condition of several design parameters. Our main results exhibit the following distinct features: (1) the state saturation, the random coupling strengths and the event-triggering mechanism are investigated within a unified framework; (2) an upper bound of the filtering error covariance matrix is established via a combination of the uses of the Taylor expansion technique and the Schur Complement Lemma; and (3) the boundedness of the filtering error covariance matrix is discussed. Therefore, the recursive filtering algorithm developed in this chapter would have both theoretical and engineering values.

Remark 8.3 *So far, an event-triggering method to recursive filtering is designed in this chapter for time-varying CNs with state saturations. Many systems existing in engineering applications are time-varying and have state saturation characteristics, thus it is very important to study on the filtering issue for time-varying CNs with state saturations. The recursive Kalman filtering method can be used for online implementation of the filter for time-varying CNs under an event-triggering transmission mechanism. The difficulties that have been overcome during the algorithm design and analysis include that (1) the analysis process of such system model is much more complicated than that of a single system model; (2) the way to design the recursive filter is challenging due to the necessity of fully considering the nonlinear saturation term $\Lambda(\cdot)$ and the event-triggering mechanism; (3) the processing method for the saturation term is critical in the boundedness analysis of the filtering error. Furthermore, it also considers the communication energy-saving problem by introducing event-triggering mechanism in this chapter.*

Remark 8.4 *The computational complexity of the developed recursive filter is $O(n^3)$ for CNs with N nodes, where n is the size of each node. In this chapter, a distinct feature of the developed filter is that an upper bound is provided on the estimation error covariance for each individual node in CNs, which means that the upper bound (which is actually a matrix) can be computed separately for each individual node during the design process. As can be seen from Theorem 8.2, the dimension of the upper-bound matrix is $n \times n$ where n is the dimension of state vectors. The computational complexity of the algorithm for node r can be represented as $O(n^3)$, which implies that the computational complexity is unrelated to the number of CN nodes (i.e. N).*

It is worth mentioning that, for the traditional state augmentation method, the dimension of the corresponding upper-bound matrix amounts to $nN \times nN$, which grows as the number of nodes increases. As such, our recursive scheme enjoys the benefits of incurring less computation burden as compared with the existing state augmentation approach.

Remark 8.5 *Compared with the existing event-triggered state estimation approach, the local design method for each node of CNs is adopted during the process of filter design in this chapter, which means that the computational complexity is unrelated to the size of the CN, thereby possessing better numerical efficiency than the conventional matrix augmentation method. Also, to some extent, the proposed scheme reduces the design conservatism while enhancing the flexibility of engineering implementation. In the context of event-triggered results (see e.g. [181] and [182]), the proposed recursive filter design method in this chapter is particularly suitable for online implementation of CNs filtering. In Theorem 8.3, the exponential boundedness is guaranteed on the filtering error (in mean square) for the addressed state-saturated CNs with random coupling strengths under event-triggering mechanism, and a parameterized relationship is established on the size of the CN as well as the design parameters λ_1, λ_2 and λ_3.*

8.3 Illustrative Examples

In this section, two simulation examples are presented to illustrate the effectiveness of the event-triggering filtering scheme developed in this chapter.

8.3.1 Example 1

Consider the CN (8.1) with four nodes where the parameters are given as follows:

$$A_{1,\ell} = \begin{bmatrix} 0.4 & 0.5 \\ 0.1 & -0.4\sin(\ell+2) \end{bmatrix}, A_{2,\ell} = \begin{bmatrix} 0.45 & 0.9+0.1\sin(\ell) \\ 0.1 & 0.4 \end{bmatrix},$$

$$A_{3,\ell} = \begin{bmatrix} 0.3 & -0.4+0.1\sin(\ell) \\ 0.3 & 0.5 \end{bmatrix}, A_{4,\ell} = \begin{bmatrix} 0.5\sin(\ell) & 0.5 \\ 0.3+0.1\sin(\ell) & 0.6 \end{bmatrix}.$$

$v_{r,\ell}$ $(r = 1,2,3,4)$ are zero-mean Gaussian white noises with covariance matrices $Q_{rv,\ell}$ and

$$Q_{1v,\ell} = \begin{bmatrix} 0.3 & 0 \\ 0 & 0.1 \end{bmatrix}, Q_{2v,\ell} = \begin{bmatrix} 0.2 & 0 \\ 0 & 0.3 \end{bmatrix},$$

$$Q_{3v,\ell} = \begin{bmatrix} 0.4 & 0 \\ 0 & 0.1 \end{bmatrix}, Q_{4v,\ell} = \begin{bmatrix} 0.6 & 0 \\ 0 & 0.2 \end{bmatrix}.$$

The nonlinear functions are chosen as

$$h(x_{r,\ell}) = \begin{bmatrix} x_{r,\ell}^{(1)} + \sin(x_{r,\ell}^{(1)} x_{r,\ell}^{(2)}) \\ 0.5x_{r,\ell}^{(2)} + \sin(x_{r,\ell}^{(1)} x_{r,\ell}^{(2)}) \end{bmatrix},$$

and the inner-coupling matrix is given by $\Gamma = 0.5I_2$. $\omega_{r,\ell}$ $(r = 1,2,3,4)$ are zero-mean Gaussian white noises with covariance matrices $Q_{r\omega,\ell} = \text{diag}\{0.5, 0.3, 0.4, 0.2\}$. The matrices $C_{r,\ell}$ and $D_{r,\ell}$ are taken as

$$C_{1,\ell} = \begin{bmatrix} 0.5 & 0.3 \\ 0.3 + 0.01\sin(\ell) & 0.2 \end{bmatrix}, C_{2,\ell} = \begin{bmatrix} 0.4 & 0.5 + 0.02\sin(\ell) \\ 0.4 & 0.2 \end{bmatrix},$$

$$C_{3,\ell} = \begin{bmatrix} 0.9 & -1.1 \\ 0.5 + 0.02\sin(2\ell) & 0.7 \end{bmatrix}, C_{4,\ell} = \begin{bmatrix} 0.8 & 0.9 + 0.01\sin(2\ell) \\ 0.4 & 0.3 \end{bmatrix},$$

$$D_{1,\ell} = \begin{bmatrix} 0.04 + 0.01\sin(\ell) \\ 0.01 \end{bmatrix}, D_{2,\ell} = \begin{bmatrix} -0.02 \\ 0.03 + \sin(2\ell) \end{bmatrix},$$

$$D_{3,\ell} = \begin{bmatrix} 0.03 + \sin(\ell) \\ 0.05 \end{bmatrix}, D_{4,\ell} = \begin{bmatrix} 0.04 - 0.01\sin(\ell) \\ -0.02 \end{bmatrix}.$$

The coupling strengths are assumed to obey a uniform distribution on the open interval $(-3, 3.5)$. Suppose that the saturation levels are $q_{1,\max} = [5\ 6]^T, q_{2,\max} = [4\ 3]^T, q_{3,\max} = [3\ 4]^T$ and $q_{4,\max} = [3\ 5]^T$. Choose the thresholds $\theta_r = 0.2$ for $r = 1,2,3,4$. The initial state values are $x_{1,0} = [-0.3\ 0.1]^T, x_{2,0} = [-0.7\ 0.2]^T, x_{3,0} = [0.5\ \ -0.2]^T$ and $x_{4,0} = [0.3\ \ -0.25]^T$. The other parameters are chosen as $\lambda_1 = 1, \lambda_3 = 1$ and $\epsilon_1 = 5$. For the value of λ_2, it is easy to set an appropriate value according to Theorem 8.3. In this example, it is chosen as $\lambda_2 = 20$.

The filter gains can be computed according to Theorem 8.2. The curves of actual states $x_{r,\ell}^{(s)}$ and their estimates $\hat{x}_{r,\ell}^{(s)}$ $(r = 1,2,3,4; s = 1,2)$ are plotted in Figs. 8.1–8.8. From Figs. 8.1–8.8, it can be seen that the designed filters have the satisfactory performance in the case of the state saturations with random coupling strengths under the event-triggering mechanism, which verifies the validity of the proposed filtering scheme in the chapter.

8.3.2 Example 2

In this example, let us study the indoor localization problem of four mobile robots with state saturations. In [71], it has been mentioned that mobile robots are often subject to state constraints (e.g. limits on position and orientation) in practical application. Furthermore, according to the kinematic equations of

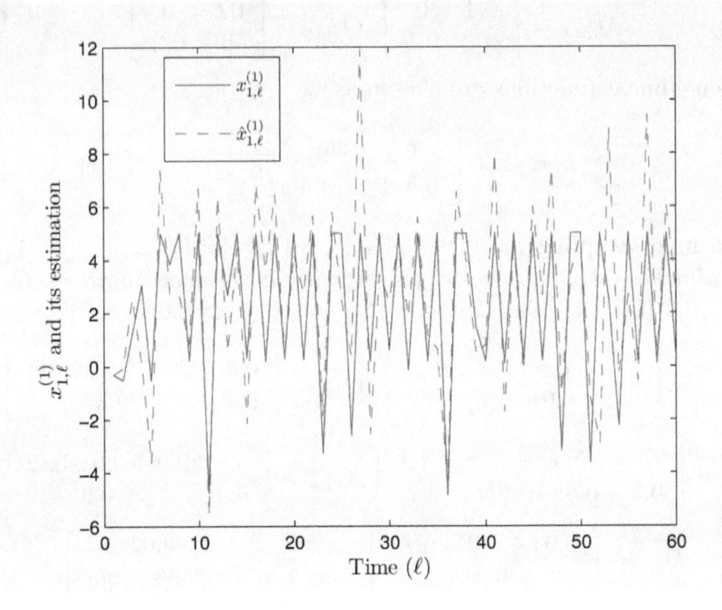

FIGURE 8.1
Actual state $x^{(1)}_{1,\ell}$ and its estimation $\hat{x}^{(1)}_{1,\ell}$.

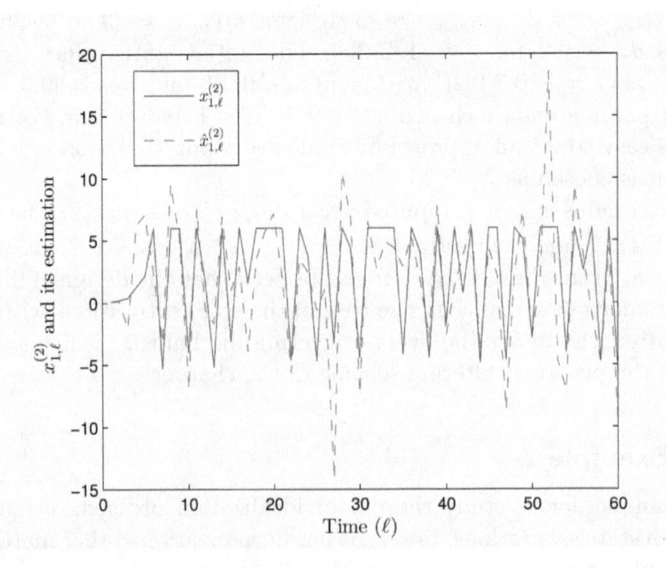

FIGURE 8.2
Actual state $x^{(2)}_{1,\ell}$ and its estimation $\hat{x}^{(2)}_{1,\ell}$.

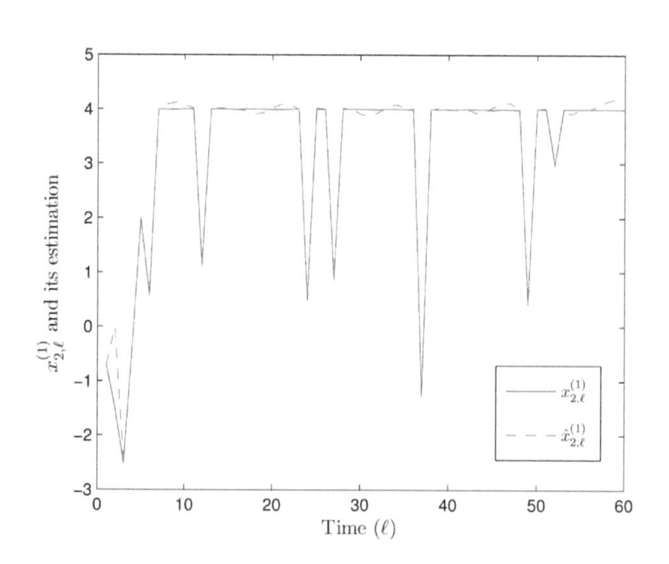

FIGURE 8.3
Actual state $x_{2,\ell}^{(1)}$ and its estimation $\hat{x}_{2,\ell}^{(1)}$.

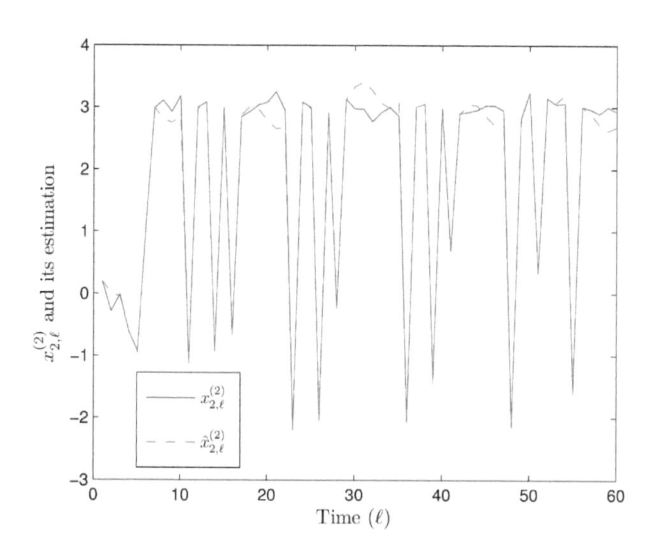

FIGURE 8.4
Actual state $x_{2,\ell}^{(2)}$ and its estimation $\hat{x}_{2,\ell}^{(2)}$.

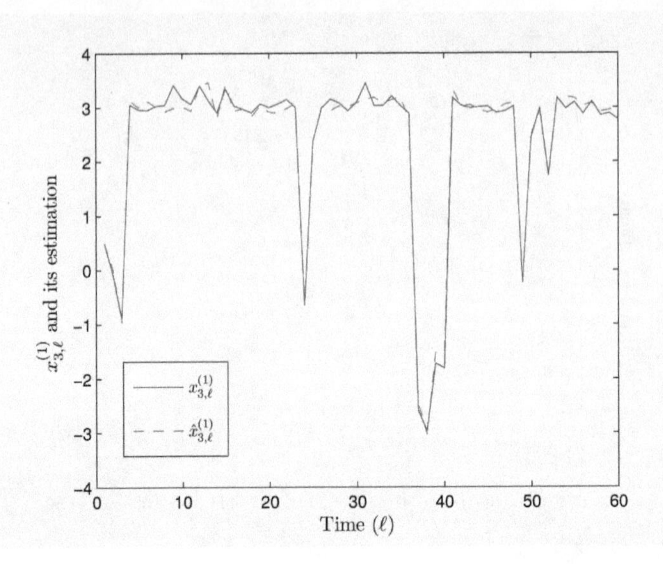

FIGURE 8.5
Actual state $x_{3,\ell}^{(1)}$ and its estimation $\hat{x}_{3,\ell}^{(1)}$.

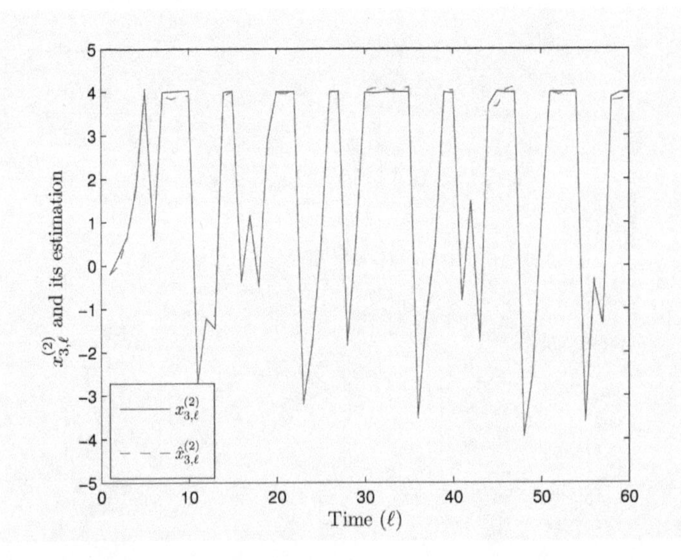

FIGURE 8.6
Actual state $x_{3,\ell}^{(2)}$ and its estimation $\hat{x}_{3,\ell}^{(2)}$.

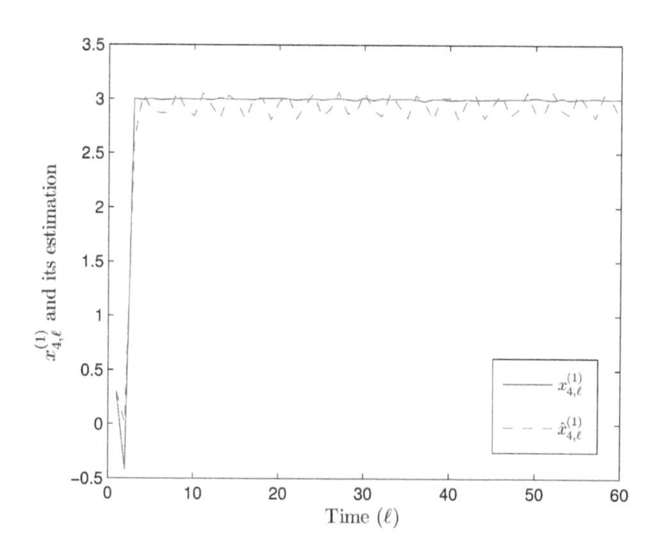

FIGURE 8.7
Actual state $x_{4,\ell}^{(1)}$ and its estimation $\hat{x}_{4,\ell}^{(1)}$.

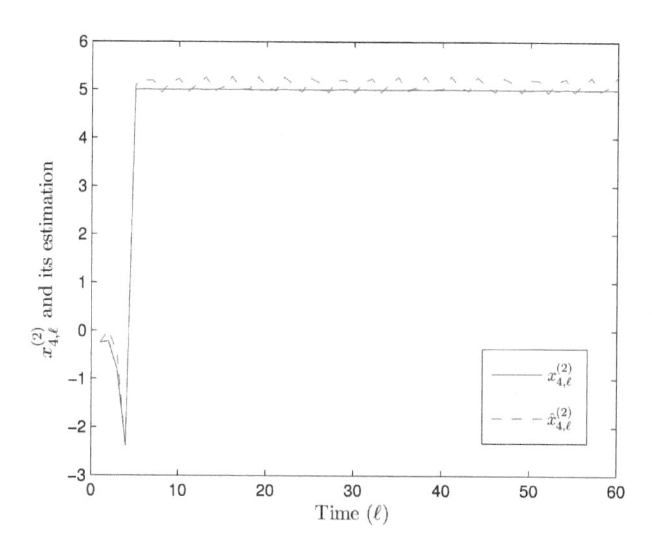

FIGURE 8.8
Actual state $x_{4,\ell}^{(2)}$ and its estimation $\hat{x}_{4,\ell}^{(2)}$.

mobile robots given in [27], we consider the following system:

$$\begin{cases} \begin{bmatrix} x_{r,\ell+1} \\ \zeta_{r,\ell+1} \\ \phi_{r,\ell+1} \end{bmatrix} = \Lambda_r \left(\begin{bmatrix} x_{r,\ell} \\ \zeta_{r,\ell} \\ \phi_{r,\ell} \end{bmatrix} + \begin{bmatrix} \varpi_{r,\ell}\cos\phi_{r,\ell} \\ \varpi_{r,\ell}\sin\phi_{r,\ell} \\ \tau_{r,\ell} \end{bmatrix} + 0.01 \sum_{s=1}^{4} \alpha_{rs,\ell} \begin{bmatrix} x_{s,\ell} \\ \zeta_{s,\ell} \\ \phi_{s,\ell} \end{bmatrix} \right) \\ \qquad + \begin{bmatrix} \omega^x_{r,\ell} \\ \omega^\zeta_{r,\ell} \\ \omega^\phi_{r,\ell} \end{bmatrix} \\ y_{r,\ell} = \begin{bmatrix} 1 & 0 & 0 \\ 0 & 1 & 0 \end{bmatrix} \begin{bmatrix} x_{r,\ell} \\ \zeta_{r,\ell} \\ \phi_{r,\ell} \end{bmatrix} + v_{r,\ell} \end{cases}$$

where $(x_{r,\ell}, \zeta_{r,\ell})$ and $\phi_{r,\ell}$ denote the position and the orientation of the r-th robot, respectively. $(\varpi_{r,\ell}, \tau_{r,\ell})$ is the velocity vector. $\omega_{r,\ell} = [\omega^x_{r,\ell}\ \omega^\zeta_{r,\ell}\ \omega^\phi_{r,\ell}]^T$ $(r = 1, 2, 3, 4)$ is a zero-mean Gaussian white noise with covariance matrix $Q_{r\omega,\ell} = \text{diag}\{0.2, 0.2, 0.2, 0.2\}$. $v_{r,\ell}$ is a zero-mean Gaussian white noise with covariance matrix $Q_{rv,\ell} = \text{diag}\{0.3, 0.3\}$ $(r = 1, 2, 3, 4)$. The saturation levels are $q_{1,\max} = [5\ 6\ 3]^T, q_{2,\max} = [4\ 3\ 2]^T, q_{3,\max} = [3\ 4\ 3]^T$ and $q_{4,\max} = [3\ 5\ 4]^T$. The other parameters are the same as those in Example 1.

Due to the limited space, only simulation results of the root-mean-square errors (RMSE) in position over four robots are shown in this chapter. At present, many existing results have not taken account of the state saturation problem for CNs in the filter design process, see e.g. [101], [133]. To illustrate the effects of the state saturation terms, Figs. 8.9–8.10 show the RMSE in position with and without considering the state saturation terms, respectively. Comparing Fig. 8.9 with Fig. 8.10, it can be observed that, even under event-triggering mechanism, the filtering performance has been much improved with consideration of the state saturation phenomenon during the filter design, and this has further confirmed the effectiveness of the proposed filtering algorithm in this chapter.

Remark 8.6 *Note that the adjustable threshold θ_r determines the triggering frequency of the information exchanges. Clearly, a smaller value of θ_r means a better filtering performance at the cost of more consumption of the bandwidth/energy resources. On the other hand, from the equation (8.26), we see that a larger θ_r leads to a larger trace of $\Psi_{r,\ell+1|\ell+1}$, which means the filtering performance is sacrificed in this case. As such, the selection of θ_r should reflect an adequate trade-off between the filtering performance and the resource consumption based on engineering practice. In addition, how to maximize θ_r with certain guaranteed filtering performance deserves further investigation, where the multi-step prediction method could be considered to further improve the filter performance.*

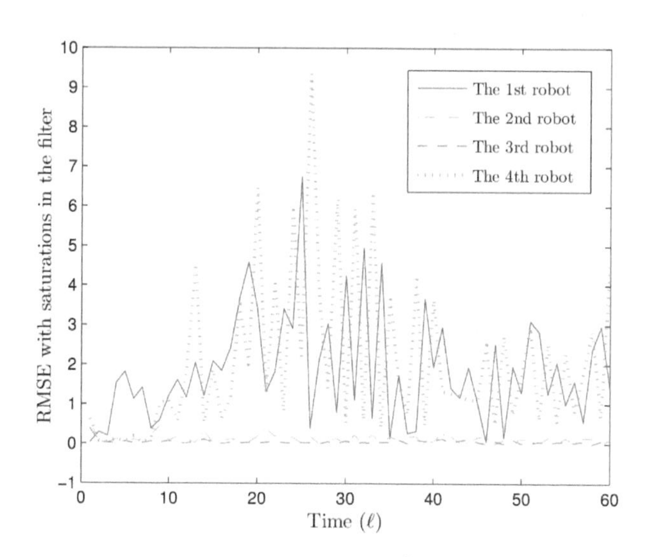

FIGURE 8.9
RMSE in position for filter when considering state saturations.

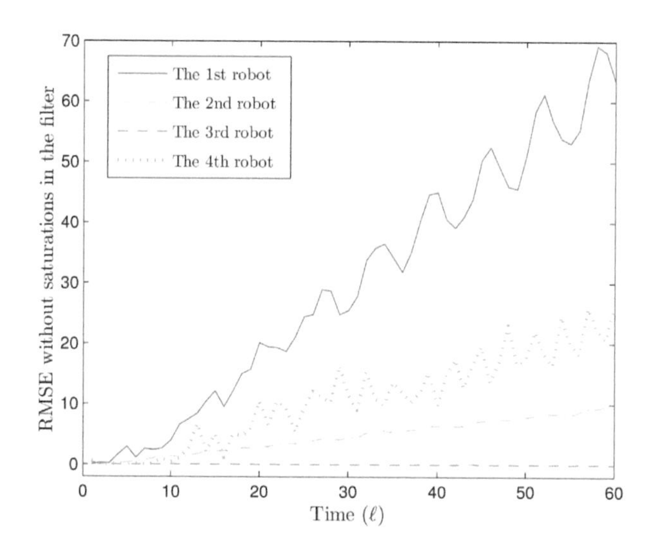

FIGURE 8.10
RMSE in position for filter without considering state saturations.

8.4 Summary

In this chapter, the event-triggering recursive filtering problem has been studied for a class of state-saturated CNs with random coupling strengths. A novel filter design scheme has been proposed for a class of time-varying CNs, where the stochastic coupling strengths have been assumed to obey the uniform random distribution. Subsequently, an upper bound of the filtering error covariance matrix has been given, and the exponential boundedness of the filtering error has been discussed. At last, a simulation example has been presented to verify the effectiveness of the proposed scheme.

9

Conclusions and Future Work

This chapter summarizes the book and points out some possible research directions related to the work done in this book.

9.1 Conclusions

The focus of this book has been placed on the filtering, estimation and control problems for nonlinear stochastic time-varying systems. In particular, several problems have been studied in detail.

- In Chapter 2, the event-based filtering problem has been dealt with for time-varying systems with fading channels, randomly occurring nonlinearities and multiplicative noise. An event indicator variable and the corresponding event-triggered scheme have been established to decide whether the measurement signal is transmitted to the filter. The event-triggered scheme has been developed on the basis of the relative error of the measurement output, and the improved stochastic Rice fading models have been used to describe the fading channels. Several unrelated random variables have been utilized in order to reflect the phenomena of state-multiplicative noises, randomly occurring nonlinearities and fading measurements. Subsequently, a novel event-triggered multi-objective controller has been designed for a class of discrete time-varying stochastic systems with randomly occurring saturations, stochastic nonlinearities and multiplicative noise. An output feedback controller has been established, which achieves the given H_∞ noise attenuation level and system state covariance constraint. Finally, the effectiveness and applicability of the developed algorithms have been demonstrated by simulation examples.

- In Chapter 3, the finite-horizon reliable H_∞ output-feedback control problem has been studied for a class of discrete time-varying systems with ROUs, RONs and measurement quantizations. A variable has been employed to quantify the actuator failures which varies in a known interval, and a single random variable has been utilized to govern the sensor failures which satisfies a certain probability distribution within the interval $[0, 1]$.

DOI: 10.1201/9781003189497-9

Bernoulli distributed white sequences with given conditional probabilities have been adopted to model the RONs and ROUs. The finite-horizon reliable control has been realized by solving a series of recursive linear matrix inequalities. On the basis of this, the current system measurement and the previous system states have been employed to control the current system state. Finally, the feasibility and effectiveness of the proposed scheme have been illustrated by a simulation example.

- In Chapter 4, the finite-horizon estimation problem has been first dealt with for a class of nonlinear time-varying systems with fading channels. The probability of fault occurrence and fading measurements have been characterized by some independent random variables. Via stochastic analysis techniques, some sufficient conditions have been obtained such that the considered dynamic system satisfies the performance constraints of the fault estimators. Second the finite-horizon fault estimation problem has been investigated for a class of nonlinear stochastic time-varying systems with randomly occurring faults. All parameters of the system have been time-varying, and the stochastic nonlinearities involved could cover several classes of very well studied nonlinearities. Two sets of Bernoulli distributed white sequences have been used to reflect the random occurrence of faults. Necessary and sufficient conditions have been established for the existence of the desired finite-horizon H_∞ fault estimators by solving coupled backward recursive Riccati difference equations. The illustrative examples have been used to highlight the effectiveness of the proposed estimation technology.

- In Chapter 5, the set-membership filtering problem has been investigated for a class of time-varying state-saturated systems with mixed time-delays under the WTOD protocol. A set of time-varying filters has been designed to obtain the estimation of the plant subject to the unknown but bounded noises and the WTOD protocol. A sufficient condition has been acquired for the designed filter to satisfy the prescribed $P(k)$-dependent constraint. Then, an optimization problem has been solved by optimizing the constraint ellipsoid of the estimation error subject to the WTOD protocol. Finally, a simulation example has been given to demonstrate the effectiveness of the proposed filter design algorithm.

- In Chapter 6, the distributed state estimation problem has been dealt with for a class of time-varying systems over sensor network with redundant channels and switching topologies. First, considering the occurrence of packet dropouts in sensor networks, the redundant channels model has been utilized to improve the reliability of data transmission, which is regulated by a set of Bernoulli-distributed random variables. The topology switching functions have been described by a Markovian chain with the help of a stochastic Kronecker delta function. By computing a set of recursive linear matrix inequalities, the distributed state estimators have been obtained, which satisfy the predetermined average H_∞ performance constraint. Second, the proposed control strategy has been extended to handle the non-fragile distributed fault estimation problem for a class of nonlinear time-varying systems with randomly occurring gain variations and randomly occurring nonlinearities over sensor networks. Third, the time-varying distributed filter has been obtained for a set of nonlinear systems over sensor networks subject to switching topologies. The RR protocol, in which all the sensors would be given identical probability to use the communication channels based on a fixed cycle, has been employed to allocate the limited bandwidth reasonably. Some parallel results have been obtained according to similar techniques. In the end, numerical simulations have been presented to highlight the obtained estimation approach.

- In Chapter 7, the variance-constrained H_∞ state estimation problem has been investigated for a class of nonlinear time-varying complex networks. The randomly varying topologies, stochastic inner couplings, nonlinearities and measurement quantization have been considered, which complicates the design of the finite-horizon state estimators. The randomly varying features of the network topology and the stochastic disturbances in the inner-coupling matrix have been described respectively by a Markov chain and a Gaussian random variable. By solving a set of RLMIs, the new finite-horizon state estimators have been obtained, which satisfy the given H_∞ performance and the estimation error covariance constraints. Furthermore, based on the output signals from partial network nodes, a novel set of time-varying robust state estimators has been designed for a class of CNs subject to randomly occurring uncertainties, multiple stochastic time-varying delays, sensor saturations under the scheduling of the random access protocol. The Markov chain has been utilized in the description of the random access protocol to schedule the transmission order of nodes. Through carrying out the stochastic analysis and inequality manipulations, sufficient conditions have been derived for ensuring the existence of the desired finite-horizon partial-nodes-based robust state estimators such that the effects from the noises on the whole estimation error dynamics are restricted below a specified H_∞ disturbance attenuation level. Finally, the effectiveness and the applicability of the proposed estimation methods have been testified by two simulation examples.

- In Chapter 8, the event-triggering recursive filtering problem has been studied for a class of state-saturated complex networks with random coupling strengths. A novel filter design scheme has been proposed for a class of time-varying complex networks, where the stochastic coupling strengths have been assumed to obey the uniform random distribution. Subsequently, an upper bound of the filtering error covariance matrix has been given, and the exponential boundedness of the filtering error has been discussed. At last, a simulation example has been presented to verify the effectiveness of the proposed scheme. In the future, we will investigate the recursive filtering problem with various network scheduling protocols, the synchronization problem for time-varying complex networks, the consensus problem based on various transmission mechanisms and the control issues for discrete-time complex networks.

9.2 Future Work

Related future research work is listed below:

- Other possible future research directions include real-time applications of the proposed reliable control theory in telecommunications, and further extensions of the present results to more complex systems with unreliable communication links, such as sampled data systems, bilinear systems and a more general class of nonlinear stochastic systems.
- Another future research direction is to further investigate the problems of nonparametric and robust sequential distributed detection for sensor networks.
- Since the security is a 'hard' performance index, the analysis in mean-square sense is more conservative for practical engineering. Therefore, the further meaningful work is to solve the security problems with probabilistic performance index in the presence of network attacks.
- An additional trend for future research is to discuss the applications of the established theories and methodologies to some practical engineering problems such as power systems and artificial intelligence systems.

Bibliography

[1] C. Ahn. Switched exponential state estimation of neural networks based on passivity theory. *Nonlinear Dynamics*, 67(1):573–586, 2012.

[2] M. Aitrami, X. Chen, and X. Zhou. Discrete-time indefinite LQ control with state and control dependent noises. *Journal of Global Optimization*, 23(3–4):245–265, 2002.

[3] A. R. Backes, D. Casanova, and O. M. Bruno. Texture analysis and classification: A complex network-based approach. *Information Sciences*, 219:168–180, 2013.

[4] S. Baromand and B. Labibi. Covariance control for stochastic uncertain multivariable systems via sliding mode control strategy. *IET Control Theory and Applications*, 6(3):349–356, 2012.

[5] M. Basin, S. Elvira-Ceja, and E. Sanchez. Mean-square H_∞ filtering for stochastic systems: application to a 2DOF helicopter. *Signal Processing*, 92(3):801–806, 2012.

[6] M. V. Basin and M. Hernandez-Gonzalez. Discrete-time filtering for nonlinear polynomial systems. *International Journal of Systems Science*, 47(9):2058–2066, 2014.

[7] M. V. Basin, A. G. Loukianov, and M. Hernandez-Gonzalez. Joint state and parameter estimation for uncertain stochastic nonlinear polynomial systems. *International Journal of Intelligent Systems*, 44(7):1200–1208, 2013.

[8] M. V. Basin and P. C. Rodriguez-Ramirez. Sliding mode controller design for stochastic polynomial systems with unmeasured states. *IEEE Transactions on Industrial Electronics*, 61(1):387–396, 2014.

[9] A. Bertrand and M. Moonen. Distributed adaptive node-specific signal estimation in fully connected sensor networks - part I: sequential node updating. *IEEE Transactions on Signal Processing*, 58(10):5277–5291, 2010.

[10] A. Bertrand and M. Moonen. Distributed adaptive node-specific signal estimation in fully connected sensor networks - part II: simultaneous and asynchronous node updating. *IEEE Transactions on Signal Processing*, 58(10):5292–5306, 2010.

[11] D. P. Bertsekas and I. B. Rhodes. Recursive state estimation for a set-membership description of uncertainty. *IEEE Transactions on Automatic Control*, 16(2):117–128, 1971.

[12] P. Bianchi, J. Jakubowicz, and F. Roueff. Linear precoders for the detection of a Gaussian process in wireless sensors networks. *IEEE Transactions on Signal Processing*, 59(3):882–894, 2011.

[13] S. Boccaletti, V. Latora, Y. Moreno, M. Chavez, and D. Hwang. Complex networks: structure and dynamics. *Physics Reports*, 424(4-5):175–308, 2006.

[14] S. Boyd, L. E. Ghaoui, E. Feron, and V. Balakrishnan. *Linear Matrix Inequalities in System and Control Theory.* Philadelphia: SIAM, 1994.

[15] R. W. Brockett and D. Liberzon. Quantized feedback stabilization of linear systems. *IEEE Transactions on Automatic Control*, 45(7):1279–1289, 2000.

[16] R. Caballero-Águila, A. Hermoso-Carazo, and J. Linares-Pérez. Optimal state estimation for networked systems with random parameter matrices, correlated noises and delayed measurements. *International Journal of General Systems*, 44(2):142–154, 2015.

[17] F. S. Cattivelli and A. H. Sayed. Diffusion strategies for distributed kalman filtering: formulation and performance analysis. In *Proceedings of Cognitive Information Processing, Santorini, Greece*, pages 36–41, 2008.

[18] F. S. Cattivelli and A. H. Sayed. Diffusion strategies for distributed Kalman filtering and smoothing. *IEEE Transactions on Automatic Control*, 55(9):1520–1533, 2010.

[19] L. Chai, B. Hu, and P. Jiang. Distributed state estimation based on quantized observations in a bandwidth constrained sensor network. In *Proceedings of the 7th World Congress on Intelligent Control and Automation, Chongqing, China*, pages 2411–2415, 2008.

[20] W. Chang, P. Chen, and C. Ku. Variance and passivity constrained sliding mode fuzzy control for continuous stochastic non-linear systems. *Neurocomputing*, 201:29–39, 2016.

[21] W. Chang and B. Huang. Variance and passivity constrained fuzzy control for nonlinear ship steering systems with state multiplicative noises. *Mathematical Problems in Engineering*, 2013:Article ID 687317, 2013.

[22] W. Che and G. Yang. Non-fragile H_∞ filtering for discrete-time systems with finite word length consideration. *Acta Automatica Sinica*, 34(8):886–892, 2008.

[23] B. Chen, K. Jamieson, H. Balakrishnan, and R. Morris. Span. An energy-efficient coordination algorithm for topology maintenance in ad hoc wireless networks. In *Proceedings of the 7th ACM International Conference on Mobile Computing and Networking, Rome, Italy*, pages 85–96, 2001.

[24] B. Chen and J. Lam. Reliable observer-based $\mathcal{H}_\infty$ control of uncertain state-delayed systems. *International Journal of Systems Science*, 35(12):707–718, 2004.

[25] B. Chen and X. Liu. Reliable control design of fuzzy dynamic systems with time-varying delay. *Fuzzy Sets and Systems*, 146:349–374, 2004.

[26] D. Chen, L. Lü, M. Shang, Y. Zhang, and T. Zhou. Identifying influential nodes in complex networks. *Physica A: Statistical Mechanics and its Applications*, 391(4):1777–1787, 2012.

[27] X. Chen and Y. Jia. Indoor localization for mobile robots using lampshade corners as landmarks: visual system calibration, feature extraction and experiments. *International Journal of Control Automation & Systems*, 12(6):1313–1322, 2014.

[28] Y. Chen, W. Yu, S. Tan, and H. Zhu. Synchronizing nonlinear complex networks via switching disconnected topology. *Automatica*, 70:189–194, 2016.

[29] M. Davoodi, N. Meskin, and K. Khorasani. Simultaneous fault detection and consensus control design for a network of multi-agent systems. *Automatica*, 66:185–194, 2016.

[30] P. Dickinson and A. Shenton. A parameter space approach to constrained variance PID controller design. *Automatica*, 45(3):830–835, 2009.

[31] D. Ding, Z. Wang, F. E. Alsaadi, and B. Shen. Receding horizon filtering for a class of discrete time-varying nonlinear systems with multiple missing measurements. *International Journal of General Systems*, 44(2):198–211, 2015.

[32] D. Ding, Z. Wang, J. Lam, and B. Shen. Finite-horizon H_∞ control for discrete time-varying systems with randomly occurring nonlinearities and fading measurements. *IEEE Transactions on Automatic Control*, 60(9):2488–2493, 2015.

[33] D. Ding, Z. Wang, B. Shen, and H. Dong. Envelope-constrained H_∞ filtering with fading measurements and randomly occurring nonlinearities: the finite horizon case. *Automatica*, 55:37–45, 2015.

[34] D. Ding, Z. Wang, B. Shen, and H. Shu. H_∞ state estimation for discrete-time complex networks with randomly occurring sensor saturations and randomly varying sensor delays. *IEEE Transactions on Neural Networks and Learning Systems*, 23(5):725–736, 2012.

[35] D. Ding, Z. Wang, B. Shen, and H. Shu. State-saturated H_∞ filtering with randomly occurring nonlinearities and packet dropouts: the finite-horizon case. *International Journal of Robust & Nonlinear Control*, 23(16):1803–1821, 2013.

[36] D. Ding, H. Dong Z. Wang, and H. Shu. Distributed H_∞ state estimation with stochastic parameters and nonlinearities through sensor networks: the finite-horizon case. *Automatica*, 48(8):1575–1585, 2012.

[37] H. Dong, X. Bu, N. Hou, Y. Liu, F. E. Alsaadi, and T. Hayat. Event-triggered distributed state estimation for a class of time-varying systems over sensor networks with redundant channels. *Information Fusion*, 36:243–250, 2017.

[38] H. Dong, N. Hou, Z. Wang, and W. Ren. Variance-constrained state estimation for complex networks with randomly varying topologies. *IEEE Transactions on Neural Networks and Learning Systems*, 29(7):2757–2768, 2018.

[39] H. Dong, Z. Wang, S. X. Ding, and H. Gao. Finite-horizon estimation of randomly occurring faults for a class of nonlinear time-varying systems. *Automatica*, 50(12):3182–3189, 2014.

[40] H. Dong, Z. Wang, S. X. Ding, and H. Gao. Finite-horizon reliable control with randomly occurring uncertainties and nonlinearities subject to output quantization. *Automatica*, 52:355–362, 2015.

[41] H. Dong, Z. Wang, S. X. Ding, and H. Gao. On H_∞ estimation of randomly occurring faults for a class of nonlinear time-varying systems with fading channels. *IEEE Transactions on Automatic Control*, 61(2):479–484, 2016.

[42] H. Dong, Z. Wang, and H. Gao. *Filtering, Control and Fault Detection with Randomly Occurring Incomplete Information.* Wiley, Chichester, UK, 2013.

[43] H. Dong, Z. Wang, D. W. C. Ho, and H. Gao. Variance-constrained H_∞ filtering for nonlinear time-varying stochastic systems with multiple missing measurements: the finite-horizon case. *IEEE Transactions on Signal Processing*, 58(5):2534–2543, 2010.

[44] H. Dong, Z. Wang, J. Lam, and H. Gao. Distributed filtering in sensor networks with randomly occurring saturations and successive packet dropouts. *International Journal of Robust and Nonlinear Control*, 24(12):1743–1759, 2014.

[45] M. Donkers and W. Heemels. Output-based event-triggered control with guaranteed L_∞ gain and improved and decentralised event-triggering. *IEEE Transactions on Automatic Control*, 57(6):1326–1332, 2012.

[46] M. C. F. Donkers, W. P. M. H. Heemels, N. van de Wouw, and L. Hetel. Stability analysis of networked control systems using a switched linear systems approach. *IEEE Transactions on Automatic Control*, 56(9):2101–2115, 2011.

[47] M. El-Koujok, M. Benammar, N. Meskin, M. Al-Naemi, and R. Langari. Multiple sensor fault diagnosis by evolving data-driven approach. *Information Science*, 259(20):346–358, 2014.

[48] N. Elia and S. K. Mitter. Stabilization of linear systems with limited information. *IEEE Transactions on Automatic Control*, 46(9):1384–1400, 2001.

[49] R. W. Eustace and B. A. Woodyatt. G. L. Merrington and A. Runacres, Fault signatures obtained from fault implant tests on an F404 engine. *ASME Transactions Journal of Engineering for Gas Turbines and Power*, 116(1):178–183, 1994.

[50] M. Farina, G. Ferrari-Trecate, and R. Scattolini. Distributed moving horizon estimation for sensor networks. In *Proceedings of the 1st IFAC Workshop on Estimation and Control of Networked Systems, Venice, Italy*, pages 126–131, 2009.

[51] G. Feng. Nonsynchronized state estimation of discrete time piecewise linear systems. *IEEE Transactions on Signal Processing*, 54(1):295–303, 2009.

[52] J. Feng, Z. Wang, and M. Zeng. Optimal robust non-fragile Kalman-type recursive filtering with finite-step autocorrelated noises and multiple packet dropouts. *Aerospace Science and Technology*, 15(6):486–494, 2011.

[53] M. Fu and L. Xie. The sector bound approach to quantized feedback control. *IEEE Transactions on Automatic Control*, 50(11):1698–1711, 2005.

[54] H. Gao and T. Chen. H_∞ estimation for uncertain systems with limited communication capacity. *IEEE Transactions on Automatic Control*, 52(11):2070–2084, 2007.

[55] H. Gao, T. Chen, and L. Wang. Robust fault detection with missing measurements. *International Journal of Control*, 81(5):804–819, 2008.

[56] H. Gao, Z. Fei, J. Lam, and B. Du. Further results on exponential estimates of Markovian jump systems with mode-dependent time-varying delays. *IEEE Transactions on Automatic Control*, 56(1):223–229, 2011.

[57] Z. Gao, T. Breikin, and H. Wang. Reliable observer-based control against sensor failures for systems with time delays in both state and input. *IEEE Transactions on Systems, Man, and Cybernetics part A: Systems and Humans*, 38(5):1018–1029, 2008.

[58] G. Garcia, S. Tarbouriech, J. M. G. da Silva, and D. Eckhard. Finite L_2 gain and internal stabilisation of linear systems subject to actuator and sensor saturations. *IET Control Theory and Applications*, 3(7):799–812, 2009.

[59] F. Gensbittel. Covariance control problems over martingales with fixed terminal distribution arising from game theory. *Siam Journal on Control and Optimization*, 51(2):1152–1185, 2013.

[60] E. Gershon. Robust reduced-order H_∞ output-feedback control of retarded stochastic linear systems. *IEEE Transactions on Automatic Control*, 58(11):2898–2904, 2013.

[61] L. E. Ghaoui and G. Calafiore. Robust filtering for discrete-time systems with bounded noise and parametric uncertainty. *IEEE Transactions on Automatic Control*, 46(7):1302–1313, 2001.

[62] F. Giulietti, L. Pollini, and M. Innocenti. Autonomous formation flight. *IEEE Control Systems*, 20(6):34–44, 2000.

[63] W. Glover and J. Lygeros. A stochastic hybrid model for air traffic control simulation. In *Seventh International Workshop on Hybrid Systems: Computation and Control, Lecture Notes in Computer Science, vol 2993, Springer, Berlin, Heidelberg*, pages 372–386, 2004.

[64] T. Gommans, D. Antunes, T. Donkers, P. Tabuada, and M. Heemels. Self-triggered linear quadratic control. *Automatica*, 50(4):1279–1287, 2014.

[65] H. Grip, A. Saberi, and X. Wang. Stabilization of multiple-input multiple-output linear systems with saturated outputs. *IEEE Transactions on Automatic Control*, 55(9):2160–2164, 2010.

[66] C. Guan, Z. Fei, Z. Li, and Y. Xu. Improved H_∞ filter design for discrete-time Markovian jump systems with time-varying delay. *Journal of the Franklin Institute*, 353(16):4156–4175, 2016.

[67] A. Gundes and H. Ozbay. Reliable decentralised control of delayed MIMO plants. *International Journal of Control*, 83(3):516–526, 2010.

[68] G. Guo, Z. Lu, and Q. Han. Control with Markov sensors/actuators assignment. *IEEE Transactions on Automatic Control*, 57(7):1799–1804, 2012.

[69] L. Guo and H. Wang. Frequency-domain set-membership filtering and its applications. *IEEE Transactions on Signal Processing*, 55(4):1326–1338, 2007.

[70] Y. Hao, J. Han, Y. Lin, and L. Liu. Vulnerability of complex networks under three-level-tree attacks. *Physica A: Statistical Mechanics and its Applications*, 462:674–683, 2016.

[71] W. He and Y. Dong. Adaptive fuzzy neural network control for a constrained robot using impedance learning. *IEEE Transactions on Neural Networks & Learning Systems*, 29(4):1174–1186, 2018.

[72] W. Heemels, R. Gorter, A. Van Zijl, P. Van den Bosch, S. Weiland, and W. Hendrix. Asynchronous measurement and control: a case study on motor synchronization. *Control Engineering Practice*, 7(12):1467–1482, 1999.

[73] T. Henningsson, E. Johannesson, and A. Cervin. Sporadic event-based control of first-order linear stochastic systems. *Automatica*, 44:2890–2895, 2008.

[74] M. Hernandez-Gonzalez and M. V. Basin. Discrete-time filtering for nonlinear polynomial systems over linear observations. In *Proceedings of the 2014 American Control Conference, Portland, OR, USA*, pages 2273–2278, 2014.

[75] M. Hernandez-Gonzalez and M. V. Basin. Discrete-time optimal control for stochastic nonlinear polynomial systems. *International Journal of General Systems*, 43(3-4):359–371, 2014.

[76] M. Hernandez-Gonzalez and M. V. Basin. Discrete-time H_∞ control for nonlinear polynomial systems. *International Journal of General Systems*, 44(2):267–275, 2015.

[77] J. Hu, Z. Wang, D. Chen, and F. E. Alsaadi. Estimation, filtering and fusion for networked systems with network-induced phenomena: new progress and prospects. *Information Fusion*, 31:65–75, 2016.

[78] J. Hu, Z. Wang, and H. Gao. Recursive filtering with random parameter matrices, multiple fading measurements and correlated noises. *Automatica*, 49(11):3440–3448, 2013.

[79] J. Hu, Z. Wang, H. Gao, and L.K. Stergioulas. Robust sliding mode control for discrete stochastic systems with mixed time delays, randomly occurring uncertainties, and randomly occurring nonlinearities. *IEEE Transactions on Industrial Electronics*, 59(7):3008–3015, 2012.

[80] J. Hu, Z. Wang, S. Liu, and H. Gao. A variance-constrained approach to recursive state estimation for time-varying complex networks with missing measurements. *Automatica*, 64:155–162, 2016.

[81] J. Hu, Z. Wang, B. Shen, and H. Gao. Gain-constrained recursive filtering with stochastic nonlinearities and probabilistic sensor delays. *IEEE Transactions on Signal Processing*, 61(5):1230–1238, 2013.

[82] J. Hu, Z. Wang, B. Shen, and H. Gao. Quantized recursive filtering for a class of nonlinear systems with multiplicative noises and missing measurements. *International Journal of Control*, 86(4):650–663, 2013.

[83] S. Hu and D. Yue. Event-based H_∞ filtering for networked system with communication delay. *Signal Processing*, 92(9):2029–2039, 2012.

[84] S. Hu, D. Yue, X. Xie, and Z. Du. Event-triggered H_∞ stabilization for networked stochastic systems with multiplicative noise and network-induced delays. *Information Sciences*, 299:178–197, 2015.

[85] H. Huang, G. Feng, and J. Cao. Robust state estimation for uncertain neural networks with time-varying delay. *IEEE Transactions on Neural Networks*, 19(8):1329–1339, 2008.

[86] S. J. Huang and G. H. Yang. Fault tolerant controller design for T-S fuzzy systems with time-varying delay and actuator faults: a k-step-fault-estimation approach. *IEEE Transactions on Fuzzy Systems*, 22(6):1526–1540, 2014.

[87] D. E. Huntington and C. S. Lyrintzis. Nonstationary random parametric vibration in light aircraft landing gear. *Journal of Aircraft*, 35(1):145–151, 1998.

[88] B. Jiang, K. Zhang, and P. Shi. Integrated fault estimation and accommodation design for discrete-time Takagi-Sugeno fuzzy systems with actuator faults. *IEEE Transactions on Fuzzy Systems*, 19(2):291–304, 2011.

[89] X. Kai, C. Wei, and L. Liu. Robust extended Kalman filtering for nonlinear systems with stochastic uncertainties. *IEEE Transactions On Systems, Man, and Cybernetics-Part A: Systems and Humans*, 40(2):399–405, 2010.

[90] M. Kalandros and L. Pao. Covariance control for multisensor systems. *IEEE Transactions on Aerospace and Electronic Systems*, 38(4):1138–1157, 2002.

[91] H. R. Karimi and H. Gao. New delay-dependent exponential H_∞ synchronization for uncertain neural networks with mixed time delays. *IEEE Transactions on Systems, Man, and Cybernetics-Part B: Cybernetics*, 40(1):173–185, 2010.

[92] H. K. Khalil. *Nonlinear Systems.* Upper Saddle River, NJ: Prentice-Hall, 1996.

[93] E. Kokiopoulou and P. Frossard. Polynomial filtering for fast convergence in distributed consensus. *IEEE Transactions on Signal Processing*, 57(1):342–354, 2009.

[94] E. Kokiopoulou and P. Frossard. Distributed classification of multiple observation sets by consensus. *IEEE Transactions on Signal Processing*, 59(1):104–114, 2011.

[95] P. Lancaster and L. Rodman. *Algebraic Riccati Equations.* Series: Oxford Science Publications, Oxford University Press, 1995.

[96] F. Li, P. Shi, X. Wang, and R. Agarwal. Fault detection for networked control systems with quantization and Markovian packet dropouts. *Signal Processing*, 111:106–112, 2015.

[97] F. Li, J. Zhou, and D. Wu. Optimal filtering for systems with finite-step autocorrelated noises and multiple packet dropouts. *Aerospace Science and Technology*, 24(1):255–263, 2013.

[98] H. Li, P. Shi, D. Yao, and L. Wu. Observer-based adaptive sliding mode control for nonlinear Markovian jump systems. *Automatica*, 64:133–142, 2016.

[99] W. Li, Y. Jia, and J. Du. Recursive state estimation for complex networks with random coupling strength. *Neurocomputing*, 219:1–8, 2017.

[100] W. Li, Y. Jia, and J. Du. State estimation for stochastic complex networks with switching topology. *IEEE Transactions on Automatic Control*, 62(12):6377–6384, 2017.

[101] W. Li, Y. Jia, and J. Du. Variance-constrained state estimation for nonlinearly coupled complex networks. *IEEE Transactions on Cybernetics*, 48(2):818–824, 2017.

[102] J. Liang, F. Sun, and X. Liu. Finite-horizon $\mathcal{H}_\infty$ filtering for time-varying delay systems with randomly varying nonlinearities and sensor saturations. *Systems Science and Control Engineering: An Open Access Journal*, 2(1):108–118, 2014.

[103] J. Liang, Z. Wang, and X. Liu. State estimation for coupled uncertain stochastic networks with missing measurements and time-varying delays: the discrete-time case. *IEEE Transactions on Neural Networks*, 20(5):781–793, 2009.

[104] J. Liang, Z. Wang, and Y. Liu. Distributed state estimation for discrete-time sensor networks with randomly varying nonlinearities and missing measurements. *IEEE Transactions on Neural Networks*, 22(3):486–496, 2011.

[105] Y.-W. Liang and S.-D. Xu. Reliable control of nonlinear systems via variable structure scheme. *IEEE Transactions on Automatic Control*, 51(10):1721–1726, 2006.

[106] D. Liu, Y. Liu, and F. Alsaadi. Recursive state estimation based-on the outputs of partial nodes for discrete-time stochastic complex networks with switched topology. *Journal of the Franklin Institute*, 355(11):4686–4707, 2018.

[107] J. Liu, Z. Gu, and S. Fei. Reliable control for nonlinear systems with stochastic actuators fault and random delays through a T-S fuzzy model approach. *Acta Mathematicae Applicatae Sinica, English Series*, 32(2):395–406, 2016.

[108] K. Liu and E. Fridman. Discrete-time network-based control under try-once-discard protocol and actuator constraints. In *Proceedings of the 2014 European Control Conference, Strasbourg, France*, pages 442–447, 2014.

[109] M. Liu, X. Cao, and P. Shi. Fault estimation and tolerant control for fuzzy stochastic systems. *IEEE Transactions on Fuzzy Systems*, 21(2):221–229, 2013.

[110] S. Liu, Z. Wang, L. Wang, and G. Wei. On quantized H_∞ filtering for multi-rate systems under stochastic communication protocols: The finite-horizon case. *Information Sciences*, 459:211–223, 2018.

[111] X. Liu, X. Yu, and H. Xi. Finite-time synchronization of neutral complex networks with Markovian switching based on pinning controller. *Neurocomputing*, 153:148–158, 2015.

[112] Y. Liu, J. Wang, and G. Yang. Reliable control of uncertain nonlinear systems. *Automatica*, 34(7):875–879, 1998.

[113] Y. Liu, Z. Wang, J. Liang, and X. Liu. Synchronization and state estimation for discrete-time complex networks with distributed delays. *IEEE Transactions on Systems, Man, and Cybernetics, Part B: Cybernetics*, 38(5):1314–1325, 2008.

[114] Y. Liu, Z. Wang, Y. Yuan, and F. E. Alsaadi. Partial-nodes-based state estimation for complex networks with unbounded distributed delays. *IEEE Transactions on Neural Networks and Learning Systems*, 29(8):3906–3912, 2018.

[115] Y. Liu, Z. Wang, Y. Yuan, and W. Liu. Event-triggered partial-nodes-based state estimation for delayed complex networks with bounded distributed delays. *IEEE Transactions on Systems, Man, and Cybernetics: Systems*, 49(6):1088–1098, 2019.

[116] Y. Long and G. H. Yang. Fault detection and isolation for networked control systems with finite frequency specifications. *International Journal of Robust and Nonlinear Control*, 24(3):495–514, 2014.

[117] J. Lu, J. Kurths, J. Cao, N. Mahdavi, and C. Huang. Synchronization control for nonlinear stochastic dynamical networks: pinning impulsive strategy. *IEEE Transactions on Neural Networks and Learning Systems*, 23(2):285–292, 2012.

[118] R. Lu, S. Chen, Y. Xu, H. Peng, and K. Xie. State estimation for complex networks with randomly varying nonlinearities and sensor failures. *Complexity*, 21(S2):507–517, 2016.

[119] Y. Luo, Z. Wang, G. Wei, F. E. Alsaadi, and T. Hayat. State estimation for a class of artificial neural networks with stochastically corrupted measurements under Round-Robin protocol. *Neural Networks*, 77:70–79, 2016.

[120] Y. Luo, Z. Wang, G. Wei, B. Shen, X. He, H. Dong, and J. Hu. Fuzzy-Logic-based control, filtering, and fault detection for networked systems: a survey. *Mathematical Problems in Engineering*, 2015:Article ID 543725, 11 pages, 2015.

[121] H. Ma and G. Yang. Fault-tolerant control synthesis for a class of nonlinear systems: sum of squares optimisation approach. *International Journal of Robust and Nonlinear Control*, 19(5):591–610, 2009.

[122] L. Ma, Z. Wang, and Y. Bo. *Variance-Constrained Multi-Objective Stochastic Control and Filtering*. John Wiley & Sons, Chichester, UK, 2015.

[123] L. Ma, Z. Wang, H. K. Lam, F. E. Alsaadi, and X. Liu. Robust filtering for a class of nonlinear stochastic systems with probability constraints. *Automation and Remote Control*, 77(1):37–54, 2016.

[124] Y. Ma, N. Ma, and L. Chen. Synchronization criteria for singular complex networks with Markovian jump and time-varying delays via pinning control. *Nonlinear Analysis: Hybrid Systems*, 29:85–99, 2018.

[125] A. Molin and S. Hirche. Structural characterization of optimal event-based controllers for linear stochastic systems. In *Proceedings of the 49th IEEE Conference on Decision and Control, Atlanta, GA*, pages 3227–3233, 2010.

[126] Y. Mostofi and M. Malmirchegini. Binary consensus over fading channels. *IEEE Transactions on Signal Processing*, 58(12):6340–6354, 2010.

[127] W. NaNacara and E. E. Yaz. Recursive estimator for linear and nonlinear systems with uncertain observations. *Signal Processing*, 62(2):215–228, 1997.

[128] Y. Niu, D. W. C. Ho, and X. Wang. Robust H_∞ control for nonlinear stochastic systems: A sliding-mode approach. *IEEE Transactions on Automatic Control*, 53(7):1695–1701, 2008.

[129] T. G. Park. Estimation strategies for fault isolation of linear systems with disturbances. *IET Control Theory and Applications*, 4(12):2781–2792, 2010.

[130] S. K. Patra and A. Ghosh. Statistics of Lyapunov exponent spectrum in randomly coupled kuramoto oscillators. *Physical Review E*, 93(3):2208–2213, 2016.

[131] C. Peng and Q. Han. A novel event-triggered transmission scheme and L_2 control co-design for sampled-data control systems. *IEEE Transactions on Automatic Control*, 58(10):2620–2626, 2013.

[132] C. Peng and T. Yang. Event-triggered communication and H_∞ control co-design for networked control systems. *Automatica*, 49:1326–1332, 2013.

[133] H. Peng, R. Lu, Y. Xu, and F. Yao. Dissipative non-fragile state estimation for Markovian complex networks with coupling transmission delays. *Neurocomputing*, 275:1576–1584, 2018.

[134] H. Ping, S. Hassan, and J. Ren. Robust adaptive synchronization of uncertain complex networks with multiple time-varying coupled delays. *Complexity*, 20(6):62–73, 2015.

[135] J. Qin, H. Gao, and W. X. Zheng. Exponential synchronization of complex networks of linear systems and nonlinear oscillators: a unified analysis. *IEEE Transactions on Neural Networks and Learning Systems*, 26(3):510–521, 2015.

[136] Y. Qin, Y. Liang, Y. Yang, Q. Pan, and F. Yang. Minimum upper-bound filter of Markovian jump linear systems with generalized unknown disturbances. *Automatica*, 73:56–63, 2016.

[137] R. Rakkiyappan and K. Sivaranjani. Sampled-data synchronization and state estimation for nonlinear singularly perturbed complex networks with time-delays. *Nonlinear Dynamics*, 84(3):1623–1636, 2016.

[138] K. Reif, S. Gunther, E. Yaz, and R. Unbehauen. Stochastic stability of the discrete-time extended Kalman filter. *IEEE Transactions on Automatic Control*, 44(4):714–728, 1999.

[139] M. Rodrigues, H. Hamdi, D. Theilliolet, C. Mechmeche, and N. BenHadj Braiek. Actuator fault estimation based adaptive polytopic observer for a class of LPV descriptor systems. *International Journal of Robust and Nonlinear Control*, 25:673–688, 2015.

[140] M. Sahebsara, T. Chen, and S. L. Shah. Optimal $\mathcal{H}_2$ filtering in networked control systems with multiple packet dropout. *IEEE Transactions on Automatic Control*, 52(8):1508–1513, 2007.

[141] A. Schaum and R. Jaquez. Estimating the state probability distribution for epidemic spreading in complex networks. *Applied Mathematics and Computation*, 291:197–206, 2016.

[142] Y. Sharon and D. Liberzon. Input to state stabilizing controller for systems with coarse quantization. *IEEE Transactions on Automatic Control*, 57(4):830–844, 2012.

[143] B. Shen. A survey on the applications of the Krein-space theory in signal estimation. *Systems Science & Control Engineering: An Open Access Journal*, 2:143–149, 2014.

[144] B. Shen, S. X. Ding, and Z. Wang. Finite-horizon H_∞ fault estimation for linear discrete time-varying systems with delayed measurements. *Automatica*, 49(1):293–296, 2013.

[145] B. Shen, X. D. Steven, and Z. Wang. Finite-Horizon H_∞ fault estimation for uncertain linear discrete time-varying systems with known input. *IEEE Transaction on Circuits and Systems-II:Express Briefs*, 60(12):902–906, 2013.

[146] B. Shen, Z. Wang, D. Ding, and H. Shu. H_∞ state estimation for complex networks with uncertain inner coupling and incomplete measurements. *IEEE Transactions on Neural Networks and Learning Systems*, 24(12):2027–2037, 2013.

[147] B. Shen, Z. Wang, and Y. S. Hung. Distributed consensus H_∞ filtering in sensor networks with multiple missing measurements: the finite-horizon case. *Automatica*, 46(10):1682–1688, 2010.

[148] B. Shen, Z. Wang, and X. Liu. A stochastic sampled-data approach to distributed H_∞ filtering in sensor networks. *IEEE Transactions on Circuits and Systems I*, 58(9):2237–2246, 2011.

[149] B. Shen, Z. Wang, and X. Liu. Bounded H_∞ synchronization and state estimation for discrete time-varying stochastic complex networks over a

finite horizon. *IEEE Transactions on Neural Networks*, 22(1):145–157, 2011.

[150] B. Shen, Z. Wang, H. Shu, and G. Wei. Robust H_∞ finite-horizon filtering with randomly occurred nonlinearities and quantization effects. *Automatica*, 46(11):1743–1751, 2010.

[151] B. Shen, Z. Wang, H. Shu, and G. Wei. H_∞ filtering for uncertain time-varying systems with multiple randomly occurred nonlinearities and successive packet dropouts. *International Journal of Robust and Nonlinear Control*, 21(14):1693–1709, 2011.

[152] H. Shen, J. H. Park, Z. Wu, and Z. Zhang. Finite-time H_∞ synchronization for complex networks with semi-Markov jump topology. *Communications in Nonlinear Science and Numerical Simulation*, 24(1-3):40–51, 2015.

[153] Q. Shen, B. Jiang, and V. Cocquempot. Adaptive fuzzy observer-based active fault-tolerant dynamic surface control for a class of nonlinear systems with actuator faults. *IEEE Transactions on Fuzzy Systems*, 22(2):338–349, 2014.

[154] S. Sheng and X. Zhang. H_∞ filtering for T-S fuzzy complex networks subject to sensor saturation via delayed information. *IET Control Theory & Applications*, 11(14):2370–2382, 2017.

[155] P. Shi, H. Wang, and C. Lim. Network-based event-triggered control for singular systems with quantizations. *IEEE Transactions on Industrial Electronics*, 63(2):1230–1238, 2016.

[156] A. Speranzon, C. Fischione, K. H. Johansson, and A. Sangiovanni-Vincentelli. A distributed minimum variance estimator for sensor networks. *IEEE Journal on Selected Areas in Communications*, 26(4):609–621, 2008.

[157] O. Sporns. The human connectome: a complex network. *Annals of the New York Academy of Sciences: The Year in Cognitive Neuroscience*, 1224(1):109–125, 2011.

[158] P. Suchomski. A reliable synthesis of discrete-time $\mathcal{H}_\infty$ control. Part I: basic theorems and J-lossless conjugators. *Control and Cybernetics*, 36(1):97–141, 2007.

[159] Y. Sun, L. Wang, and G. Xie. Average consensus in networks of dynamic agents with switching topologies and multiple time-varying delays. *Systems & Control Letters*, 57:175–183, 2008.

[160] A. Sundaresan and P. K. Varshney. Location estimation of a random signal source based on correlated sensor observations. *IEEE Transactions on Signal Processing*, 59(2):787–799, 2011.

[161] R. Szewczyk, E. Osterweil, J. Polastre, A. Mainwaring M. Hamilton, and D. Estrin. Habitat monitoring with sensor networks. *Communications of the ACM*, 47(6):34–40, 2004.

[162] P. Tabuada. Event-triggered real-time scheduling of stabilizing control tasks. *IEEE Transactions on Automatic Control*, 52(9):1680–1685, 2007.

[163] Y. Tang, H. Gao, and J. Kurths. Distributed robust synchronization of dynamical networks with stochastic coupling. *IEEE Transactions on Circuits and Systems-I: Regular Papers*, 61(5):1508–1519, 2014.

[164] J. Teng, H. Snoussi, and C. Richard. Decentralized variational filtering for target tracking in binary sensor networks. *IEEE Transactions on Mobile Computing*, 9(10):1465–1477, 2010.

[165] A. Testolin, M. Piccolini, and S. Suweis. Deep learning systems as complex networks. *Journal of Complex Networks*, page https://arxiv.org/abs/1809.10941 (accessed on 25 January 2019).

[166] E. Tian, D. Yue, and C. Peng. Reliable control for networked control systems with probabilistic actuator fault and random delays. *Journal of the Franklin Institute*, 347(10):1907–1926, 2010.

[167] E. Tian, D. Yue, T. Yang, Z. Gu, and G. Lu. T-S fuzzy model-based robust stabilization for networked control systems with probabilistic sensor and actuator failure. *IEEE Transactions on Fuzzy Systems*, 19(3):553–561, 2011.

[168] S. Tong, Y. Li, and T. Wang. Adaptive fuzzy decentralized output feedback control for stochastic nonlinear large-scale systems using DSC technique. *International Journal of Robust and Nonlinear Control*, 23(4):381–399, 2013.

[169] Y. Tong, L. Zhang, M. V. Basin, and C. Wang. Weighted H_∞ control with D-stability constraint for switched positive linear systems. *International Journal of Robust and Nonlinear Control*, 24(4):758–774, 2014.

[170] C. Trabelsi, O. Bilaniuk, Y. Zhang, D. Serdyuk, S. Subramanian, J. F. Santos, S. Mehri, N. Rostamzadeh, Y. Bengio, and C. J. Pal. Deep complex networks. *In Proceedings of the 6th International Conference on Learning Representations, Vancouver, BC, Canada*, page https://arxiv.org/abs/1705.09792 (accessed on 10 January 2019).

[171] V. Ugrinovskii and E. Fridman. A Round-Robin type protocol for distributed estimation with H_∞ consensus. *Systems & Control Letters*, 69:103–110, 2014.

[172] R. J. Veillette. Reliable linear-quadratic state-feedback control. *Automatica*, 31:137–143, 1995.

[173] R. J. Veillette, J. V. Medanic, and W. R. Perkins. Design of reliable control systems. *IEEE Transactions on Automatic Control*, 37(3):290–304, 1992.

[174] B. Wang, X. Meng, and T. Chen. Event based pulse-modulated control of linear stochastic systems. *IEEE Transactions on Automatic Control*, 59(8):2144–2150, 2014.

[175] F. Wang, J. Liang, Z. Wang, and F. E. Alsaadi. Robust synchronization of complex networks with uncertain couplings and incomplete information. *International Journal of General Systems*, 45(5):589–603, 2016.

[176] G. Wang, J. Cao, and J. Lu. Outer synchronization between two nonidentical networks with circumstance noise. *Physica A: Statistical Mechanics and Its Applications*, 389(7):1480–1488, 2010.

[177] J. Wang, H. Zhang, Z. Wang, and H. Liang. Local stochastic synchronization for Markovian neutral-type complex networks with partial information on transition probabilities. *Neurocomputing*, 167:474–487, 2015.

[178] L. Wang, Z. Wang, T. Huang, and G. Wei. An event-triggered approach to state estimation for a class of complex networks with mixed time delays and nonlinearities. *IEEE Transactions on Cybernetics*, 46(11):2497–2508, 2016.

[179] X. Wang. Complex networks: topology, dynamics and synchronization. *International Journal of Bifurcation and Chaos*, 12(05):885–916, 2002.

[180] X. Wang. *Event-Triggering in Cyber-Physical Systems*. Ph.D. dissertation of The University of Notre Dame, Notre Dame, 2009.

[181] Y. Wang, H. Karimi, and Y. Yan. An adaptive event-triggered synchronization approach for chaotic Lur'e systems subject to aperiodic sampled-data. *IEEE Transactions on Circuits and Systems II: Express Briefs*, 66(3):442–446, 2019.

[182] Y. Wang, Y. Xia, C. Ahn, and Y. Zhu. Exponential stabilization of Takagi-Sugeno fuzzy systems with aperiodic sampling: an aperiodic adaptive event-triggered method. *IEEE Transactions on Systems, Man, and Cybernetics: Systems*, 49(2):444–454, 2019.

[183] Z. Wang, H. Dong, B. Shen, and H. Gao. Finite-horizon H_∞ filtering with missing measurements and quantization effects. *IEEE Transactions on Automatic Control*, 58(7):1707–1718, 2013.

[184] Z. Wang, J. Lam, L. Ma, Y. Bo, and Z. Guo. Variance-constrained dissipative observer-based control for a class of nonlinear stochastic

systems with degraded measurements. *Journal of Mathematical Analysis and Applications*, 377(2):645–658, 2011.

[185] Z. Wang, Y. Liu, and X. Liu. H_∞ filtering for uncertain stochastic time-delay systems with sector-bounded nonlinearities. *Automatica*, 44(5):1268–1277, 2008.

[186] Z. Wang, B. Shen, and X. Liu. H_∞ filtering with randomly occurring sensor saturations and missing measurements. *Automatica*, 48(3):556–562, 2012.

[187] Z. Wang, Y. Wang, and Y. Liu. Global synchronization for discrete-time stochastic complex networks with randomly occurred nonlinearities and mixed time-delays. *IEEE Transactions on Neural Networks*, 21(1):11–25, 2010.

[188] Z. Wang, G. Wei, and G. Feng. Reliable H_∞ control for discrete-time piecewise linear systems with infinite distributed delays. *Automatica*, 45:2991–2994, 2009.

[189] G. Wei, L. Wang, and F. Han. A gain-scheduled approach to fault-tolerant control for discrete-time stochastic delayed systems with randomly occurring actuator faults. *Systems Science & Control Engineering: An Open Access Journal*, 1:82–90, 2013.

[190] C. Wen, Z. Wang, T. Geng, and F. E. Alsaadi. Event-based distributed recursive filtering for state-saturated systems with redundant channels. *Information Fusion*, 39:96–107, 2018.

[191] C. Wen, Z. Wang, Q. Liu, and F. E. Alsaadi. Recursive distributed filtering for a class of state-saturated systems with fading measurements and quantization effects. *IEEE Transactions on Systems, Man, and Cybernetics: Systems*, 48(6):930–941, 2018.

[192] J. Wu, H. Deng, Y. Tan, and D. Zhu. Vulnerability of complex networks under intentional attack with incomplete information. *Journal of Physics A: Mathematical and Theoretical*, 40(11):2665–2671, 2007.

[193] L. Wu, W. Zheng, and H. Gao. Dissipativity-based sliding mode control of switched stochastic systems. *IEEE Transactions on Automatic Control*, 58(3):785–791, 2013.

[194] Y. Xiao, Y. Cao, and Z. Lin. Robust filtering for discrete-time systems with saturation and its application to transmultiplexers. *IEEE Transactions on Signal Processing*, 52(5):1266–1277, 2004.

[195] L. Xie, C. de Souza, and M. Fu. $\mathcal{H}_\infty$ estimation for discrete time linear uncertain systems. *International Journal of Robust Nonlinear Control*, 1(2):111–123, 1991.

[196] Y. Xu, R. Lu, H. Peng, K. Xie, and A. Xue. Asynchronous dissipative state estimation for stochastic complex networks with quantized jumping coupling and uncertain measurements. *IEEE Transactions on Neural Networks and Learning Systems*, 28(2):268–277, 2017.

[197] Y. Xu, R. Lu, and P. Shi. Finite-time distributed state estimation over sensor networks with round-robin protocol and fading channels. *IEEE Transactions on Cybernetics*, 48(1):336–345, 2018.

[198] X. G. Yan and C. Edwards. Nonlinear robust fault reconstruction and estimation using a sliding mode observer. *Automatica*, 43(9):1605–1614, 2007.

[199] X. G. Yan and C. Edwards. Robust decentralized actuator fault detection and estimation for large-scale systems using a sliding mode observer. *International Journal of Control*, 81(4):591–606, 2008.

[200] F. Yang and Y. Li. Set-membership filtering for discrete-time systems with nonlinear equality constraints. *IEEE Transactions on Automatic Control*, 54(10):2480–2486, 2009.

[201] F. Yang and Y. Li. Set-membership filtering for systems with sensor saturation. *Automatica*, 45(8):1896–1902, 2009.

[202] G. Yang and J. Wang. Robust nonfragile Kalman filtering for uncertain linear systems with estimation gain uncertainty. *IEEE Transactions on Automatic Control*, 46(2):343–348, 2001.

[203] G. Yang, J. Wang, and Y. Soh. Reliable $\mathcal{H}_\infty$ controller design for linear systems. *Automatica*, 37(5):717–725, 2001.

[204] G. H. Yang, J. Lam, and J. Wang. Reliable controller design for nonlinear system. In *Proceedings of the 35th IEEE Conference on Decision and Control, Kobe, Japan*, pages 112–117, 1996.

[205] G. H. Yang and H. Wang. Fault detection for a class of uncertain state-feedback control systems. *IEEE Transactions on Control Systems Technology*, 18(1):201–212, 2010.

[206] X. Yang, J. Cao, and J. Lu. Synchronization of Markovian coupled neural networks with nonidentical node-delays and random coupling strengths. *IEEE Transactions on Neural Networks and Learning Systems*, 23(1):60–71, 2012.

[207] L. Yao, H. Wang J. Qin, and B. Jiang. Design of new fault diagnosis and fault tolerant control scheme for non-Gaussian singular stochastic distribution systems. *Automatica*, 48:2305–2313, 2012.

[208] Y. I. Yaz and E. E. Yaz. On LMI formulations of some problems arising in nonlinear stochastic system analysis. *IEEE Transactions on Automatic Control*, 44(4):813–816, 1999.

[209] J. You, S. Yin, and H. Gao. Fault detection for discrete systems with network-induced nonlinearities. *IEEE Transactions on Industrial Informatics*, 10(4):2216–2223, 2014.

[210] K. You, W. Sun, M. Fu, and L. Xie. Attainability of the minimum data rate for stabilization of linear systems via logarithmic quantization. *Automatica*, 47(1):170–176, 2011.

[211] K. You and L. Xie. Minimum data rate for mean square stabilization of discrete LTI systems over lossy channels. *IEEE Transactions on Automatic Control*, 55(10):2373–2378, 2010.

[212] W. Yu, G. Chen, Z. Wang, and W. Yang. Distributed consensus filtering in sensor networks. *IEEE Transactions on Systems, Man, and Cybernetics, Part B (Cybernetics)*, 39(6):1568–1577, 2009.

[213] D. Yue, J. Lam, and D. W. C. Ho. Reliable h_∞ control of uncertain descriptor systems with multiple time delays. In *IEE Proceedings - Control Theory and Applications*, 150: 557–564, 2003.

[214] D. Yue, E. Tian, and Q. Han. A delay system method for designing event-triggered controllers of networked control systems. *IEEE Transactions on Automatic Control*, 58(2):475–481, 2013.

[215] D. Zhang, W. Cai, L. Xie, and Q. Wang. Non-fragile distributed filtering for T-S fuzzy systems in sensor networks. *IEEE Transactions on Fuzzy Systems*, 23(5):1883–1890, 2014.

[216] D. Zhang, H. Su, S. Pan, J. Chu, and Z. Wang. LMI approach to reliable guaranteed cost control with multiple criteria constraints: The actuator faults case. *International Journal of Robust and Nonlinear Control*, 19(8):884–899, 2009.

[217] J. Zhang, A. K. Swain, and S. K. Nguang. Simultaneous robust actuator and sensor fault estimation for uncertain nonlinear Lipschitz systems. *IET Control Theory & Applications*, 8(14):1364–1374, 2014.

[218] K. Zhang, B. Jiang, and P. Shi. Fast fault estimation and accommodation for dynamical systems. *IET Control Theory & Application*, 3(2):189–199, 2009.

[219] K. Zhang, B. Jiang, and M. Staroswiecki. Dynamic output feedback fault tolerant controller design for Takagi-Sugeno fuzzy systems with actuator faults. *IEEE Transactions on Fuzzy systems*, 18(1):194–201, 2010.

[220] K. Zhang, B. Jiang, K. Zhang, B. Jiang, P. Shi, and J. Xu. Analysis and design of robust H_∞ fault estimation observer with finite-frequency specifications for discrete-time fuzzy systems. *IEEE Transactions on Cybernetics*, 45(7):1225–1235, 2015.

[221] L. Zhang, Z. Ning, and Z. Wang. Distributed filtering for fuzzy time-delay systems with packet dropouts and redundant channels. *IEEE Transactions on Systems, Man, and Cybernetics: Systems*, 46(4):559–572, 2016.

[222] W. Zhang, Y. Huang, and H. Zhang. Stochastic H_2/H_∞ control for discrete-time systems with state and disturbance dependent noise. *Automatica*, 43(3):513–521, 2007.

[223] Q. Zhao and J. Jiang. Reliable state feedback control system design against actuator failures. *Automatica*, 34(10):1267–1272, 1998.

[224] Z. Zheng and J. Haddad. Distributed adaptive perimeter control for large-scale urban road networks with delayed state interconnections. *IFAC-PapersOnLine*, 50(1):5295–5300, 2017.

[225] M. Zhong, D. Zhou, and S. X. Ding. On designing H_∞ fault detection filter for linear discrete time-varying systems. *IEEE Transactions on Automatic Control*, 55(7):1689–1695, 2010.

[226] Q. Zhou, P. Shi, S. Xu, and H. Li. Observer-based adaptive neural network control for nonlinear stochastic systems with time delay. *IEEE Transactions on Neural Networks and Learning Systems*, 24(1):71–80, 2013.

[227] L. Zou, Z. Wang, and H. Gao. Observer-based H_∞ control of networked systems with stochastic communication protocol: The finite-horizon case. *Automatica*, 63:366–373, 2016.

[228] L. Zou, Z. Wang, H. Gao, and X. Liu. State estimation for discrete-time dynamical networks with time-varying delays and stochastic disturbances under the Round-Robin protocol. *IEEE Transactions on Neural Networks and Learning Systems*, 28(5):1139–1151, 2017.

[229] L. Zou, Z. Wang, Q.-L. Han, and D. H. Zhou. Ultimate boundedness control for networked systems with Try-Once-Discard protocol and uniform quantization effects. *IEEE Transactions on Automatic Control*, 62(12):6582–6588, 2017.

Index